Jürgen Kranz Joseph Rotem (Editors)

Experimental Techniques in Plant Disease Epidemiology

With 34 Figures

AF793740

Springer-Verlag Berlin Heidelberg New York
London Paris Tokyo

Editors

Prof. Dr. Jürgen Kranz

Tropeninstitut Phytopathologie, Schottstr. 2, 6300 Gießen, FRG

Prof. Dr. Joseph Rotem

Department of Plant Pathology, Volcani Center
Agricultural Research Organization, 50–250 Bet Dagan, Israel

ISBN-13: 978-3-642-95536-5 e-ISBN-13: 978-3-642-95534-1
DOI: 10.1007/978-3-642-95534-1

Library of Congress Cataloging-in-Publication Data.
Experimental techniques in plant disease epidemiology / J. Kranz and J. Rotem, editors. p. cm.
Includes bibliographies and index.
ISBN-13: 978-3-642-95536-5
1. Plant diseases—Epidemiology. 2. Epidemiology—Research-Methodology. I. Kranz, Jürgen.
II. Rotem, J. (Joseph), 1920–
SB731.E98 1987 632′.3—dc 19 87-18752 CIP

Softcover reprint of the hardcover 1st edition 1988

Typesetting: Hagedorn, Berlin;
2131/3020-543210

Preface

Following the publications of Van der Plank, the subject of epidemiology has over the last few years been increasingly treated in books by other authors, who often emphasize the statistical, mathematical and modeling aspects. While this is stimulating and necessary for the development of this new branch in plant pathology, such an emphasis might create the impression that mathematical treatment is the raison d'être for epidemiologists. Mathematics in its various applications doubtless holds a position similar to that of chemistry in many other areas of plant pathology and the computer is also an important tool. Nevertheless despite the undisputed influence of mathematics and computer on the thinking approach to research problems of epidemiologists, epidemiology remains an experimental discipline. Experiments provide the data bases for models and in turn help to test them. It is from this mutal interplay of theory and empirics that epidemiology derives its scientific thrust and charm.

Although it is still far from being complete and well established, the theoretical side of epidemiology has attained a certain consolidation. Students and researchers in epidemiology today can choose from a broad spectrum of ideas, concepts, and methods for their research. Experimental methods in epidemiology are, however, less advanced. This book is therefore intended to help in the development of a methodology able to provide a choice of adequate methods widely acceptable amongst epidemiologists for varying applications and objectives. Experimental methods in epidemiology range from the design of appropriate field and growth chamber experiments, measurement of disease, pathogens, weather factors and host parameters to the analysis of, for instance, race dynamics and development of fungicide resistance, or the identification of variables useful in forecasting and crop loss assessment. Virus diseases and nematodes, although not essentially different in their epidemiology from other etiological groups, have their own pecularities to be dealt with. The same applies to geographical aspects and long-distance dispersal. All these different applications are treated in this book by authors well experienced in the appropriate fields.

Epidemiology as a science and as the interface between laboratory research and practical crop protection in the field must not only produce good, reliable, and realistic information from experiments, but also to derive from this information generally or specifically applicable concepts or tactics and strategies. Models and comparative epidemiology are tools to achieve these ends. Three chapters which place epidemiological experiments into the context of models and concepts therefore conclude this volume.

This book is thus an introduction to readers, particularly novices in epidemiology, to solutions of their experimental problems with the technical means at hand. The authors not only describe the state of the art. They often reveal the "tricks of the trade", present and future problems involved in various approaches and techniques, and they provide ample lists of relevant references. With this guidance we hope for further improvements in the methodology for experimentation in epidemiology.

Gießen and Rehovot, August 1987

J. Kranz
J. Rotem

List of Contributors

Ajai	Space Applications Center, Indian Space Research Organization, Ahmedabad 380053, India
Aust, H. J.	Institut für Mikrobiologie, TU Braunschweig, Mendelssohnstr. 4, D-3300 Braunschweig, Federal Republic of Germany
Barker, K. R.	Department of Plant Pathology, North Carolina State University, Raleigh, NC 27695, USA
Berger, R. D.	Plant Pathology Department, University of Florida, Gainesville, FL, 32611, USA
Campbell, C. L.	Department of Plant Pathology, North Carolina State University, Raleigh, NC 27695, USA
Dekker, J.	Laboratory of Phytopathology, Agricultural University, 6709 PD Wageningen, The Netherlands
Ellis, M. A.	Department of Plant Pathology, The Ohio State University, Ohio Agricultural Research and Development Center, Wooster, OH, 44691, USA
Friesland, H.	Deutscher Wetterdienst, Zentrale Agrarmeteorologische Forschungsstelle, Bundesallee 50, D-3300 Braunschweig, Federal Republic of Germany
Gillespie, T. J.	Department of Land Resource Science, Ontario Agricultural College, University of Guelph, Guelph, Ontario N1G 2W1, Canada
Gilligan, C. A.	Department of Applied Biology, University of Cambridge, Pembroke Street, Cambridge, CB 2 3DX, UK
Hau, B.	Phytopathologie und Angewandte Entomologie, Tropeninstitut, Justus-Liebig-Universität, Schottstr. 2, D-6300 Gießen, Federal Republic of Germany
Irwin, M. E.	University of Illinois and Illinois Natural History Survey, Urbana, USA
James, T. D. W.	Department of Environmental Biology, Ontario Agricultural College, University of Guelph, Guelph, Ontario N1G 2W1, Canada
Johnson, K. B.	Department of Plant Pathology, University of Minnesota, St. Paul, MN 55108, USA
Kranz, J.	Phytopathologie und Angewandte Entomologie, Tropeninstitut, Justus-Liebig-Universität, Schottstr. 2, D-6300 Gießen, Federal Republic of Germany
Madden, L. V.	Department of Plant Pathology, The Ohio State University, Ohio Agricultural Research and Development Center, Wooster, OH, 44691, USA
Nagarajan, S.	Division of Mycology and Plant Pathology, New Delhi 110012, Indian Agricultural Research Institute, India
Noe, J. P.	Department of Plant Pathology, North Carolina State University, Raleigh, NC 27695, USA
Raccah, B.	Department of Virology, Agricultural Research Organization, The Volcani Center, Bet Dagan 50250, Israel
Reynolds, K. M.	Department of Plant Pathology, Ohio State University, Wooster, OH 44691, USA
Rotem, J.	Department of Plant Pathology, Agricultural Research Organization, The Volcani Center, Bet Dagan 50250, Israel
Schrödter, H.	David-Mansfeld-Weg 7, D-3300 Braunschweig, Federal Republic of Germany

Seem, R. C.	Department of Plant Pathology, New York State Agricultural Experiment Station, P.O. Box 462, Geneva, NY 14456, USA
Sutton, J. C.	Department of Environmental Biology, Ontario Agricultural College, University of Guelph, Guelph, Ontario N1G 2W1, Canada
Teng, P. S.	Department of Plant Pathology, University of Minnesota, St. Paul, MN 55108, USA
Weltzien, H.	Institut für Pflanzenkrankheiten, Nußallee 9, D-5300 Bonn, Federal Republic of Germany
Welz, G.	Phytopathologie und Angewandte Entomologie, Tropeninstitut, Justus-Liebig-Universität, Schottstr. 2, D-6300 Gießen, Federal Republic of Germany

Contents

Introduction (J. Kranz and J. Rotem) 1

I General Techniques 5

Chapter 1. **Experiments and Procedures in Epidemiological Field Studies** (H. J. Aust and J. Kranz) 7

1 Introduction 7
2 Planning and Conduct of Field Experiments 8
  2.1 Formulation of the Hypothesis and Selection of Variables 9
  2.2 Single-Plot Field Experimentation 9
  2.3 Sampling in Single-Plot Experiments 10
  2.4 Data Acquisition and Handling 11
  2.5 Experimental Error 12
3 Combination of Field and Growth Chamber Experiments in Studying Epidemiological Problems 13
  3.1 Examples for the Combination of Field and Growth Chamber Experiments 15
    3.1.1 Manipulating the Pathogen 15
    3.1.2 Manipulating the Host — The Use of Trap Plants 15
    3.1.3 Manipulating the Environment 16
  3.2 Conclusions 16
4 References 16

Chapter 2. **Techniques of Controlled-Condition Experiments** (J. Rotem) 19

1 Introduction 19
2 The Controlled Habitat 20
  2.1 Light 20
  2.2 Temperature 20
  2.3 Humidity 20
  2.4 Free Leaf Moisture 21
  2.5 Relation of the Controlled Habitat to the Field Habitat 22
3 The Choice of Hosts and Pathogens 22
  3.1 The Ideal Host 22

3.2 The Ideal Pathogen 23
3.3 Relation of the Reaction of Hosts and Pathogens Under Controlled Conditions to Their Reaction in the Field 23
4 Types of Controlled-Condition Experiments 23
4.1 Monocyclic Experiments: Inoculation, Infection and Colonization, Sporulation 23
4.1.1 Inoculation 24
4.1.2 Infection and Colonization 25
4.1.3 Sporulation 26
4.2 Polycyclic Experiments 27
4.3 Auxiliary Experiments 28
4.4 Relation of Results Obtained Under Controlled Conditions to Results Obtained in the Field 28
5 Preparation of Data for Simulation 29
6 Epilogue 30
7 References 31

II Measurement and Their Analysis 33

Chapter 3. **Measuring Plant Disease** (J. Kranz) 35

1 Introduction 35
2 About Terms 35
3 Sources of Error in Disease Assessment 38
3.1 Individual Errors 38
3.2 Object-Dependent Errors 39
4 Sampling Techniques for Disease Assessment 39
5 Techniques used in Measuring Disease 42
5.1 Actual Measurement of Disease Intensity 42
5.2 Use of Keys, Standard Diagrams and Classes of Disease Intensities 43
5.3 Use of Incidence Values to Estimate Severity Values 45
5.4 Automatic Measurement and Remote Sensing 45
5.5 Measurement of Disease Intensities Through Spore Catches 46
6 Specific Applications of Disease Measurement 46
6.1 Crop Measurement 46
6.2 Assessment of Several Diseases in the Same Sampling Unit 47
6.3 Disease Assessment for Models and Forecasts 47
6.4 Differences Between Disease Assessment on Shoot and Root Diseases 47
6.5 Some Implications in Disease Assessment for Crop Losses 48
6.6 Measuring Disease in Surveys 49
7 References 49

Chapter 4. **The Measurement and Analysis of the Effects of Crop Development on Epidemics** (R. C. Seem) 51

1 Introduction 51

2 The Dynamic Interaction of Host and Pathogen 52
2.1 The Pathogen Affects the Host 52
2.1.1 Plant Growth Dynamics and Loss of Photosynthetic Tissue 53
2.1.2 Growth Stimulation Caused by Disease 53
2.2 The Host Affects the Pathogen 53
2.2.1 Host Susceptibility 53
2.2.2 New Tissue 55
2.2.3 Plant-Altered Environment 55
3 Measurement and Analysis of Host Development 55
3.1 Physiological Time 56
3.2 Host Phenology 56
3.3 Measurements for Growth Analysis 57
4 Measuring the Dynamic Interaction 59
4.1 Measuring Time- and Space-Dependent Susceptibility 59
4.2 Measuring Disease in the Presence of Changing Host Tissue 60
4.2.1 Increases in the Amount of Susceptible Tissue 60
4.2.2 Decreases in the Amount of Susceptible Tissue 61
5 Analysis of the Dynamics Interaction 62
5.1 Analysis of Changing Susceptible Host Tissue 62
5.2 Analysis of Changing Asymptotes 63
5.3 Analysis by Simulation 63
6 Considerations and Modifications of Techniques for the
Interaction of Perennial Crops 64
6.1 Survival Strategy of Perennials 64
6.2 Size of Perennials 64
6.3 Time-Dependent Susceptibility of Perennials 65
6.4 Sampling and Assessment of Perennials 65
7 Summary and Conclusions 66
8 Acknowledgments 66
9 References 67

Chapter 5. **Quantitative Assessment of Inoculum Production, Dispersal, Deposition, Survival and Infectiousness in Airborne Diseases** (J. Rotem) 69

1 Introduction 69
2 Production, Dispersal and Buildup of Inoculum 69
2.1 Trapping Spores in the Air 70
2.2 Collecting Spores from Leaves 72
3 Survival and Infectiousness 74
3.1 Survival of Overseasoning Inoculum 75
3.2 Survival of Inoculum in the Season 76
3.3 Survival of Spores on Plants Under Controlled Conditions 77
3.4 Survival of Detached Spores Under Controlled Conditions 78
3.5 Techniques for Testing the Effects of Radiation 78
4 Testing Inoculum Potential with Potted Trap Plants 79
5 Epilogue 81
6 References 82

Chapter 6. **Analysis of the Spatial Pattern of Soilborne Pathogens**
(C. A. Gilligan) . . . 85

1 Introduction . . . 85
2 Objectives and Requirements for the Analysis of Spatial Pattern of Soilborne Pathogens . . . 86
2.1 Metameters for Measurement of Spatial Pattern . . . 86
2.2 Definable Entities for Inoculum, Infection and Disease . . . 87
2.3 Sampling Units and Sampling Method . . . 87
3 Analysis of Unmapped Data . . . 89
3.1 Goodness-of-Fit to Probability Functions . . . 89
3.2 Indices of Aggregation . . . 92
4 Analysis of Mapped Data . . . 93
4.1 Analysis of Dichotomous Data . . . 94
4.1.1 Transects . . . 94
4.1.2 Transects Across Diffuse Clumps of Disease . . . 95
4.1.3 Disease Mosaics . . . 95
4.1.4 Two-Dimensional Lattice Maps . . . 95
4.2 Analysis of Discrete and Continuous Variables . . . 96
5 References . . . 97

Chapter 7. **Electronic Monitoring and Use of Microprocessors in the Field**
(J. C. Sutton, T. J. Gillespie and T. D. W. James) . . . 99

1 Introduction . . . 99
2 Environmental Data Acquisition Systems . . . 99
2.1 Data Recording . . . 99
2.2 Data Communication . . . 102
3 Monitoring Weather Variables . . . 104
3.1 Temperature . . . 104
3.2 Environmental Water . . . 105
3.2.1 Atmospheric Humidity and Water Potentials . . . 105
3.2.2 Liquid Water on Living Foliage and Fruits . . . 106
3.2.3 Wetness of Dead Tissues . . . 108
3.2.4 Rain . . . 109
3.3 Wind . . . 109
3.4 Irradiance . . . 110
4 Monitoring Soil Variables . . . 110
4.1 Soil Temperature . . . 110
4.2 Soil Water . . . 110
5 Monitoring Sites . . . 112
6 References . . . 112

Chapter 8. **The Analysis of Weather Factors in Epidemiology**
(H. Friesland and H. Schröedter) . . . 115

1 Introduction . . . 115
2 Some Practical Hints for the Monitoring of Weather Factors . . . 116

3 Techniques in Use for the Analysis of Biometeorological Data 118
3.1 Temperature 119
3.1.1 Mean Temperatures and Daily Course of Temperature 119
3.1.2 Frequency of Temperatures 119
3.1.3 Temperature Sums 120
3.1.4 Temperature-Response-Functions 122
3.2 Humidity 123
3.2.1 Relative Humidity 123
3.2.2 Duration of Wetness 124
3.3 Temperature-Humidity Combination 125
3.4 Precipitation 126
3.5 Wind and Turbulence 127
3.6 Light and Radiation 128
4 Estimation of Missing Micrometeorological Data 129
5 Epilogue 132
6 References 132

III Special Topics 135

Chapter 9. **The Analysis of Effects of Control Measures on the Development of Epidemics** (R. D. Berger) 137

1 Introduction 137
2 Epidemiological Principles of Control 137
3 Analysis of the Effect of Sanitation 138
3.1 The Sanitation Ratio 138
3.2 An Alternative Hypothesis for the Effect of Sanitation 139
3.3 Methods to Test the Effect of Sanitation 140
3.4 Other Considerations of Sanitation on the Epidemic 140
4 Analysis of Effects of Control Measures on Epidemic Rate 141
4.1 Methods 141
4.2 Analysis of Disease Progress 142
4.3 Disease Progress After a Single Application of a Fungicide 147
4.3.1 The Theoretical Disease-Progress Curve 147
4.3.2 The Actual Disease-Progress Curve 147
4.3.3 Prediction of the Increase in Disease 149
5 Epilogue 149
6 References 150

Chapter 10. **How to Conduct and Measure Fungicide Resistance** (J. Dekker) .. 153

1 Introduction 153
2 Measurement of Resistance 154
2.1 Principles 154
2.2 Use of Artificial Media 155
2.3 Use of Living Plant Material 156
3 Estimation of Resistance Risk 158

3.1 Emergence of Resistance on Agar Medium 158
3.2 Fitness and Buildup of Resistance in the Field 158
3.3 Use of Models 159
4 Detection and Monitoring of Resistance in the Field 159
4.1 Sampling 159
4.2 Handling and Processing of Samples 160
4.3 Recording and Interpretation of Results 161
5 Epilogue 161
6 References 161

Chapter 11. **Analysis of Virulence in Pathogen Populations** (G. Welz) 165

1 Introduction 165
2 Methods of Virulence Analysis 166
2.1 Sampling the Population 166
2.1.1 Sampling Techniques 166
2.1.2 Sample Size 167
2.1.3 Frequency and Timing of Sampling 168
2.2 Testing the Sample 168
2.2.1 Design of Virulence Tests 168
2.2.2 Reading Infection Types or Counting Colonies 170
2.2.3 Measuring Agressiveness 171
2.3 Analyzing the Data 172
2.3.1 Virulence Frequencies and Associations 172
2.3.2 Race Designation 173
2.3.3 Measuring Virulence and Race Dynamics 173
2.3.4 Measuring Pathogen Diversity 174
3 References 176

Chapter 12. **Analysis of Epidemiological Components in Yield Loss Assessment** (P. S. Teng and K. B. Johnson) 179

1 Introduction 179
2 Components Analysis in Yield Loss Assessment 180
2.1 Analysis of the Disease Progress Curve (DPC) 180
2.2 Relating DPC Components to Empirical Loss Models 181
3 Components Analysis in Crop Loss Assessment 185
3.1 The Problem of Scale and Aggregation 186
3.2 Crop-Pest Loss Models 186
4 Concluding Remarks 187
5 References 188

Chapter 13. **How to Develop Plant Disease Forecasters** (L. V. Madden and M. A. Ellis) 191

1 Introduction 191

2 Fundamental Forecasting Systems 192
2.1 Infection 192
2.1.1 Experimental Procedures 193
2.1.2 Data Analysis 194
2.1.3 Other Physical Factors and Nonconstancy 195
2.1.4 Vectors 195
2.1.5 Inoculum Level 196
2.2 Other Disease Cycle Components 196
3 Empirical Forecasting Systems 196
3.1 Single Season Predictions 197
3.1.1 Qualitative Development of Forecasters Based on Weather 197
3.1.2 Qualitative Development of Forecasters Based on Disease Intensity 197
3.1.3 Quantitative Development of Forecasters 198
3.2 Multiple Season Predictions 201
3.2.1 Qualitative Development 201
3.2.2 Quantitative Development 201
3.3 Combining Fundamental and Empirical Forecasting Systems 203
4 Field Testing 203
5 Implementation — Delivery and Operation 204
6 Summary and Conclusions 206
7 References 206

Chapter 14. **Techniques for Studying Aphid-Borne Virus Epidemiology**
(B. Raccah and M. E. Irwin) 209

1 Introduction 209
2 Detecting Plant-Pathogenic Viruses in Plant and Vector Hosts 210
2.1 Visual Inspection of Infected Plants 210
2.2 Laboratory Procedures for Detecting Viruses in Plants and Vectors 210
2.3 Indexing for Disease Progress in the Field 212
2.4 Indexing Incoming Infection in the Field 213
3 Sources of Plant-Pathogenic Viruses 214
4 Assaying for Vector Potential 214
4.1 Laboratory Assays 215
4.2 Field Live Assays 216
5 Vector Potential 217
5.1 Vector Abundance and Activity 217
5.2 Monitoring Movement and Host-Seeking Activity 218
6 Epilogue 220
7 References 220

Chapter 15. **Techniques in Quantitative Nematology**
(K. R. Barker and J. P. Noe) 223

1 Introduction 223

2 Nematode Community characteristics 223
2.1 Spatial and Temporal Population Patterns 224
2.2 Sampling 225
3 Role of Laboratory and Greenhouse Research 226
4 Use of Artificial Infestations in Fields 226
5 Utilizing Natural Field Infestation 227
5.1 Exploitation of Contagious Spatial Patterns 227
5.2 Means of Establishing Ranges of Population Densities 229
5.3 Following an Epidemic over Prolonged Periods 230
6 Modeling Approaches for Nematode-Induced Epidemics 231
6.1 Simple Population Models 231
6.2 Collecting Data to Derive Models 232
6.3 Formulating the Models 233
6.4 Validation 234
7 Epilogue 234
8 References 234

Chapter 16. **Use of Geophytopathological Information** (H.C. Weltzien) 237
1 Introduction: Definitions — Objectives 237
2 Data Collection 237
3 Data Documentation 239
4 Ecological Interrelationships 239
5 Forecasting 241
6 References 242

Chapter 17. **Monitoring and Mapping Long-Distance Spread of Plant Pathogens** (S. Nagarajan and Ajai) 243
1 Introduction 243
2 Factors that Facilitate Long-Distance Dispersal and Deposition 243
3 Methodology of Monitoring 244
3.1 Ground Surveys 244
3.2 Monitoring Dispersal from the Air 245
3.2.1 Monitoring the Synoptic Situation 245
3.2.2 Use of Weather Satellites 246
3.2.3 Validation of Disease by LANDSAT 246
3.3 Future Trends in Aerial Reconnaissance 247
4 References 248

IV The Synopsis of Experimental Results 251

Chapter 18. **Modeling Epidemics of Root Diseases and Development of Simulators** (C.L. Campbell, K.M. Reynolds, and L.V. Madden) 253
1 Introduction 253

2 Scope, Purpose, and Types of Models in Plant Disease Epidemiology 253
3 Systems Analysis in Plant Pathology 255
4 Simulation Modeling and General System Theory 255
5 Simulation Models for Root Diseases 257
5.1 Primary Infection 258
5.2 Colonization and Inoculum Production 260
5.3 Effects of Colonization on Plant Growth and Yield 261
5.4 Secondary Infection 262
6 Interfacing Epidemiological Models with Crop and Single Plant Growth Models 263
7 References 264

Chapter 19. **Modelling Epidemics of Polycyclic Foliar Diseases and Development of Simulators** (B. Hau) 267

1 Introduction 267
2 Modelling of Polycyclic Epidemics 267
2.1 Objectives of Modelling Epidemics 267
2.2 Steps in Modelling Epidemics 268
2.2.1 Restriction of Models 268
2.2.2 Approaches to Modelling 269
2.2.2.1 The Analytic Approach 269
2.2.2.2 The Synthetic Approach 270
2.2.3 Experimentation 270
2.2.4 Development of Models 271
2.2.5 Testing of Models 272
2.2.6 Implementation and Application of Models 273
3 Combination of Models 274
3.1 Combination of Crop Growth and Plant Disease Models 274
3.2 Combination of Epidemiological Models 274
4 Concluding Remarks 275
5 References 275

Chapter 20. **The Methodology of Comparative Epidemiology** (J. Kranz) 279

1 Introduction 279
2 Some Basic Aspects of Comparative Epidemiology 280
2.1 The General Framework 280
2.2 Terminology 280
2.3 Published Evidence as Data Base for Comparison 280
3 Comparative Experiments 281
3.1 The Scope 281
3.2 The Criteria 282
3.2.1 Comparison of the Pathosystems and Population Levels 282
3.2.2 Comparison of Epidemics, Their Behaviour and Patterns 283
3.2.3 Comparison of Soilborne Diseases and their Epidemics 284
3.2.4 Comparison at the Community Level 284

4 Tools and Procedures in Comparative Epidemiology 285
4.1 Data Reduction and Curve Approximation 285
4.2 Use of Coefficients and Multivariate Statistical Methods 286
4.3 Use of Computers in Comparison 287
5 Conclusions 287
6 References 289

Subject Index 291

Introduction

J. Kranz and J. Rotem

This volume is devoted to the description of experimental techniques but it is not merely a source book. The choice of methods described and the evaluation of their specific merits is influenced by the character of the various chapters and attitudes of their authors. Therefore, some chapters are relatively more factual, while others also include ideas that may lead to adaptations of a technique needed. In addition, various authors have personal preferences or opinions in relation to specific techniques or approaches. These differences are evident in chapters in which similar techniques are mentioned by different authors. For instance, Rotem in Chapter 2 and Madden and Ellis in Chapter 13 describe methods used in infection tests in growth chambers. Rotem, apparently spoiled by having many chambers at his disposal, does not describe a situation where two chambers only had to suffice for testing the effect of several temperatures. For Madden and Ellis this is apparently a situation actually encountered and they describe ways of meeting such a situation. In this and other cases, different approaches complement each other. For instance, Campbell, Reynolds and Madden describe methods and approaches to modelling root diseases (Chapter 18) and Hau complemented the subject by discussing modelling of foliage diseases (Chapter 19). Discussing experimental procedures in the field, Aust and Kranz (Chapter 1) refer the reader to experiments made under controlled conditions, while Rotem in Chapter 2, describing experiments made under these conditions, stresses the necessity for field work. All these individual approaches and treatments are accepted as long as no undue overlaps – and some gaps – occur. Also, no particular attempt is made to unify the terminology used by authors in the book. We even have left terms used unaltered which are inappropriate in plant pathology. An example is colonization (for incubation or pathogenesis), a purely mycological term with no notation of disease and damage (fungi may colonize inorganic and organic matter, dead or alive).

Descriptions of various techniques have been arranged in four sections I to IV with self-explanatory titles. Section I, General Techniques, starts with Chapter 1 by Aust and Kranz, who discuss some techniques and approaches used in the field, where the epidemics occur and, as mentioned before, advise the researcher to conduct part of his study under controlled conditions. In Chapter 2, Rotem describes the techniques used under controlled conditions, e.g. pertaining to certain causal relationships, and discusses their role in studies in an open habitat, stressing the importance of keeping the field in mind when planning a controlled-condition study.

In Section II on Measurements and Their Analyses, Chapter 3 by Kranz deals with the highly important and often inadequately performed measuring of diseases. Kranz discusses the relative advantages of measuring severity versus measuring incidence, and sampling, to ensure the accuracy and precision to be aimed at with such measurements.

Some specific applications of disease assessment are referred to. He also stresses the need of measuring the crop itself, a topic further discussed by Seem in Chapter 4. The reason for Seem's treatise is the dynamic interaction between host and pathogen. The chapter discusses the relative virtues of various methods of measuring the host: direct measurements of its foliage, evaluations based on phenological events or derived from special simulation programs, etc. Seem also discusses special problems connected with measuring perennial plants.

An entirely different aspect of measuring is treated by Rotem in Chapter 5. It deals with evaluating the amounts of inoculum, its dispersal, infectiousness and survival. These parameters are most difficult to quantify and in many studies they are but roughly estimated. The techniques described in this chapter do not eliminate the need of estimation but, with careful interpretation of results, might reduce the error. This chapter, emphasizing aspects of above-ground diseases, is followed by Chapter 6, written by Gilligan. It is devoted to techniques used in measuring distribution of soilborne diseases and their inoculum. Gilligan discusses cases in which the inoculum is difficult to distinguish or count and of diseases, the distribution of which is regular, random or clustered. For theses diseases it is most important to adopt the most suitable method of collecting, plotting and analyzing data, and various approaches to these tasks are discussed.

The next two chapters in this section deal with meteorology. In Chapter 7 Sutton, Gillespie and James describe the recent achievements in measuring weather factors with electronic instruments. Discussing the action and the use of thermistors, thermocouples, artificial leaves and other sensors, the authors provide the reader with a choice for selecting the proper instrument for measuring microclimatic variables. They also discuss the advantages of monitoring the weather inside and outside the experimental plot. In Chapter 8 the analysis of the measured weather factors is discussed by Friesland and Schrödter. They evaluate the usefulness of biometerological parameters and the most adequate techniques for their derivation and use. As agricultural meterologists with great experience in plant disease epidemiology, they reconcile aspects from either side and thus supplement the preceding chapter by ensuring professional standards in handling weather data.

Section III deals with Special Topics not necessarily connected with each other, but all dealing with definite objectives in epidemiological experimentation. The first two chapters are on disease control. In Chapter 9, Berger deals with analysis and presents, the results of control experiments. Included in his chapter is a computer program for the calculation of the area under disease progress curve. Chapter 10, written by Dekker, is devoted to experiments for the detection of pathogens' resistance to fungicides and their analysis. Dekker stresses that only the employment of generally accepted methods makes it possible to compare results obtained by various researchers. Leading the reader step by step through the subject of fungal resistance, he describes methods for defining resistant strains and measuring their behaviour with and without the presence of fungicides. In a very similar approach Chapter 11 by Welz proposes concepts and methods to study resistance of plants and the effects of host-plant resistance on the dynamics of epidemics and the selection processes involved.

Another special topic of experimentation, treated by Teng and Johnson (Chapter 12), are studies on the impact diseases, and their epidemics, have on the yield of crops. Among other aspects of loss assessment, they discuss the problems in the estimation of losses caused simultaneously by several diseases in large areas. They claim that the only realistic way to understand and predict the effect of such diseases on yield is by coupling disease and crop models. Coupling of such models is also mentioned in Chapter 3, 4, 18 and 19.

Chapter 13 by Madden and Ellis logically follows the one on loss assessment: Forecasts are based on (at least some concept) of economic damage thresholds, which, in turn, have to have damage coefficients (loss equivalents) as an essential component. The authors describe methods employed in the developing of forecasts. Stressing that a usable forecast must be as simple as possible and thoroughly validated, the authors discuss the value of some of its parameters and the ways of quantifying their level in experimentation. Although preferring forecasts derived from experimentation, Madden and Ellis state that empirical systems may be as functional as more elaborate models.

The following two chapters of Section III deal with epidemics caused by viruses and nematodes. In Chapter 14, Raccah and Irwin discuss experiments due to the complexity of viral epidemics and the methods of their research. They stress, amongst other topics, the identification and the influence of vectors. Although the chapter deals only with the epidemiology of virus diseases transmitted by aphids, it contains useful hints also for studying epidemiology of other virus diseases.

Problems similar in part to experiments with soilborne diseases (Chapter 6) are discussed by Barker and Noe in relation to the epidemiology of nematodes (Chap. 15). Among other aspects, the authors discuss peculiarities of sampling and specific merits of experiments performed in various habitats, e.g. laboratory, greenhouse and field. Barker and Noe advise how to lay down an experiment with nematodes and establish various population densities. They further describe procedures for modelling of nematode epidemics which relates to Chapter 18.

Amongst the special topics for experimentation in epidemiological research there are finally two chapters, which offer methods for studies on the macroscale. In Chapter 16 Weltzien describes the collection, documentation and use of geophytopathological information, and the relation of the collected data to ecological conditions. This is followed by Chapter 17 in which Nagarajan and Ajai review techniques used for evaluation of the long-distance spread of diseases. A quantitative assessment of such dispersal is difficult if possible at all. Therefore, the evaluation is done by a number of indirect methods: ground surveys of the diseases, the assumed dispersal of spores in specific synoptic situations and indications provided by weather and resource satellites.

Section IV of the book. 'The Synopsis of Experimental Results' is devoted to modelling and the use of experimental data in comparative epidemiology. With these concluding chapters the book refers to the overriding goal of epidemiology. Not the mass of data derived from experiments will justify its research, but rather its ability to formulate general rules, concepts and principles applicable to both the explanation of phenomena, and the design of advanced control tactics and strategies. Models and comparison are important research tools to these ends. Hence, in the following chapters authors warn that data collected from the literature may not be appropriate for modelling and advocate experimentation. They explain how experimentation and mathematical modelling interact and depend on each other.

In Chapter 18, Campbell, Reynolds and Madden discuss modelling of root diseases, and in Chapter 19 Hau deals with modelling of foliage diseases. The two chapters complement each other. Campbell et al. discuss the virtues of simplicity versus complexity in models, describe techniques used in developing models of root diseases, and problems of interfacing epidemiological models with those of crop development. Hau likewise advocates the combination of disease and crop simulators. He also discusses the synthetic versus the analytic approach to building the models.

Finally, in Chapter 20 Kranz discusses how, in the framework of comparative epidemiology, the methods described in the preceding chapters can be employed to conceptualize in a meaningful way. He presents comparative epidemiology as having analytic and synoptic functions in the search for general principles in the epidemiology of plant diseases, and in the test of the validity of proposed hypotheses and theories.

I General Techniques

Chapter 1

Experiments and Procedures in Epidemiological Field Studies

H. J. Aust[1] and J. Kranz[2]

1 Introduction

A major objective of epidemiology is to study plant diseases and their population dynamics in natural environments. Consequently, the field is the basic laboratory of the epidemiologist (Kranz 1972). Indeed, in order to attain quantitative data on the components of the disease triangle or square, the use of field experiments is indispensable (Aust and Bretschneider-Herrmann 1982). But the study may be accelerated when supplemented by the use of greenhouse, growth chamber and computer facilities.

We can divide epidemiological field experiments into the following two main groups: (1) those established for studying the effect of as many variables as possible (holistic) of host, pathogen and environment on disease development and (2) those in which only few key variables are studied to explain most of the effects governing the epidemic (meristic). The decision whether to employ the holistic or the meristic approach depends not only on the purpose of the study and/or the research facilities, but also on the financial, instrumental and staff situation. In some cases the holistic approach may have the character of a highly complex experiment if a problem arises which cannot be solved by one research worker alone. The epidemiologist then may, for instance, join with a biologist and an agrometeorologist to conduct the experiments. The holistic approach can also be the more suitable method if a plant disease has been recently discovered or has not yet been well studied. But this approach may then have the character of an experiment in which only the key variables of the larger number checked are screened. An experienced epidemiologist may also be able to reduce the whole number of variables a priori to only a limited number of variables which need to be studied. In both the latter cases the holistic approach becomes a meristic one.

Single-plot experiments, either incomplete (meristic) or complete (holistic), are suitable for studies within the disease triangle (host-pathogen environment). If, however, effects of human interference on the pathosystem are investigated within the disease square of adequate size, then for each different treatment (fungicides cultivars, plant density, etc.), at least one single plot is needed for each treatment and has to be equipped accordingly. Once hypotheses have emanated from such single-plot experiments, standard designs are necessary to test them.

[1] Institut für Mikrobiologie, TU Braunschweig, Mendelssohnstr. 4, D-3300 Braunschweig, FRG
[2] Phytopathologie und Angewandte Entomologie, Tropeninstitut, Justus-Liebig-Universität, D-6300 Gießen, FRG

Standard design (e.g. randomized block) field experiments are chosen to test various hypotheses, for instance, for the development of forecasts, for studying the effect of resistance genes or for checking the effect of fungicides on the dynamics of an epidemic. In these cases only few variables are usually measured and must components of the disease square may be ignored.

Finally, it should be noted that in this chapter only experimentation with air borne diseases is considered. Campbell et al. (Chap. 18) and Gilligan (Chap. 6), both in this volume, refer to aspects of field experiments with soilborne diseases.

In addition, practically all the following chapters have some relevance for field experimentation and its combination with a controlled environment.

2 Planning and Conduct of Field Experiments

For the planning of field experiments as described above, the first step is to collect all available, relevant information of the following categories:

1. Information that consolidates the hypothesis and defines the objective adequately.
2. Information that helps to select the relevant variables, their dimensions, etc.
3. Information about methods and instruments needed for the experiment, including those which must be adapted or newly developed.

The next step should be the experimental design and the operational plan for conducting and evaluating the experiment. During this stage, field record forms will have to be designed and tested for both data acquisition and compatibility with recording and evaluation (see Sect. 2.4).

Quality and validity of the results obtained in a field experiment depend on whether or not the planning was thorough and consistent with the objective. Then, it is important to consider that field experiments last longer, are often laborious and costly and are not as easily repeated as are many growth chamber or laboratory studies.

Field experimentation as described here requires continual measurement and observation in the plot. Most meteorological data should be recorded hourly (see Sutton et al., Chap. 7). From such records all other meteorological parameters or variables can be derived. Once planned, all measurements must be conducted strictly according to a time schedule. Various methods of spore trapping may be employed (see Rotem, Chap. 5). Data that is conveniently stored on magnetic tape or discs may later be utilized for any particular need. It is essential that measurements prone to subjective error, like visual disease assessment, always be done by the same person, preferably by the researcher himself (see Kranz, Chap. 3). For the measurement and recording of the host properties (see Seem Chap. 4). After the experiment is terminated an immediate, though often preliminary, evaluation of the collected data should follow in order to improve the design, methods, instrumentation and variables (number and kind) to be used in the next experiment. This implies that a check should be done to see whether the results fit the hypothesis or whether altered field experiements should be conducted, with or without auxiliary studies in the laboratory, greenhouse and growth chamber (see Rotem, Chap. 2). Depending on the results, the hypothesis may also be reconsidered and modified, if necessary.

2.1 Formulation of the Hypothesis and Selection of Variables

Information and concepts needed to formulate the hypothesis on which the experiment is based may come from the literature, a priori considerations, observations, pilot trials or from previous greenhouse, growth chamber or laboratory experiments. Often a discussion with experienced colleagues may be helpful. If it is difficult to formulate the hypothesis clearly, a preliminary or pilot experiment is the appropriate approach. In this case as many variables as possible are included, since at the beginning of the experiment one may not know the most relevant variables (see Mogk 1974). The great number of variables can be accomodated by adequately designed record forms as well as computers, and can be analyzed with a number of statistical methods, e.g. multivariate analysis (Butt and Royle 1974). For the definition of variables to be used in holistic field experiments we also refer to Butt and Royle (1974). With meristic field experiments the choice and definition of variables is less difficult, since as a result of preceding experiments the researcher has clearer ideas about which additional variables should be incorporated. He, however, must make the decision early in the project not only on which variables to choose, but also on their precise definition and method of measurement. More often than not, the practical accessibility of the variable and its suitability for future monitoring may be more important in determining its selection than its actual biological relevance.

2.2 Single-Plot Field Experimentation

Most field experiments involve one or more plots of the host crop under study and a varying number of instruments. The size and shape of experimental plots varies with the subject being specifically studied. For instance, much larger plots are needed to study the effects of cultural factors (e.g. methods of irrigation) than to study the impact of specific cultivars on race dynamics (see Chap. 10 and 11).

Strict instructions must define the manner in which regular, precise and possibly numerous measurements are to be conducted in order to ensure that essential variables of host, pathogen, disease and, sometimes, human interference are appropriately measured as much under natural conditions as possible. All measurements should be done consistently at a specific time and under the same conditions. Randomization, replication, proper sample size, etc. must be ensured within the plots. An example for such a design is given in Fig. 1.1 which deals with studies on the coffee-rust pathosystem in Kenya (Becker 1976). When appropriate, trap plants (see Sect. 3.1.2) should be used and placed in the plot, either randomly or systematically.

A sufficiently large plot size should ensure that as many sampling units as needed are present and that they are free from interference of border effects, undesirable disease gradients (e.g. interplot interference; see Paysour and Fry 1983) and irregular microclimatic factors, etc. The plot should be representative of all aspects that are to be studied and be subject to regular disease development. Instruments needed for meteorological measurements may be of the standard type, e.g. thermohygrographs, or may be based on computerized electronic devices (see Sutton et al., Chap. 7). The kind of equipment employed depends on the objective of the project and financial means. Financial constraints, however, must not always be a disadvantage, since they may stimulate inventive scientists to create new solutions with simple but effectively working instruments.

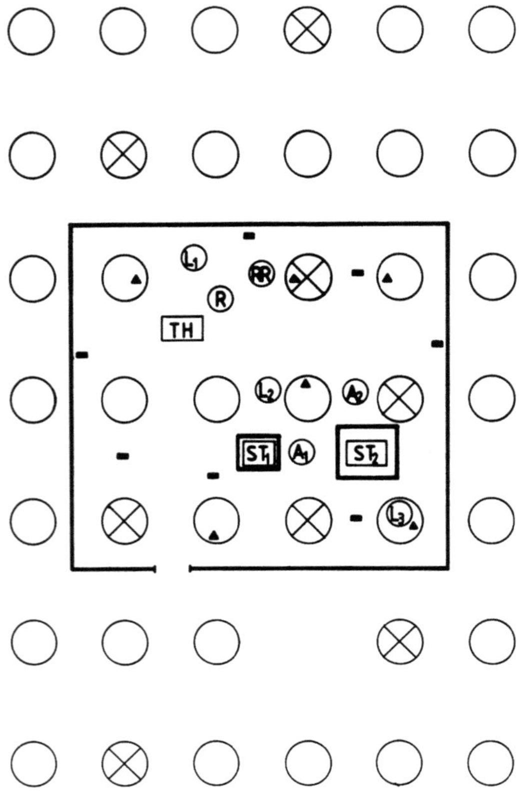

Fig. 1.1 Arrangement of instruments for conducting epidemiological experiments in the pathosystem of coffee rust in Kenya (Becker 1976). *Abbreviations: ST_1* Spore trap 1.25 m above the ground; *ST_2* spore trap 2.50 m above the ground; *A* anemometer; *TH* thermohygrograph; *R* rain gauge; *RR* rain gauge with recorder; *L* leaf wetness recorder; ■ slide trap; ▲ rain water trap; *X* trees with tagged branches for recording disease and host development

2.3 Sampling in Single-Plot Experiments

Sampling techniques (see also Chap. 3 and 4) have to ensure representative measurements at a minimum cost while attaining the precision required in all types of field experimentation. Whether sampling should be randomized systematically or stratified depends on spatial patterns of pathogen distribution and also on the growth habit of the host. If the pathogen is distributed uniformly and host growth is also uniform, a randomized selection of sampling units is appropriate. If one or both of them are not uniform a stratified sampling or a systematic determination of typical plants as sampling units is preferable.

Phenometric measurements and observations on the host are made using sampling units, tagged or untagged. They may include every newly formed leaf, internode, etc. Tagging ensures a follow-up of host and disease development throughout the lifetime of the host plant. In some studies this is essential to minimize experimental error. In others it may be preferable to avoid interference with ongoing host development. In such cases one should randomly or systematically select new sampling units on every date of recording, particularly if sampling units have to be removed for measurements in destructive sampling.

Of particular importance are sampling intervals. At least five measurements should be strived for within disease prone periods (i.e. during the epidemic). Also velocity of disease progress during the progressive phase of the epidemic may lead to rather short intervals between measurements (see also Berger, Chap. 9).

Normally, with well-organized and consistent sampling and recording, a large quantity of data can be obtained with comparatively small investment of time.

2.4 Data Acquisition and Handling

Holistic field experiments and in many cases meristic field experiments, as well, also tend to yield bulky and complex data. Hence, when planning an epidemiological field experiment attention should also be given to this aspect. A biometrician may be consulted and field record forms may be designed to facilitate recording of data. As data acquisition with hand-held computers becomes a more common procedure, appropriate and previously tested programs must be available before the onset of field experiments.

The decision also has to be made where to record: in the field or (with destructive sampling) in the laboratory. The latter is usually more convenient but may lead to greater experimental error than recordings of the same tagged plants at each date in the field.

Recording of data can be done with record sheets with or without an assistant (who does the writing) or using a dictaphone. All these procedures require subsequent recording, which may prove to be a bottleneck of epidemiological field work. Record forms should be designed to facilitate recording and leave space for entries, which may be coded. Figures as entries are preferable to letters or signs. Record forms should permit some flexibility to allow recording of variables which were not conceived before the beginning of the experiment. For details of the design of field record forms see Mogk (1974).

Sometimes preliminary measurements, e.g. length and width of lesion size, must be noted. Later, after calculating lesion size with a special function, these values are eventually entered. With hand-held computers or, where applicable, data acquisition with microcomputers (e.g. graphic tablets, image analysis), this step can be saved and thus the procedure becomes less error prone. The disadvantage of this 'mobile' data acquisition can be limited flexibility for new or unusual events. Also with hand-held computers it may be more efficient when two people record the data, one for doing the measurements and the other for recording. For the handling and evaluation of data a number of commercial programs (e.g. MSTAT, SPSS, dBASE III, Plotit) are also available for microcomputers. Some are quicker for recording, others are more feasible for corrections and rearrangement of data.

Correction of data, or at least checks on plausibility and completeness are important operations before data are statistically processed. This can be facilitated by identical format and coding amongst the various data sets of a research program.

Data evaluation can be executed in four steps. First, the variables are described with their means, standard deviations, minima, maxima, etc. and tested for normality, homogeneity of variances, etc. Second, tabulations, scattergrams or similar methods are used to obtain a general view of possible relationships and results. Third, data is transformed, and fourth, the appropriate suitability of data and multistatistical methods is chosen. Multivariate methods (e. g. multivariate linear regressions, discriminant analysis), non-linear methods and meaningful non-parametric methods are of particular interest to the epidemiologist. For details, we refer to modern books on statistics and some of the following chapters in this volume (Madden and Ellis, Chap. 13 and Berger, Chap. 9, for the analysis of disease progress curves; Friesland and Schrödter, Chap. 8, for the analysis of weather factors; and Teng and Johnson, Chap. 12, for interplot interference).

2.5 Experimental Error

As is common in scientific experiments we may have experimental errors deriving from a variety of sources, e. g. edaphic, topographic or nutritional factors. When dealing with biological objects it is mainly the variability of single individuals which causes errors. These occur if samples are not properly chosen or if there is an insufficient number of replicates, or an uneven distribution of disease.

Van der Plank (1963) coined the term "representational error" to describe the error when plots differ from the fields which they are meant to represent. Representational errors occur in most standard designs, experiments which include untreated plots. For example, the development of powdery mildew in sprayed cereal plots is much affected by the proximity to unsprayed crop (Jenkyn and Bainbridge, 1974). Differences in disease gradients between similarly treated plots can mainly be related to the variance of wind direction and the treatments that affected the neighbouring plots (e. g. Jenkyn et al. 1979). In planning an experiment with adjacent but differently treated plots (e. g. with different cultivars or fungicides), the researcher should anticipate various interference phenomena: a positive interference that may lead to an underestimation of treatment effect (Van der Plank 1963), or a negative interference that may lead to an overestimation of the treatment effect (James et al. 1976). Problems with such errors arise in any field experiment in which there is a large variability in the amount of disease. According to James et al. (1976), it is easier to make experimental designs to exclude positive interference than those to exclude negative interference (Chap. 12). In order to reduce negative interference in the reference plots (i. e. to limit inoculum loss), the experiment would have to be increased to such a size that it could not be accomodated in a normal-sized experimental area. Therefore, Jenkyn (1977) proposed that if land area is limited, the problems might be solved by using systematic rather than random designs. In this context it is worthwhile to mention that guard rows between plots also have a minor effect on the transfer of inoculum from plot to plot (also, see Welz, Chap. 11).

According to Burleigh and Loubane (1984), one can conclude that optimal plot size in epidemiological field experiments is a function of the spore-dispersal characteristics of the pathogens used. Therefore, the area for unrestricted development of a pathogen with wind dispersed spores is larger than the area needed for a pathogen disseminated by splash dispersal or even soilborne pathogens. Only a separation of the plots by at least a plot width of disease-free crop reduces the transfer of inoculum (Jenkyn 1977). Recently, Paysour and Fry (1983) developed a model for planning field experiments with aerially disseminated pathogens. This model can be used to select appropriate plot size and spacings.

3 Combination of Field and Growth Chamber Experiments in Studying Epidemiological Problems

Holistic and meristic field experiments as described in the foregoing sections may have one disadvantage, and that is the complexity of interactions between host, pathogen, environment and man. This complexity is further complicated by the variability of the host's susceptibility, the variability of the pathogen's virulence and the variability of

Table 1.1. Advantages and disadvantages of field and growth chamber experiments

Advantages of field experiments	Disadvantages of field experiments
1. Field experiments ensure better understanding of regulatory and interactive processes of entire epidemics. 2. Information derived from phenomena that result from interactions between host and pathogen populations under natural conditions are more relevant for the estimation of parameters and therefore applicable in the sense of comparative epidemiology to similar events of other epidemics. 3. Selection of the acting cultural, key environmental and disease factors is less arbitrary, reducing the danger of deducing artifactual conclusions. 4. Results of field experiments may be directly used for practical solutions. 5. Field experiments normally require low capital investment.	1. Results of field experiments are often descriptive and yield bulky data. Therefore, computer facilities are necessary for handling data and evaluation is needed to pinpoint the key factors. 2. The causal relations between host, pathogen and environment, which govern the course of an epidemic, are often difficult to determine because too many factors act simultaneously. 3. From the results obtained from field experiments in one season it is sometimes difficult to judge how the disease may behave in another season, since levels of inoculum, susceptibility of the host and climatic conditions may vary. 4. Considerable work and money is often wasted when the disease fails to appear. 5. Repetition of field experiments often has to be postponed to the next year.
Advantages of growth chamber experiments	**Disadvantages of growth chamber experiments**
1. Growth chambers enable the manipulation of many factors when running experiments simultaneously in several chambers, thus elucidating the causal relationships between the elements of the studied patho-system. 2. The manipulated factors include not only meteorological factors, but also controlled, dose-dependent influence such as inoculum or different host nutrition as well as temporal factors such as duration of moisture periods. 3. Most growth chamber experiments are quickly completed and repeated. 4. Execution of experiments is not affected by the outdoor season and by the availability of crop and disease.	1. Experiments are related to the reaction of individual plants and small pathogen populations. 2. Peculiarities of crop microclimates are missing because of the attributes of growth chambers. Therefore, each growth chamber represents a special habitat, but is also more likely to represent an artefact. 3. Results of growth chamber experiments must be checked with respect to their application to field conditions. 4. Growth chambers require high initial investment and later additional costs for their maintenance.

Table 1.2 Application of epidemiological studies conducted in the field and in growth chambers

Studied in the field	Studied in growth chamber
1. Epidemic patterns as affected by edaphic and topographic conditions. 2. Effects of sowing dates, cultural practices, control treatments and age-conditioned host susceptibility on disease development. 3. Patterns of disease development in host populations as influenced by climatic conditions. These include the effect of rain, wind and spells of unusually hot, dry, cool and humid weather. The influence of these meteorological events on host development, formation of microclimate in the crop and buildup of inoculum, as well as on the susceptibility of cultivars which in turn depend on resistance genes and age. 5. Study of special epidemiological processes, e.g. spore dispersal, buildup of inoculum and spatial spread of disease, not possible in growth chambers. 6. Interactions such as competition or compensation phenomena with other diseases or pests evaluated frequently in non-quantitative ways. 7. Development and verification of forecasting criteria, simulation models and results obtained from studies in controlled environments. 8. Creation of ideas to be studied in growth chambers.	1. Only age-conditioned, susceptible growth stages of the host suit the growth chamber experiment. The results obtained are quantitative but may not represent the patho-system in the field when the nutritional age-conditioned susceptibility of the host is also influenced by the metabolic sink of the fruiting organs. 2. Possibility for assessing the influence of the factors of temperature, humidity and dew on individual plants or plant organs of various ages during phases of infection, colonization and sporulation; but excluding the effects of quantity or differences of rain, radiation and wind in the field. Elucidation of discrete factors which in the field interact with each other (e.g. temperature, radiation, darkness, rain, humidity, dew, luxurious development of the plant canopy, microclimate). 3. Testing the effect of pathogen races including those not yet present in the field. 4. Screening for the susceptibility of cultivars, most easily performed with small plants (e.g. seedlings), to one or several races of pathogen. 5. Verification of interactions like competition or compensation phenomena occurring within the studied pathosystem. 6. Establishment of data and parameters needed for building simulation models. Verification of phenomena observed in the field but that are not well understood. 7. Creation of ideas to be studied in the field.

meteorological factors. For example, in the field nearly all meteorological factors act simultaneously on the host as well as on the pathogen. Since these factors are also intercorrelated, it is sometimes difficult to judge which individual factor exerts the greatest influence on the epidemic. Here, growth chambers (Chap. 2) enable one factor to be varied while the others are kept constant. However, a growth chamber represents only another habitat and plants as well as pathogens may adapt themselves to the conditions of this habitat (Aust and Bretschneider-Herrmann 1982). Growth chambers have proved to be useful in parameter identification (see also Chap. 13), but estimation of parameter values must be confirmed by field experiments if not done before. To solve this dilemma a combination of these two experimental methods (field/growth chamber) is feasible. This approach unites in a convenient way the advantage of controllable environmental conditions obtained with the growth chamber with the advantage of acquiring relevant quantitative data with the field experiment. In order to clarify this subject the main applications of growth chamber and field experiments with their advantages and disadvantages are summarized in the Tables 1.1 and 1.2. More detailed and technical aspects will be dealt with in Chapter 2.

3.1 Examples for the Combination of Field and Growth Chamber Experiments

Epidemiologists may approach their problems with combined field and growth chamber research facilities intuitively. We here propose an approach which, in the future, will help epidemiologists to tackle epidemiological problems in a more consistent way.

3.1.1 Manipulating the Pathogen

Reproducing the pathogen normally does not require much space. This can be done with facultative parasites in petri dishes in the laboratory or for obligate parasites on living plants in a controlled environment. The inoculum produced in such a way can be easily transferred to the field to inoculate the host plants growing in a natural environment.

It depends on the purpose of the experiment as to whether the scientist inoculates whole plants or single organs of the plants in the field. For checking latent periods or lesion growth it is, for instance, worthwhile to mark and inoculate single leaves in different positions, or if information is necessary about how a specific level of disease creates a certain yield loss, one can spray several plots in the field with defined levels of inoculum. One prerequisite is that the disease does not yet exist in the crop or only at very low levels. Another prerequisite is the existence of favourable weather conditions.

This approach has several advantages as indicated in Tables 1.1 and 1.2. First, the researcher has no uncertainties about whether or not the disease will appear in this season; second, the scientist really knows exactly when the pathogen reached the leaf surface with a defined dose of inoculum on the plants in the field or on single plant organs. Hence, the researcher obtains more detailed information concerning the pathogen (e.g. infection, incubation periods, sporulation) in relation to host development and meteorological factors in the field.

3.1.2 Manipulating the Host — The Use of Trap Plants

In the same way in which the pathogen is produced in the laboratory or in a growth chamber, it is possible to grow host plants in a greenhouse or a growth chamber and to transfer them later to the field. For instance, every trap-plant technique uses this approach. Royle (1978) utilized this technique for a daily check of leaf infections of hop powdery mildew in a hop garden, while Aust et al. (1983) used the same method for counting the number of barley powdery mildew leaf infections in the field. Trap plants are also useful for checking incubation periods of airborne pathogens (Jenkyn 1973; Aust 1981; Aust et al. 1978). In all these cases the introduced host plants are submitted to natural spore dispersal and the natural environment of the crop. In this manner the researcher obtains information about dose-dependent influences like the number of infections per cm^2 leaf surface or time factors like incubation periods.

An additional step would be to grow host and pathogen separately in growth chambers, inoculate the host plants in the laboratory and immediately transfer them to the field. Such a design was chosen to check the number of infections on plants in the shade as compared to those that developed under natural meteorological factors in the sun. In this case host and pathogen were manipulated and only the variability of the natural environment acted on the exposed plants. These few examples may suffice to again demonstrate the benefits of combining field and growth chamber experiments.

3.1.3 Manipulating the Environment

Environment commonly is the only one of the three components of the disease triangle that is not always modifiable in the field. However, in some studies exactly this, the change of the environment of host or pathogen in the field, must be monitored. In this case host development, dispersal patterns of airborne pathogens and their development in the field may be affected. To check for these factors not only measurements of the meteorological factors are necessary, but also the use of trap plants is indispensable.

For instance, barriers caused by vertical screens or rows of bushes or trees may change the microclimate behind them, depending on wind direction and sun irradiation. Deposition of spores behind such barriers can be much higher as compared to deposition in the open field (Schrödter 1952; Kranz and Aust 1979), whereby higher numbers of deposited spores on the plants behind the barrier together with a more favourable microclimate can lead to higher disease development on the plants in the immediate vicinity of the barrier. Another approach for changing the natural environment would be to use horizontal screens as are sometimes utilized to cultivate tobacco (Waggoner et al. 1959). Such screens may prevent wetting of the leaves by dew or short periods of rain. In the case of powdery mildew this results in better conditions for spore dispersal and a favourable microclimate on the plants under the screen, finally leading to a higher infection rate of the plants under the screen (Kranz and Aust 1979). This approach is another example in which combining field and growth chamber experiments enables the researcher to obtain information on the influences of changed meteorological factors, as well as on how changed microclimate affects disease development in the field.

3.2 Conclusions

Studies conducted using a combination of field and growth chamber facilities are a compromise between growth chamber and field experiments. The combined field-growth chamber approach is designed to reduce the variability of one or two components of the disease triangle, thus leading to experimental results of field experiments which are easier to interpret than those of conventional field experiments. The combined approach also gives quantitative data and/or better insight into the causal relationship of a pathosystem. The proposed designs can be modified in many other ways, depending on the objective, the pathosystem used and the ingenuity of the experimenter.

4 References

Aust HJ (1981) Über den Verlauf von Mehltauepidemien innerhalb des Agro-Ökosystems Gerstenfeld. Acta Phytomed 7; 76 pp.

Aust HJ. Bretschneider-Herrmann B (1982) Nutzen und Grenzen biologischer Forschung unter kontrollierten Umweltbedingungen mit besonderer Berücksichtigung phytopathologischer Fragestellungen. Angew Bot 56: 201–217

Aust HJ, Hau B, Mogk M (1978) Wirkung von Temperatur und Konidiendichte auf die Variabilität der Inkubationszeit des Gerstenmehltaues. Z Pflanzenkr, Pflanzenschutz 85: 581–585
Aust HJ, Hau B, Kranz J (1983) EPIGRAM-a simulator of barley powdery mildew. Z Pflanzenkr, Pflanzenschutz 90: 244–250
Becker SM (1976) Untersuchungen zur Verbreitung der Uredosporen von *Hemileia vastatrix* Berk. und Br. in Kenia. Ph D thesis Univ Gießen
Burleigh JR, Loubane MJ (1984) Plot size effects on disease progress and yield of wheat infected by *Mycosphaerella graminicola* and barley infected by *Pyrenophora teres*. Phytopathology 74: 545–549
Butt DJ, Royle DJ (1974) Multiple regression analysis in the epidemiology of plant disease. In: Kranz J (ed) Epidemics of plant diseases. Springer Berlin Heidelberg New York, pp. 78–114
James WC, Shih CS, Hodgson WA, Callbeck LC (1976) Representational errors due to interference in field experiments with late blight of potato. Phytopathology 66: 695–700
Jenkyn JF (1973) Seasonal changes in incubation time of *Erysiphe graminis* f.sp. *hordei* Ann Appl Biol 73: 15–18
Jenkyn JF (1977) Interference between plots as a source of error in experiments with *Erysiphe graminis* Pestic Sci 8: 428–429
Jenkyn JF. Bainbridge A (1974) Disease gradients and small plot experiments on barley mildew. Ann Appl Biol 76: 269–279
Jenkyn JF, Bainbridge A, Dyke GV, Todd AD (1979) An investigation into inter-plot interactions, in experiments with mildew on barley, using balanced designs. Ann Appl Biol 92: 11–28
Kranz J (1972) Einige Voraussetzungen für die Planung und Durchführung von Feldversuchen in der Epidemiologie. Z Pflanzenkr, Pflanzenschutz 79: 573–581
Kranz J, Aust HJ (1979) Schatten als epidemiologische Einflußgröße beim Gerstenmehltau. Z Pflanzenkr, Pflanzenschutz 86: 533–545
Mogk M (1974) Automatic data processing in analysis of epidemics. In: Kranz J (ed) Epidemics of plant diseases. Springer Berlin Heidelberg New York, pp. 55–77
Paysour RE, Fry WE (1983) Interplot interference: a model for planning field experiments with aerially disseminated pathogens. Phytopathology 73: 1014–1020
Royle DJ (1978) Powdery mildew of hop. In: Spencer DM (ed), The powdery mildews. Academic Press, London, pp 381–409
Schrödter H (1952) Untersuchungen über die Wirkung einer Windschutzpflanzung auf den Sporenflug und das Auftreten der *Alternaria*-Schwärze an Kohlsamenträgern. Angew Meteorol 1: 154–158
Van der Plank JE (1963) Plant diseases: epidemics and control. Academic Press of London
Waggoner PE, Boyd AB, Reifswyder WE (1959) The climate of shade. Conn Agric Exp Stn New Haven Bull

Chapter 2

Techniques of Controlled-Condition Experiments

Joseph Rotem[1]

1 Introduction

When cultivation of specific crops was expanded beyond their traditional habitats, diseases, previously thought to be confined to specific environments, were found to appear also in the new, apparently adverse habitats. This phenomenon usually results from intensive cropping techniques which render originally adverse habitats more favourable to disease development. However, experiments under controlled conditions (CC) showed that some pathogens possessed hitherto unknown potentials that helped them to thrive under the new conditions. For instance, CC tests demonstrated that development of *Phytophthora infestans* is facilitated by dryness at a specific stage of its life cycle (Rotem and Cohen 1974). Other CC tests revealed that some pathogens which had been known to require prolonged wettings for infection succeeded in semi-arid habitats due to their ability to use several short wettings interrupted by dry intervals (Bashi and Rotem 1974). Only CC tests supply data on temperature and wetting requirements in discrete phases of the pathogen's life cycle. Data from CC experiments assist simulation and forecasting programs and rationalize approaches to chemical and cultural control. Such contributions of CC experiments are well known und their merits need no further elucidation. Therefore, instead of repeating the advantages of studies in CC, we stress in this chapter the shortcomings encountered in such studies.

Controlled conditions may be created in greenhouses, but due to the latter's size and construction, the conditions are less accurate than in smaller instruments. CC instruments attempt to simulate environmental conditions in the field. Technology permits simulation of all field parameters, but the investment required, limits the instruments available to simulation of some parameters only. Factors like radiation, wind and microclimate differ in CC and in the field. For this and other reasons CC experiments are valid in the study of some problems but not of others. They indicate what *may*, but not necessarily what *must* happen in the field. Therefore, as stressed frequently throughout this chapter, results obtained under CC should be verified, or at least supplemented, by field trials (see Aust and Kranz, Chap. 1).

[1] Department of Plant Pathology, Agricultural Research Organization. The Volcani Center, Bet Dagan 50250, Israel

2 The Controlled Habitat

Construction and operation of the controlled-condition instruments are described by the makers. Here, we point to some problems which are not usually mentioned in commercial pamphlets.

2.1 Light

Due to the problem of heating by incandescent lamps, most growth chambers are equipped with fluorescent bulbs. Some brands of fluorescent lamps emit also the red part of the spectrum, but they are expensive and most researchers prefer to use the regular, red-deficient fluorescent tubes in combination with incandescent lamps that provide the red. Light intensity in growth chambers is sometimes still measured in foot candles, despite the irrelevance of this measure to most biological phenomena in plants. It is advisable to express light intensity in more appropriate units such as micro-Einsteins that measure the visible part of the spectrum. Even in the most sophisticated growth chambers light intensity is weaker than in the open and interferes with the conduct of specific trials. In less sophisticated instruments, present in most laboratories, the low light intensity restricts the duration of experiments through its effect on plants. When light is a factor of importance, trials must be conducted in greenhouses, even though controlling temperature and humidity in them is less accurate.

2.2 Temperature

Temperature in a growth chamber is an easily controlled factor. Sensors for controlling temperature (as well as humidity) are located at one site and respond to the conditions existing at the site. In addition to the precision of the thermo- and humidistats, the efficiency of control depends also on air movement throughout the chamber. The rather insensitive thermographs supplied with many growth chambers, draw a straight line when the actual deviation from the set air temperature may be 1 °C or more. The more accurate the instrument used to measure the temperature, the greater are the deviations from the set temperature. Deviations of 1 °C in the air are acceptable because, due to the mass of plant tissue, its temperature fluctuates less than that of the air. In our measurements a fluctuation of 1 °C in the air corresponded to a fluctuation of 0.5° and 0.25 °C on the leaf surface in light and dark periods respectively. Fluctuations in leaf-surface temperature of plants kept in transport plastic bags acting as moist chambers (see Sect. 2.3), ranged from 0.1° to 0.2 °C in the dark. During the light period these plastic bags created a greenhouse effect, and the leaf temperature exceeded by up to 10 °C the temperature recorded by thermographs not kept under plastic. Under such conditions the temperature should be measured (and, if possible, controlled) by thermocouples clipped to leaves inside plastic bags, and the temperature outside these bags should be fixed accordingly.

2.3 Humidity

In most cases humidity in growth chambers is expressed in terms of relative humidity (RH), but some researchers claim that the vapour pressure deficit in bar units is a more relevant measure. Relative humidity can be calibrated according to the dew-point principle, but this is more complicated and less used than control with dry and wet bulb thermometers or RH humidistats. Most growth chambers fail to maintain low RH, especially at low temperatures, and high RH at high temperatures. Precise, commercial humidity chambers are available and enable the researcher to conduct specific experiments. Small boxes equipped with precise control of humidity (e.g. Hartmann et al. 1982; Leach 1980) are less expensive than the commercial chambers and are suitable for specific purposes. Most growth chambers not equipped with humidity control, maintain a relatively stable RH of 50 to 70%. This is so because at low temperatures, which normally induce high RH, the cooling system works for longer periods and removes excess humidity, and at high temperatures the RH is decreased by heat. The range of 50–70% RH is suitable for most experiments in which humidity is not an experimental factor. This is a common situation, because free leaf moisture rather than relative humidity is the critical factor in specific phases of the life cycle of most pathogens.

2.4 Free Leaf Moisture

Free leaf moisture may be created by atomizers or foggers, by a dew chamber and by spraying plants with water and enclosing them in plastic bags. Each of these methods has its specific advantages. Atomizing plants with water inside a box (e.g. a wooden frame with plastic walls) is easy, inexpensive and good for semi-quantitative inoculations. The disadvantage of this method lies in an uneven deposition of drops on various leaves. A better deposition of drops is obtained in dew chambers. These chambers have a water bath at the bottom, cooled walls and/or ceilings and one or more benches for plants above the water bath. Cooled walls and/or ceiling induce the escape of long-wave radiation from the foliage. This escape lowers the foliage temperature below that of the air and induces condensation of water vapour on leaves. The air and plant temperature results from the difference between the heated bath and the cooled walls. The greater this difference, the quicker is the formation of dew and the bigger and denser are its drops. For specific purposes we may prefer small or big drops or a barely distinguishable film of water. Dew chambers of various sizes are available and their specification is described by the manufacturers. Many researchers use homemade dew chambers which are less convenient but work just as well as manufactured ones (e.g. Clifford 1973). The chaepest dew chamber is a metal box with a heated bath at its bottom. This box is placed in a cool room at 3–5 °C. In such a system the wall and ceiling temperature is that of the surrounding room. The air and plant temperature and the amounts of dew are determined by the bath temperature only.

In some cases plants placed in dew chambers may not be uniformly covered by dew. This is due mainly to somewhat different temperature of plants placed close to the cooled walls and plants situated in the centre. In addition, leaves close to each other interfere with the escape of long-wave radiation and, thus, with the formation of dew. Therefore, some of the leaves should be removed, plants should not be placed too close to walls and they should not be crowded. Plants placed on several benches, one above another, also interfere with the uniform escape of long-wave radiation and affect the formation of dew. A more uniform coverage by dew drops is obtained by using only one bench. These are serious disadvantages but the convenience of work with dew chambers is so tempting that most researchers prefer not to use plastic bags unless especially accurate results are required.

Spraying plants with water and covering them with plastic bags secures a more uniform deposition of water on the leaves. Disadvantages of this method are discussed in Section 4.1.

2.5 Relation of the Controlled Habitat to the Field Habitat

As a model is a simplified representation of reality, so the controlled environment is a simplified representation of the field environment. The main difference between the CC and the field habitat stems from deficient radiation in the controlled habitat and its effects on crop microclimate. The difference obtained in the field between the temperature of the upper leaf side (exposed to the sun) and the lower side (protected from the sun), and between the outer and inner leaves in the plant canopy, is normally absent in the controlled habitat. Some CC instruments are equipped with programmers and cycle the environmental conditions between given minima and maxima of temperature and humidity. The most sophisticated instruments have computerized control programs for simulation of various patterns of weather during the experimental period. However, most instruments work under constant conditions, or under rapidly changing levels between constant day and night temperatures, RH and light. The absence of transient conditions in the mornings and evenings, and the absence of high winds and rain, preclude a straight relation of environmental effects produced in the controlled habitat to those present in the field. These differences should be taken into account when interpreting results obtained in CC experiments.

3 The Choice of Hosts and Pathogens

3.1 The Ideal Host

In this respect ideal is the host in which morphological and physiological properties provide for the best use of the CC habitat. Due to low light intensity in most growth chambers and limitation of space in all CC instruments, we prefer to work with small and rapidly growing plants, seeded and grown in CC before use in experiments. If the plants are to be used when bigger or older, they have to be grown in greenhouses or in the open, where their development may be less uniform than of plants grown in CC. Low light intensity in most CC restricts the duration of experiments, but some hosts are more tolerant than others. In general, an experimental period of 2–3 weeks is tolerated by most hosts. Some host species do not fit the CC techniques due to their large size, their sensitivity to weak light or the delayed appearance of symptoms.

Limitation of space in CC restricts the number of replicates, and plant uniformity is therefore extremely important. Many CC trials fail because the experimental differences between replicates are greater than those between treatments. It is always better to throw away the test plants that deviate form the norm than to obtain uncertain results.

3.2 The Ideal Pathogen

Due to the limitations of light and space in most CC, it is easier to work with pathogens that react quickly to environmental treatments and produce easily distinguished symptoms. It is more difficult, and sometimes impossible, to work with pathogens that react slowly and/or fail to produce clear symptoms. To such pathosystems belong many soilborne pathogens which are studied mainly in various kinds of Wisconsin tanks (e.g. Steele 1967) under the semi-controlled conditions of greenhouses. The greenhouse, rather than the growth chamber habitat, also suits pathogens that cause cankers in plants in general, and in trees in particular.

3.3 Relation of the Reaction of Hosts and Pathogens Under Controlled Conditions to Their Reaction in the Field

The predisposition to disease of plants grown in CC may differ from that of plants grown in the field (see Chap. 1). Cultivation in small pots speeds up ageing, and the physiological age of plants grown in this way differs from that of plants grown in the field. Potted plants also produce less yield, and this prevents application of the CC techniques to studies of crop losses.

When the researcher is tempted to obtain quick results, and considers the limitation of space in CC habitats, he tends to work with small and young, rather than with larger and adult plants, despite a possible difference in the reactions of plants of various ages. Therefore, in selected treatments, results obtained with small test plants have to be verified in tests with older plants grown in bigger pots. Nevertheless, there are some plant species, like small grains, which are successfully treated in CC when relatively large and old. Even this does not absolve the researcher from the necessity to verify in the field the reactions of test plants previously observed in CC. For more details on the subject of plant susceptibility and the significance of field versus CC experiments, see Populer (1978).

4 Types of Controlled-Condition Experiments

For didactic reasons, we may divide the CC experiments into three classes. The first class includes the monocyclic experiments that cover only one cycle, from inoculation to subsequent sporulation, and in most cases only one phase of the pathogen's life cycle. But epidemics in the field are polycyclic processes because they consist of recurrent cycles of disease. With specific pathosystems we can conduct polycyclic experiments in growth chambers, and this is the second class of studies carried out under CC. The third class of CC experiments are tests designed to introduce a quantitative element into a study executed mainly in the field. These tests are especially valuable in the verification of some phenomena observed in the field but not subject to quantitative assessment in its complex habitat.

4.1 Monocyclic Experiments: Inoculation, Infection and Colonization, Sporulation

The techniques employed in monocyclic experiments are well known. In the following we discuss measures that maximize the probability of success in such experiments. The use of highly uniform test plants has already been mentioned in Section 3.

4.1.1 Inoculation

An uneven deposition of inoculum on test plants may result in a greater difference between replicates of the same treatment than between treatments exposed to various experimental variables.

In the field, spores of only some pathogens are normally dispersed in drops of water. However, despite the airborne character and relatively low concentration of most inocula in the field, inoculation in CC is most frequently made with inoculum suspended in water, and with a highly concentrated inoculum. In CC, inoculation with dry inoculum is possible only with spores like rusts and powdery mildews, which are produced in quantity and easily removed when dry. Spores of such pathogens may be inoculated with the aid of settling tower (e.g. Aslam and Schwarzbach 1980; Eyal et al. 1968; Kulik and Asai 1961; Mortensen et al. 1979). Because of the soft and sticky nature of powdery mildew spores, researchers used a settling tower with a fan that blew the spores from sporulating plant material (Kirby and Frick 1963). The inoculated plants may be wetted immediately, or after an intermediate exposure to dryness, as happens in the field. Inoculation with dry inoculum is difficult with pathogens, the spores of which are firmly attached to the sites of their formation. To detach these spores, shaking or rubbing the spore-bearing leaf or culture in water is the most practical method of preparing inoculum. Settlement of spores on the bottom of the spore-suspension container can be slowed down by addition of 0.1% water agar or an emulsifier like Tween 20. The latter also prevents clumping of spores. Spraying the inoculum on test plants until the drops run off is the easiest but least uniform method of inoculation. More accurate is the use of a turntable. Plants rotating on the turntable are sprayed with a water suspension of spores. The turntable technique suits also inoculation with dry inoculum (e.g. Brown and Fittler 1981; Mortensen et al. 1979). Futher improvement in the uniformity of deposited spores is achieved by inoculation with mechanical inoculators. These instruments are activated by constant air pressure and release a standard volume of water suspension of spores on the inoculated plants (Andres and Wilcoxson 1984) or on a given target area of each leaf (Schein 1964). Some researchers prefer to use automatic micropipette dispensers that release uniform by size drops on discrete targets. Others transfer these drops with a pencil-like rubber stick (J. Kranz, personal communication) or with filter paper discs dipped in a suspension of spores. Each technique has its specific advantages. The micropipette, the pencil-like stick and the filter paper methods produce small lesions, facilitate several replicate inoculations on each leaf and save inoculum. The filter paper method is apparently the best to secure infection, but the number of spores deposited on the inoculated target is difficult to count. The mechanical inoculators have the advantage of covering a relatively large portion of the leaf surface. This is especially important in diseases which do not spread beyond the site covered by inoculum.

For specific purposes we can collect dry inoculum of spores firmly attached to the sites of their formation. They are sucked onto a mounted membrane filter attached to a vacuum pump. The filter is moved close to the surface of the spore-bearing leaf. The membrane with spores may be exposed to various environmental conditions before being cut into small pieces and placed on a wet leaf to induce infection. However, the membrane is never evenly covered with spores, and this rather complicated method is semi-quantitative at best. Semi-quantitative results are obtained also when such membranes, or potted plants acting as spore traps, are placed next to fruiting bodies which release spores when wetted.

4.1.2 Infection and Colonization

Wetting periods, wetting period temperatures and inoculum doses are the main factors studied in the process of infection. The researcher should be aware of the fact that the level of each of these factors may modify the action of other factors and change the ranges of optimum and marginal temperatures and wet periods (Aust et al. 1980; Rotem 1978). In the study of infection, inoculated plants are kept under various conditions for the duration of the wetting period only.

Wetting conditions are obtained in dew chambers or in plastic bags as described in Section 2. With either technique, but especially with plants covered with plastic bags, considerable time may pass until the temperature of the enclosed plant is adjusted to that of the growth chamber. To shorten this period, the plants to be inoculated the next day are kept overnight in growth chambers. The same is done with plastic bags, already wetted inside; and with the water to be used for the suspension of the inoculum.

After termination of the wet period, plants removed from dew chambers or plastic bags may actually remain wet for an additional time. This period can be shortened to about 15 min by exposing the wet plants to wind created by fans. However, some plant species (e.g. cucumbers and potatoes) are sensitive to quick drying, especially after having been exposed to prolonged wetting. After conclusion of the wet period, the test plants are incubated under standard conditions in a growth chamber.

In some pathosystems, infection (as well an sporulation) can be induced not necessarily by one, relatively long wet period, but by several, short wet periods interrupted by periods of dryness. The interrupted, wetting-period regimes are effective only with pathogens in which the germinating spores can withstand transient conditions of dryness (Bashi and Rotem 1974). In such study it is advisable to test also the survival of spores as described in Chapter 5.

Evaluation of infection is usually based on symptoms and must therefore be postponed until the symptoms appear. In specific cases, such as rusts and powdery mildews, infection is evaluated by the appearance of spores on the inoculated target, rather than of symptoms. In some downy mildews, such as those on onions or peas, symptoms do not appear before the pathogen sporulates. In such cases the plant first undergoes normal incubation procedures; it is then placed under moist conditions to induce sporulation, the intensity of which indicates the degree of infection. In hypersensitive cultivars, necrotic symptoms are evidence of a failure of infection. This should be confirmed by the absence of spores after leaves with necrotic lesions are kept in moist chambers.

The effect of environmental factors on colonization is studied with plants exposed to the same wetting-period conditions, then removed and incubated under various environmental conditions in growth chambers. Temperature rather than humidity is the main factor studied, and its effect is evaluated according to symptom development. However, symptoms do not necessarily indicate the colonizing ability of the pathogen. In many pathosystems high incubation temperature may inhibit the pathogen, but enhance the necrotic reaction. In contrast, low temperature may inhibit lesion development despite satisfactory development of the pathogen. In both cases the effect of the temperature on colonization is indicated by the area from which spores emerge, rather than by lesion size per se.

4.1.3 Sporulation

Sporulation is sometimes tested with detached leaves which may fail to represent the pathosystem. My preference is to use whole plants. With pathogens that sporulate under wet conditions, these conditions are created in dew chambers or under plastic bags, as described in Section 2. *Phytophthora infestans* on potatoes produces more spores on plants kept in dew chambers than on those covered by plastic bags, but this does not necessarily apply to other systems. In dew chambers as well as in plastic bags, sporulation is affected by the size of drops that cover the leaf, and large drops are usually associated with less sporulation. In many systems spores are produced at high humidity, near saturation [e.g. some downy mildews (Cohen and Rotem, in press)]. Sporulation of rusts and powdery mildews does not require free leaf moisture and, at least in powdery mildews, is inhibited by water.

Some studies are aimed at estimating the number of produced spores as affected by factors (incubation temperature, light, lesion development, etc.) acting on the pathosystem before the spores are actually formed. Other studies are aimed at determining the latent period, viz. the time between inoculation and sporulation, or at estimating the effects of wetting periods (or RH), light and temperature during and not before, the period of the actual formation of spores. Effects of these factors on sporulation are easily studied with pathogens that sporulate in wetness. It is difficult with pathogens like rusts that sporulate in dryness, and in which sporulation is a prolonged and continuous process. In some pathogens, like species of *Alternaria*, the process of sporulation requires two wetting periods in darkness interrupted by a dry and light period. These species, and possibly also others, sporulate under interrupted wetting-period regimes (Rotem et al. 1978).

Most CC studies estimate sporulation in plants exposed to wet conditions for a single wetting period. This is an artificial situation, because in the field plants are exposed to recurrent periods of wetting and produce successive batches of spores. It is impossible to calculate the number of spores likely to be produced throughout the process of disease, by the number of spores formed in a single wetting period. This is so because more spores per lesion per each wetting period are produced in plants exposed to only one such wetting than to recurrent wettings.

The recurrent wetting experiment is difficult to conduct. Its duration is conditioned by temperature regimes, because high temperatures shorten, and low temperatures extend the period over which the system sporulates. Some researchers estimate daily production of spores by using spore traps kept in growth chambers. This technique may reflect the sporulation of pathogens that produce easily detached and dispersing spores, such as oidia of powdery mildews (e.g. Aust and Kranz 1974), but not of pathogens that produce firmly attached spores. Strong winds with gusts are needed for efficient dispersal of these spores. Since such winds are normally not present in growth chambers, a direct estimation of the number of produced spores is the preferred technique. We have to evaluate the number of spores formed after each wet night without damaging the foliage. This is simple with spores of rusts and powdery mildews that are detached by gentle shaking, but difficult with pathogens in which the spores can be separated in quantities only by shaking the leaves in water. In these cases we have to divide the test plants into several groups. After each wet night one of these groups is removed from further experimentation. The leaves are detached, shaken and the spores are counted. Spores on the remaining plants (i.e. plants to be evaluated after the following nights) are washed down with streams of water, or blown off with air. Despite this procedure, some spores remain attached to leaves. The number of spores retained on leaves is counted in samples of plants removed from further experimentation. The number of spores produced daily is the number counted after each wet night minus the number retained on plants on the previous day.

Some researchers count spores with the aid of Coulter counters. Use of this instrument is reliable only with small, uniform and preferably round spores present in samples with a minimum of dust or leaf particles. In relatively clean samples, estimation of the number of spores in often done according to the optical density of the spore suspension in a spectrophotometer. The wavelength measured should fit the investigated species. Values for the standard curve are determined by actual counts with a hemacytometer. With rusts, the number of spores is often assessed indirectly by weighing the samples or measuring their volume in a narrow tube. The number of spores in other pathosystems is often too small, or the samples are too dirty, to use these methods and the spores have to be counted in a microscope. Hemacytometers are widely used but give erroneous measurements with large (e.g. *Helminthosporium* spp.) spores. In our studies with large spores, we preferred the method of filtration: part of the spore suspension is filtered through an 8-μ membrane filter, on which the spores are counted (Bashi and Rotem 1975).

In CC, as well as in the field, large differences are found between the number of spores produced on various lesions, as described in Section 2.2. In many cases the more accurate the method of evaluation (i.e. counting of spores), the greater the differences between individual replicates and the less valid the statistical differences between treatments. Evaluation of sporulation according to a visual key of grades produces more uniform and statistically better results and leaves the researcher, the referee and the reader happy. However, we prefer the statistically problematic counting, because it reflects the nature of sporulation in the field.

4.2 Polycyclic Experiments

Epidemics induced in growth chambers (mini epidemics) make it possible to study concurrent and recurrent phases of the pathogen's life cycle, and the reaction of the host, from the first appearance of disease until the destruction of plants. Mini epidemics are induced in growth chambers programmed for gradual change between given minima and maxima of temperature and humidity. Environmental cycles include also 'dew' periods, during which the plants are covered by fine drops of water released from humidifiers situated next to plants; or 'rain' periods simulated by sprinklers attached to the ceiling. When dew is simulated, the wet period is at night and, if possible, the circulation of air in the growth chamber is kept to a minimum.

The mini epidemic starts when the chamber is filled with healthy plants and with a source of inoculum e.g. one plant with a few lesion. It is, in our practice, to introduce the lesioned plant for a period of one dewy night, for production of spores; and to leave it for the following windy day, for spore dispersal. Dispersal of spores is assisted by fans. The dispersed spores land on healthy plants, and infect them during the subsequent wet night. These cycles are repeated daily (e.g. Rotem and Cohen 1974). Mini-epidemics caused by powdery mildews, which sporulate and infect in dryness, can be induced without the wet periods at night.

The production of mini-epidemics is more complicated than appears from this description. Most commercial chambers are not equipped with instruments controlling cycles of temperature, humidity, dew, rain and air movement. They have unsealed, electrical connections (e.g. in lamp housing) which, in wet periods, may cause short circuits and endanger the researcher. They require installation of auxiliary instruments (e.g. fans) and technical skill of the researcher. On top of these difficulties, only certain pathosystems suit this kind of research. Mini-epidemics succeed with pathogens that reproduce rapidly, easily disperse their spores, require short periods of incubation and destroy the crop before the plants start to suffer from the low light intensity prevailing in growth chambers. The pathosystems with which we succeeded were downy and powdery mildews of cucurbits and potato late blight, which destroyed the hosts in less than 4 weeks (e.g. Rotem and Cohen 1974); the period of about 6 weeks required for such work with powdery mildew of tomatoes and peppers made mini-epidemic studies difficult. We failed to achieve mini-epidemics of *Alternaria solani* in potatoes, which required a long experimental period due to the inadequate dispersal of its firmly attached spores and the slow buildup of inoculum.

Another use of CC is to mimic courses of weather events as measured in the field and shich appear to have an effect on the epidemic under study. This requires equipment regulated by means of computer programs.

4.3 Auxiliary Experiments

These tests complement experiments carried out mainly in the field. Section 5 in Chap. 5, which discusses the use of potted trap plants in the study of inoculum potential in the field, exemplifies the role of dew and growth chambers for completion of the work. In the field, these trap plants are exposed to various experimental conditions. A quantitative evaluation of results is possible only when trap plants removed from various exposures in the field are incubated under standard conditions in growth chambers until the appearance of symptoms.

4.4 Relation of Results Obtained Under Controlled Conditions to Results Obtained in the Field

Airborne diseases are those diseases most frequently studied in CC. In the field airborne diseases are transferred by inoculum that, in most cases, is dispersed during the dry daytime, is highly diluted during its flight, is unevenly deposited on healthy leaves and is exposed to environmental hazards before subsequent rain or dew induces germination and infection. However, in the field, the mass of healthy leaves and produced inoculum, and the recurrence of all phases of the pathogen life cycle, compensate for the unevenness of inoculation and the resulting infections are relatively uniform. In CC we secure uniform infections by quantitative inoculations, but are not able to simulate all processes that occur in the field. Concentrated inoculum is used instead of diluted inoculum and, except for rust and powdery mildew diseases, the test plants are inoculated with a water suspension of spores not exposed to environmental hazards between the time of their formation and infection. Due to these differences, the results obtained in CC identify the factors of importance and indicate the relative difference between various factors rather than their actual quantitative effect in the field.

The study of sporulation is more difficult than the study of infection. Few studies compared the amount of spores produced in CC and in the field under relatively similar conditions. We found that sporulation of downy mildews on tobacco and peas and of potato late and early blights was more prolific in the field than in CC. Such differences suggest that caution must be exercised in the interpretation of results obtained in CC, and that there is a need to verify results obtained in CC in specific field trials.

The mini-epidemics induced in growth chambers reflect the action of several basic factors affecting epidemics in the field, but not the effect of radiation. Auxiliary experiments with potted plants exposed in the field before or after being treated in CC can improve our understanding of phenomena occurring in the field. These simple auxiliary tests often change a descriptive observation made in the field into an experimental study. For more information on the combination of growth chamber and field experiments in the solution of epidemiological problems, see Aust and Kranz, Chapter 1.

5 Preparation of Data for Simulation

There is no substitute for data obtained in CC tests for computing specific parts of simulation programs (Kranz and Hau 1980). This is the case of data quantifying the effect of wetting periods and temperatures on infection. Even so we are not always sure whether these data should be derived from tests carried out under fluctuating temperatures that mimic the temperature regimes in the field, or whether several tests at various constant temperatures can be used to fit any particular situation in the field; or whether CC tests should quantify the events of compensation (Aust et al. 1980; Rotem 1978) or whether these events too can be calculated in the program.

Tests in CC can provide the data needed to compute the effects of submarginal temperatures on the survival of a pathogen which is already established in the host tissue. The latent period under various environmental and host conditions is easily studied in CC and difficult to study in the field. Less suitable for use in simulation are CC data that quantify the number of produced spores. As discussed in this section and in Section 2.2, sporulation in the field may be more prolific than in CC. The infectious period, viz. the number of days during which the system sporulates, can be measured accurately in CC when the measure aims at establishing the time factor only; and is less accurate when it aims also at measuring the number of produced spores. In these cases we cannot rely only on data derived from tests in CC and should provide the simulation program with data obtained also from a direct measurement in the field.

The following are phenomena for which the CC technique fails to provide reliable data for use in simulation. The phase of spore dispersal is difficult to quantify in CC. Data on survival of the dispersing spores should derive mainly from the field, since survival is strongly affected by sunlight. However, in all field experiments, auxiliary tests in CC may help in the preparation of data.

In all cases the CC tests should be designed for the specific purpose of simulation, and should measure the events tested in terms used in the computer programs. Some simulation program use historical data from tests not designed for this particular purpose and may fail to simulate epidemics. For instance, a hypothetical study in the past quantified the relative effect of various temperature on infection without aiming at the use of these data in simulation programs. The researcher infected his plants with a massive dose of inoculum, and evaluated the results in percentage of infected leaves. The techniques employed were good for establishing the relative effects of treatments, but did not provide measures needed for simulation. In using such historical data the program analyst may be tempted to dismiss the effect of massive inoculum, and to transform the percentage figures used in the original evaluation into units that fit programming. Such a procedure may result in a relatively minor mistake in computing one cycle of the simulated epidemic. The mistake may grow to large proportions when computed for a whole epidemic, and to catastrophic proportions when sporulation instead of infection is the transformed measure. More about the use of experimental data for the development of simulators, and the problems involved in Chapters 18 and 19 by Campbell et al. and Hau.

6 Epilogue

Some epidemiological phenomena, such as dispersal of spores, are best studied in the field, and CC experiments will have an indicative rather than a quantitative value. Other phenomena, such as the discrete effects of specific factors, are best studied in CC. Sometimes the idea to study a given phenomenon in CC derives from inconclusive observations previously made in the field, or from obscure, strange or contradictory phenomena discovered in the analysis of a field epidemic. In these cases, only the easily manipulated habitat of CC instruments makes it possible to determine the role of specific factors, to search for some missing link or to look for a logical explanation of the process observed in the field. Studies in CC help us to separate the effects of concurrently acting factors and to study the phenomena of competition, interaction or compensation which are masked by the complexity of the field habitat. Recent progress in epidemiology results in part from such studies. However, some phenomena discovered in CC may have a significant effect only in the simple habitat of the growth chamber. The same phenomena may have only a minor influence in the field, where their value may be diminished by the action of other factors. This situation is exemplified by two cases: (1) CC experiments showed that ageing tomato plants are quite resistant to *Stemphylium botryosum* f. sp. *lycopersici*. However, in the field, epidemics of *Stemphylium* blight in tomatoes occur only in ageing and resistant plants, due to the scarcity of inoculum when the plants are young and its abundance in the old field at the end of the growing season. (2) In CC trials, early blight developed better in nutrient-deficient than in well-fed potatoes. No such difference was found in the field, because the amount of nutrients in the soil sufficed for regular early blight development in the non-fertilized plots. These examples show that some phenomena found in CC are of negligible value in the field. In general, an epidemiological study should combine CC and field trials, as pointed out by Aust and Kranz in Chapter 1. However, the relative ease of conducting experiments in CC tempts many researchers to neglect the field. The researcher then becomes a CC addict. Sometimes he produces papers which are correct from the methodological viewpoint but contribute little to epidemiology. Occasionally, this author has been one of these sinners, and he warns his colleagues and himself: in planning CC experiments, always remember the field.

7 References

Andres MW, Wilcoxson RD (1984) A device for uniform deposition of liquid-suspended urediospores on seedling and adult cereal plants. Phytopathology 74: 550–552

Aslam M, Schwarzbach E (1980) An inoculation technique for quantitative studies of brown rust resistance in barley. Phytopathol Z 99: 87–91

Aust HJ, Bashi E, Rotem J (1980) Flexibility of plant pathogens in exploiting ecological and biotic conditions in the development of epidemics, In: Palti J, Kranz J (eds) Comparative epidemiology. Pudoc, Wageningen, pp 46–56

Aust HJ, Kranz J (1974) Eine automatische Sporenfalle für den Gebrauch in Klimaschränken. Angew Bot 48: 267–272

Bashi E, Rotem J (1974) Adaptation of four pathogens to semi-arid habitats as conditioned by penetration rate and germinating spore survival. Phytopathology 64: 1035–1039

Bashi E, Rotem J (1975) Sporulation of *Stemphylium botryosum* f. sp. *lycopersici* in tomatoes and of *Alternaria porri* f. sp. *solani* in potatoes under alternating wet-dry regimes. Phytopathology 65: 532–535

Brown JF, Fittler JF (1981) A quantitative method of inoculating plants with uniform densities of fungal spores. Aust Plant Pathol 10: 51–53

Clifford BC (1973) The construction and operation of a dew-simulating chamber. New Phytol 72: 619–623

Cohen Y, Rotem J (1987) Sporulation of foliar pathogens, In: Pegg GF, Ayers PG (eds) plant infecting fungi. Cambridge Univ Press, pp 314–333

Eyal Z, Clifford BC, Caldwell RM (1968) A settling tower for quantitative inoculation of leaf blades of mature small grain plants with uredospores. Phytopathology 58: 530–531

Hartmann H, Sutton JC, Thurtell GW (1982) An apparatus for accurate control of atmospheric water potentials in studies of foliar plant pathogens. Phytopathology 72: 914–916

Kirby AHM, Frick EL (1963) Greenhouse evaluation of chemicals for control of powdery mildews. I. A method suitable for apple and barley. Ann Appl Biol 51: 51–60

Kulik MM, Asai GN (1961) Use of a portable inoculation tower in laboratory, greenhouse and field tests of fungicides to control rice blast. Plant Dis Rep 45: 907–910

Kranz J, Hau B (1980) Systems analysis in epidemiology. Annu Rev Phytopathol 18: 67–83

Leach CM (1980) An apparatus for precise control of humidity, temperature, air flow and light in spore discharge studies. Phytopathology 70: 189–191

Mortensen K, Green GJ, Atkinson J (1979) A method for uniform infection of seedling and adult cereal plants by *Puccinia graminis* f. sp. *tritici*. Phytopathology 69: 420–423

Populer C (1978) Changes in host susceptibility with time. In: Horsfall JG, Cowling EB (eds) Plant disease, advanced treatise, vol 2. Academic Press, London, pp 239–262

Rotem J (1978) Climatic and weather influences on epidemics. In: Horsfall JG, Cowling EB (eds) Plant disease, advanced treatise, vol 2. Academic Press, London, pp 317–337

Rotem J, Cohen Y (1974) Epidemiological patterns of *Phytophthora infestans* under semi-arid conditions. Phytopathology 64: 711–714

Rotem J, Cohen Y, Bashi E (1978) Host and environmental influence on sporulation in vivo. Annu Rev Phytopathol 16: 83–101

Schein RD (1964) Design, performance, and use of a quantitative inoculator. Phytopathology 54: 509–513

Steele AE (1967) A constant temprature bath for pot-grown plant. Plant Dis Rep 51: 171–173

II Measurement and Their Analysis

Chapter 3

Measuring Plant Disease

J. Kranz[1]

1 Introduction

The measurement of disease intensity plays the same key role for plant pathology as does diagnosis. Without quantification of disease no studies in epidemiology, no assessment of crop losses and no plant disease surveys and their applications would be possible. Disease assessment is also needed for many other applications in plant pathology, such as screening for resistance and fungicides. It expresses the effects of various treatments or factors on disease in experiments, and disease control.

Under measurement we also include assessment methods used to estimate disease intensities. Except for precise laboratory or field studies, disease intensities must be estimated, unless some equipment (see pp. 45–46) can ensure precise measurements. These methods have varying degrees of sophistication, accuracy and precision. For field studies they should be easy and quick in use for a wide range of conditions but also adequately reliable and reproducible, accurate and precise. This is not always achieved and for more refined studies the presently available methods of measurement are often unsatisfactory. Reasons for this are discussed in a number of publications to which we refer in this chapter (Horsfall and Barratt 1945; Large 1966; Kranz 1970; Amanat 1977; Berger 1981; Sherwood et al. 1983; Seem 1984; Forbes and Jeger 1987).

In disease assessment we distinguish two distinct components: (1) disease assessment proper, and (2) sampling techniques. Both components imply different aspects. In the context of this chapter we shall deal first with assessment, then with a description of techniques employed and, finally with some specific applications of them. We cannot, however, give here an answer to every problem which may arise.

2 About Terms

Disease intensity can be expressed either as incidence (or frequency) or severity. *Incidence* is the percentage of diseased plants or plant parts in the sample (or population), irrespective of their individual severity. Disease *severity* is the percentage of the relevant host tissue or organ covered by symptoms (or lesions) of the disease. Severity results from the number and size of lesions. These two components of severity may change independently during disease progress. When intensity values are divided by 100 they become the y ($0 < y \leqq 1$) values as used in infection rates.

[1] Phytopathologie und angew. Entomologie, WZ Tropeninstitut, Justus-Liebig-Universität, 6300 Gießen, FRG

The choice between evaluation of disease according to its *severity* or incidence depends largely on the type of disease and on the objective of disease assessment. Disease severity appears to be more appropriate for rusts, downy and powdery mildews, leaf spots and similar diseases. If not stated otherwise, each severity value comprises both the number of lesions and the total of their area. In some cases it may be essential to differentiate between the symptom area proper and the chlorosis, necrosis or leaf shedding caused by the pathogen indirectly. In specific cases there is a difference between disease severity and destruction which may result from disease – accelerated senescence of the host (Seem 1984). For crop loss assessment such a distinction may be irrelevant but it may be important for purely epidemiological purposes. Hence, when measuring disease it should be made clear whether disease and/or destruction is referred to.

Evaluation of disease itensity according to its *incidence* is suitable for most diseases in the early stages of their epidemics. But it applies mainly to disease which affects entire plants, e.g. systemic virus diseases, wilts, smuts or spots on fruits, if one lesion makes the latter unfit for sale. Disease incidence can be assessed by simple counting of diseased vs healthy sampling units (see p. 45) and thus is less cumbersome and error prone than assessment of disease severity. Sampling units may, however, be defined diseased only if disease severity is beyond a certain level (e.g. physiological damage threshold) and then assessed as disease incidence. Figures for disease intensity should be unconfounded, or, as Large (1966) stated "...results must be easily communicable to others...". Disease

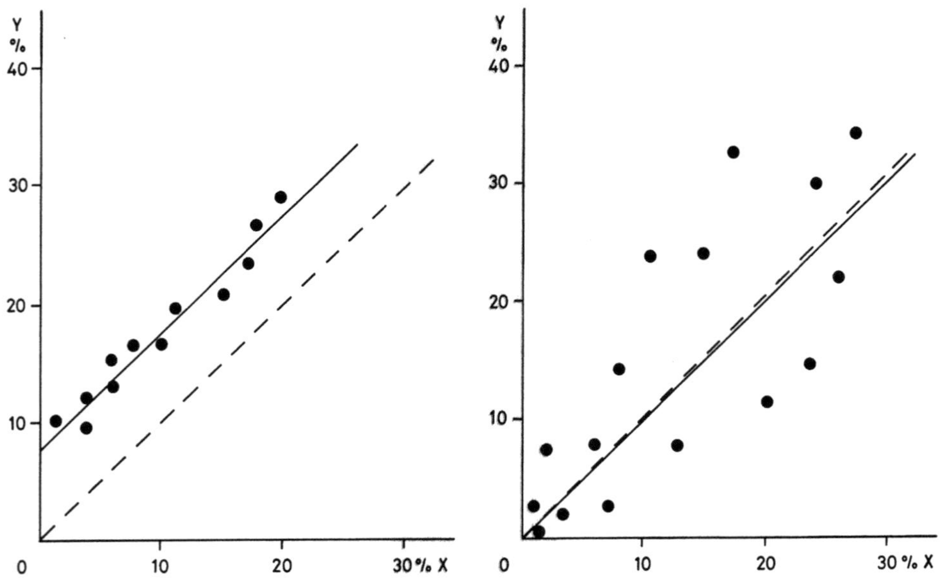

Fig. 3.1 Examples for accuracy and precision in disease assessment. The *abscissa* gives the true, the *ordinate* the estimated severities in percent; the correct estimates for both assessors should have been on the *broken line*. The person on the *left* made inaccurate but (within his concept) very precise estimates, whereas mean estimates of the one on the *right* were highly accurate but imprecise (Amanat 1977)

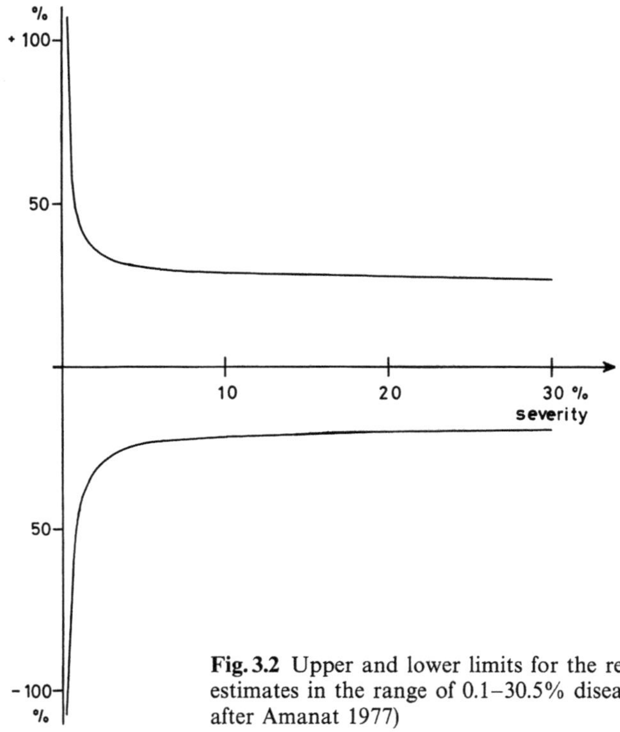

Fig. 3.2 Upper and lower limits for the relative deviations of permissible estimates in the range of 0.1–30.5% disease severity (see text; readjusted after Amanat 1977)

indices are often a nuisance because it is hard to understand what they mean in quantitative terms. Terms like slight, moderate, severe disease are obsolete nowadays. The same applies to the arbitrary, not quantitatively defined grades, e.g. 1–9.

Two terms often used in measuring disease intensities are accuracy and precision. Accuracy describes the degree of conformity with a given standard or truth, whereas precision is the degree of refinement in measurement (Fig. 3.1). To characterize the individual estimations, i.e. assessments made by one person on a series of intensities, a regression line $y = a + b\,x$ can be fitted to the estimates. If the parameter a is more or less equal to 0 and b equals 1, then the estimates coincide with the expected values. In such a case, the person has estimated the diseased area of the organ (e.g. leaf) accurately. For instance, if the true (i.e. measured) value is 30%, we accept estimates between 25 and 35% to be accurate. For a disease severity of 5% the permissible error ranges from 3.75 to 7% (Fig. 3.2). The interval is not symmetric for the smaller value because the estimates below the exact value are limited by 0. Even such an accuracy is not sufficient for a good estimation, because it does not account for the variation around the regression line (Hau and Kranz 1979). The coefficient of determination r^2 is used as a measure of the scatter around the regression line; it is independent of a and b. Estimates are precise if r^2 is close to 0.95. Tests for precision, i.e. repeated estimates of the same intensity level by one person, may also be done by means of the coefficient of variation ($C = 100\,\sigma\,n - 1/x$), while the test for accuracy is done by the chi-square fit.

3 Sources of Error in Disease Assessment

Disease assessment proper comprises physiological perception of stimuli and its psychological processing and evaluation. Perception in our case is the human ability to distinguish by visual acuity the diseased proportions of plant surface. The psychological processing and evaluation of stimuli is made against the background of some concepts or standards. One concept is the decimal system. The other is expectation based on experience and/or knowledge. Apart from personal optimism or pessimism the evaluation may be conditioned by the reputation of a given disease of being mild or severe.

An essential information for realistic disease assessment, for the design of standard diagrams and for an appropriate training is the maximum disease severity per sampling unit that occurs in the field. Here, tremendous abuse prevails. In measurements (not estimates!) of five leaf-disease pathosystems Kranz (1977) found that during disease progress only up to 10% of all leaves had more than 37% disease severity. This level incidently is the same as the upper limit for wheat stem rust in the modified Cobb scale (Large 1966). However, at the peak of a powdery mildew and a rust epidemic, 27 and 43% of the remaining leaves respectively, had disease severities between 37 and 100%. Even then, the majority of disease severities per plant in non-systemic diseases rarely have higher disease severities than, perhaps, 30% per plant, and that this level may rather be lower if averaged for a whole field. There are, however, exceptions: Fields can be fully destroyed by diseases such as potato late blight, some downy and powdery mildews, where it becomes impossible to distinguish between lesions and dead tissue with no pathogen present.

A specific though common feature of destruction due to disease is shedding of affected organs. This makes disease assessment usually more complicated, particularly as plant organs may also shed due to other factors. Also, the actual disease severity on a leaf or fruit may vary greatly on the organs shed. With plants or cultivars which tend to react with shedding, either records have to be kept of individual organs or they have to be collected in special recipients like framed nettings. They then may be rated for their disease severity. For crop loss assessment this distinction might not be needed as long as a disease is the sole cause for the drop. With very small leaves, e.g. asparagus, the dry weight of collected leaves might be rated against the total dry weight of entirely healthy ferns of a given age.

3.1 Individual Errors

Estimators usually tend to overestimate. This is particularly pronounced in the lower range of disease severities (Kranz 1970), and often exceeds actual values up to two or three times. But there are estimators who underestimate in lower ranges and overestimate higher values, and vice versa. Sherwood et al. (1983) found that their scorers differentiated among frequencies of leaf spots more readily than among their sizes. Another problem amongst estimators are divergences between their tendencies for accurate and precise estimates (Fig. 3.1). Only 9% of the inexperienced estimators were both precise and accurate (Amanat 1977).

Estimators unintentionally rely on the decimal system, as they prefer certain values to others (Koch and Hau 1980). These 'preferred values' comply with the Weber-

Fechner Law. Indeed, all tests seem to support Horsfall and Barratt (1945) that estimates of disease severities follow the logarithmic function according to this law and increase linearly as the stimuli increase logarithmically. As suggested by Baird and Noma (1978, cited by Hebert 1982), there are three more types of stimulus-response curves vs linear, exponential and power functions. Hence, linearity of response should be tested before designing scales and standard diagrams.

3.2 Object-Dependent Errors

The quality of disease estimates can be affected by factors different from the physiological stimuli and psychological responses. The additional factors are mainly object-dependent and include (1) complexity of the sampling unit, (2) size and shape of lesions and (3) colour and number lesions in the sampling unit. Also, light conditions and leaf wetness can lead to deviating estimates.

Variability in estimates and the tendency to overestimate was found to be lower when the sampling unit was the individual leaf, instead of the entire branch assessed as a whole (Amanat 1977). Three-dimensional samples have been particularly difficult to estimate. Therefore, precision of disease assessment can be increased by choosing the smallest possible sampling unit, such as leaf or fruit. Errors increase also in samples with more numerous lesions and with darker shades of colour. Overestimations decrease with increased intensity and decreasing number of spots (Sherwood et al. 1983). Hence, with larger and fewer lesions the estimates are more precise (Amanat 1977).

The points mentioned above have to be considered in the training for disease assessment. Training is essential, particularly if several estimators work in a joint program. But training alone may not eliminate all sources of error, even when using standard diagrams (as did Sherwood et al. 1983). Therefore, also trained scorers should test their estimates frequently by means of exactly measured specimens or standard diagrams (see Sect. 5.2), which emphasize low levels of disease severity and different spot sizes.

4 Sampling Techniques for Disease Assessment

As plant pathologists usually deal with populations, appropriate sampling is the second component of disease assessment to ensure its accuracy. All estimates of disease thus have to be related to an appropriate sampling unit of the crop, such as plant part, plant or plant population (e.g. plots, fields, farms, regions, countries). For a sample to be representative for a crop the spatial distribution of disease at the time of sampling has to be observed (for details, see Poushinsky and Basu 1984; as well as Seem, Gilligan, Berger, Dekker, Welz and Barker and Noe, Chaps. 4, 6, 9, 10, 11 and 15, for specific aspects and purposes).

The degree of accuracy and precision required in disease assessment varies in relation to the objective of a particular study. It certainly must be highest in epidemiological growth chamber or field experiments and lowest in large-scale surveys. In some epidemiological studies the population of lesions may even be divided into lesions of different size or age (see Sect. 5).

The decision to use random, systematic, stratified or multiple-stage *sampling techniques* is influenced by the spatial distribution of a disease in the crop which obviously may change during the season (see Chap. 4). Diseased plants, or lesions, may be distributed randomly, or may be clustered (aggregate), regular or systematic according to the soil-, wind- or vectorborne nature of diseases. Hence, the choice of sampling techniques requires adequate knowledge (or pilot studies) of the disease and its behaviour in space and time. For the most appropriate sampling technique see, for instance, Cochran (1963). In most cases a systematic sampling is recommended, e.g. walk in a diagonal or 'W'-pattern across a plot or field and take x sampling units at y metres distant along the path. For decision making in disease control sequential sampling is often appropriate (for above examples, see Boivin and Sauriol 1984).

Sampling units can be defined in compliance with the objective, and, in some cases with sampling dates. Hence, if yield loss due to rusts is highly correlated with disease severity on flag leaves, these leaves are the choice sampling unit. In other cases the sampling unit is the entire plant, the stem basis, etc. They may either be tagged or chosen according to the sampling technique adopted as 'destructive' samples at each sampling date. With the latter method, however, only the disease intensity at the time of sampling can be assessed. If information on the full scope of disease progress is wanted sampling units have to be tagged. This would ensure that disease severity on units already gone can be kept on record and be plotted accordingly; plants, plots or fields (but see Sect. 5.4). Sometimes collective sampling units, e.g. dry weight of shed needles of coniferous plants, may be used.

An adequate *sample size* ensures a desirable level of accuracy. For practical reasons an optimum sample size (Chochran 1963; Analytis and Kranz 1972; Seem 1984) is chosen. This size depends on the required accuracy and precision for the objective in relation to available funds, personnel and technology. Peculiarities of a given disease affect both the choice of sampling unit and the determination of its size and affect frequency as well as the technique of sampling. Sample size and the criterion used for assessment of disease intensity may change in the course of an epidemic (see Sect. 6.4).

To accurately determine whether lesions are clustered or whether the infection sites are randomly dispersed, a larger sample size would be required for disease severity than for disease incidence. In each case the sample size needed would depend on the degree of clustering (Campbell and Pennypacker 1980). The latter authors sampled plants infected by soilborne pathogens per quadrat (incidence) and the number of lesions per quadrat (severity) and calculated a variance-to-mean ratio. When this ratio was not significantly different from unity, it indicated a Poisson distribution of infected plants (or lesions), i.e. an even or a purely random dispersion of sampled plants or lesions. A ratio significantly greater than unity could indicate that the lesions are clustered or aggregated and follow a negative binomial distribution. However, this may be different for assessments of incidence and severity. The value of the 'k' or 'dispersion' parameter of the negative binomial given an indication of the degree of agreegation present. Values of k approaching zero indicate extreme aggregation, whereas clustering decreases with k closer to infinity (Campbell and Pennypacker 1980). With a chi-square analysis we can test the goodness of fit of expected to observed frequency values for both distributions.

Adams et al. (1985) sampled five diseases of potato tubers and established experimentally the sample variance after transformation of percent values to logits. For some of these diseases a variance ($V = \sigma^2/n$) of 0.01 confidence limit of 95% was chosen as giving acceptable accuracy for the incidence, with different quantitative scores for some of the diseases. With the above level of variance 15–20 samples of 50 tubers each might generally be needed for assessments of differences amongst diseases.

The detection of very low disease incidence (presence or absence of disease) very often requires stratified sampling in favour of finding the disease (Seem et al. 1985). Based on the assumption that the disease (e.g. grapevine downy mildew) at its initial stage is randomly (Poisson) distributed, which must not always apply, one equation is needed to estimate when the disease is present in sampling units at a level equal to or greater than a given threshold value. Using the upper boundary of the 99% confidence interval for n as sample size, it can be calculated with the following formula:

$$n = \ln(1-\alpha)/\ln(1-(\hat{P}_I - 2.58\,\Sigma))\,, \qquad (3.1)$$

where α = significance level (0.95 or 0.99);
$\hat{P}_I$ = proportional of diseased units;
$\sum = \sqrt{P_I(1-\hat{P}_I)/t}$ standard error of $\hat{P}_I$;
t = total number of units searched.

Books on statistics (e.g. Cochran 1963) offer other equations.

Another, though rough estimate of sample sizes can be obtained from pilot sampling (Fig. 3.3). Disease intensities are estimated from a number of sampling units and the

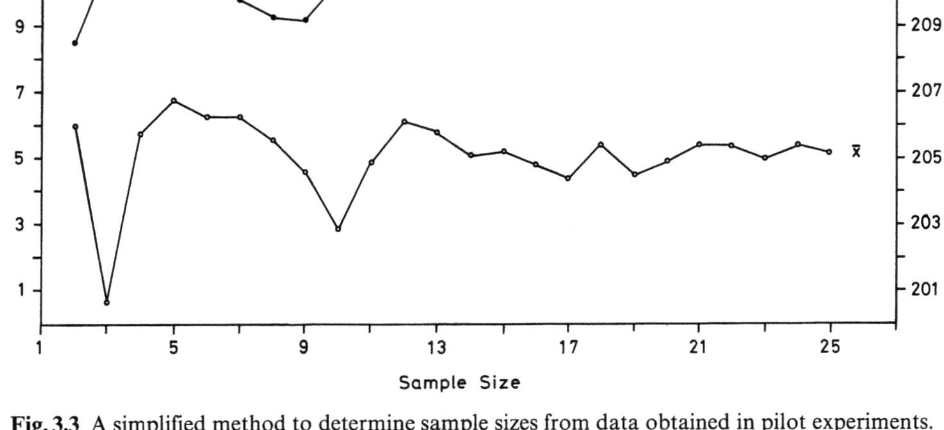

Fig. 3.3 A simplified method to determine sample sizes from data obtained in pilot experiments. The mean ($\bar{x}$) and standard deviation (*SD*, *s*) are computed for each new sampling unit entered from the data set obtained. When variation in both curves becomes more or less steady (e.g. from 13 onwards) a sample size may be chosen according to the desired precision

running means and their corresponding standard deviations are plotted over the abscissa on which the number of units entered are shown. Usually both curves flatten after a certain number of entries. When both curves are sufficiently stabilized over a particular number of units on the abscissa (e.g. 15, in Fig. 3.3), one may chose this as the sample size.

Frequency of sampling will greatly depend on the course and time span of the disease progress curve. As a rule of thumb one needs at least five points of measurements for establishing a reliable curve. Sampling may be done at equidistant time intervals or at defined growth stages of the host (Chap. 4, Sect. 3.2). Berger (1981) suggests sampling performed in accordance with (the highly variable) latent periods. On the other hand, loss assessment models (see Teng and Johnson, Chap. 12) may require only one assessment of disease severity in a particular growth stage of the host.

A *sampling plan* should be drawn up with instructions on all the above elements when disease assessment on a large scale is to be done. For sequential sampling, a useful tool in decision making, a decision table or upper and lower decision lines have to be computed from pilot studies (see Boivin and Sauriol 1984).

5 Techniques used in Measuring Disease

In epidemiology disease intensity expresses the size of the population of lesions. The rate of their increase describes the dynamics of an epidemic as a reflection of past events (or states). The age composition of lesions, such as between infectious lesions and lesions which are not yet or no longer infectious is important because of the pathogen's potential to propagate and its effect on the possible future dynamics of an epidemic. The infectious stage of lesions may be different with biotropic and necrotrophic pathogens (see Chap. 5). Disease intensities per sample, plot, field, etc. are usually computed as arithmetic means with the 0-severity of healthy sampling units included. If, however, the objective of the experiment or study requires the exclusion of 0-values, i.e. disease severity is for diseased plants etc. only, then disease incidence should also be quoted. The effect of changing the amount of susceptible host tissue is referred to by Seem (Chap. 4).

Apart from proper sampling and training, disease assessment could be improved or facilitated by: (1) measurement of lesion sizes, (2) use of standard diagrams or class scales of disease intensities, (3) use of incidences to estimate severities, (4) automatic measurement and (5) remote sensing.

5.1 Actual Measurement of Disease Intensity

Counting of lesions and/or measurement of their size is practical only in research projects. If lesion size is regular, lesions per leaf (or other organs) may be counted, their diameter (or length and width) measured and the total surface of lesions calculated. Lesion surface is then related to the leaf surface and expressed as disease severity in percent. If lesions differ in size they can be assigned to classes of different sizes and can be computed accordingly. Columns for these classes may be provided for in field record forms. Such a recording, however, requires coefficients which have to be established beforehand by exact measurements (e.g. ellipsoids, circles). Image analysis with a microcomputer may be used in destructive sampling (see also Sect. 5.4).

5.2 Use of Keys, Standard Diagrams and Classes of Disease Intensities

Standard diagrams (e.g. James 1971) in the first place are most useful tools for the "calibration" of the estimators' eyes, and for training. For their design, however, a realistic range of severity values per sampling unit and "preferred values" from the decimal system should be adopted. For many diseases the range of disease severity below 40% could be divided into several classes between 0.1% and 40% with only one class beyond this level. Actual design of such visual aids is based on precise measurement of lesions along a calibration curve or line within this range, or beyond (Fig. 3.4). Measurements of lesion size can be done by means of microcomputers or other methods, e.g. the one described by Sherwood et al. (1983). As the shade of colour affects estimates, coloured, standard diagrams specific for a given disease would be advantageous particularly if the age of the lesions indicates their sporulation intensity.

Standard diagrams may also be used for scaling disease intensities. Values shown in standard diagrams (Fig. 3.5) may be chosen as lower and upper limits of classes, or medians, arithmetic or geometric means. Estimates are assigned either for classes or to these means which are multiplied with their frequencies to directly obtain the mean of the sample.

The number of classes influences the accuracy of the estimate if the frequency distribution is skewed. For very discriminative work in epidemiology a maximum of seven classes in addition to a zero class seems to be optimum. The 12 classes of the Horsfall and Barratt scale (HB scale) certainly overdo the case. Berger (1981), who discusses merits and disadvantages of this scale, feels that many upper classes are superfluous without having an appropriate number of classes in the range near zero. He proposes some amendments of this scale in addition to the ELANCO tables (Redman et al. 1968; cited by Berger 1981) and suggests another table for converting ratings into

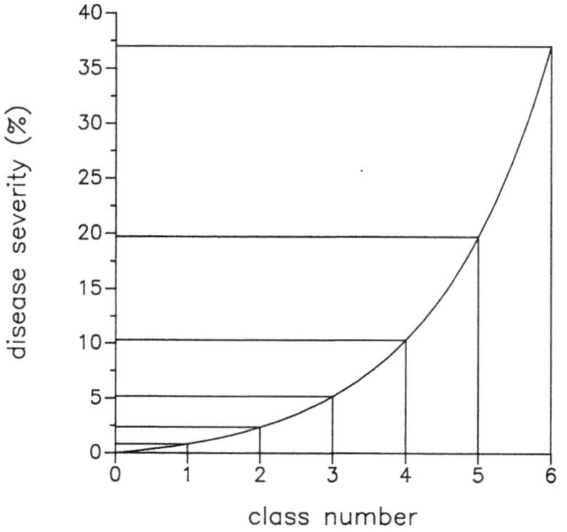

Fig. 3.4 An example for the deviation of six classes in the range from 0 to 37% based on the Weber-Fechner Law. Here, upper class limits are determined; class means would be just halfway between both class limits on the abscissa

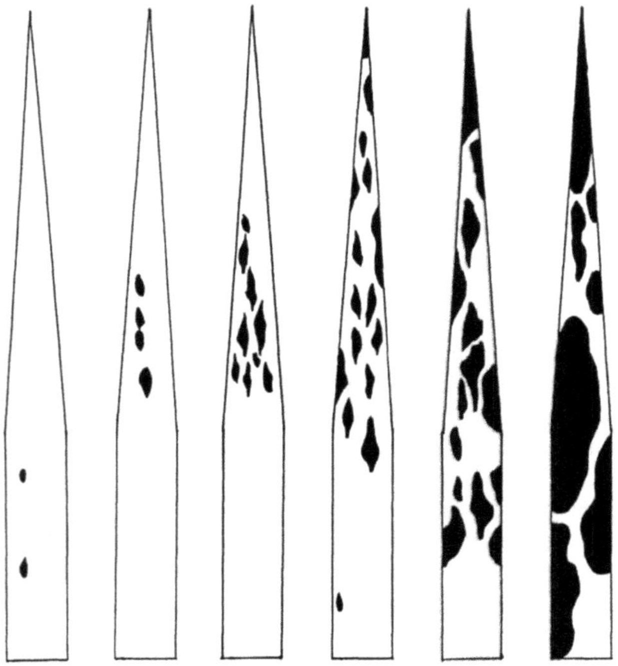

Fig. 3.5 A standard diagram for *Septoria* leaf blotch on wheat based on the Weber-Fechner Law. Values given here are the class means (Ellmer and Jörg, unpublished)

percentages. Hebert (1982) doubts whether the logarithmic function, which is underlies the HB scale, is really a valid basis for disease assessment (see p. 39).

Results obtained by inexperienced estimators indicate that their ability to assign disease severities to their appropriate classes is not great: of 203 estimates only 15.8% were assigned to their proper class, 23.2% to lower and 61.1% to higher ones (Amanat 1977). Sherwood et al. (1983) are also pessimistic about improvements achieved by the use of classes. Direct estimates, which are somewhat less cumbersome, can also ensure good results, provided the estimator is trained by means of standard diagrams.

Standard diagrams and disease assessment scales may be used more successfully for ranking or classification of diseased material, e.g. screening for resistance. Problems arise when different degrees of disease severity are present on various sides of a leaf or fruit. For cooperative programs standard diagrams will be of great use to ensure concordance between estimators.

Standard diagrams may also be developed for other facets, e.g. degree of defoliation, deformation and vigour of plant growth. Large (1966) distinguished between field keys and keys for research purposes. A model for field keys is the one designed by the British Mycological Society (1947) for potato late blight. This key has been successfully used for a number of decades (Table 3.1). It combines incidence and severity and is simple

Table 3.1 Field key for potato late blight *(Phytophthora infestans)*[a]

Percentage	description
0	No disease observed.
0.1	Only a few scattered plants affected, not more than 1 or 2 spots in 12-yd. radius.
1	Up to 10 spots per plant, or general, slight spotting.
5	About 50 spots per plant; up to 1 in 10 leaflets in 10 infected.
25	Nearly every leaflet with lesions, plants still retaining normal form; plants may smell of blight, field looks green although every plant is affected.
50	Every plant affected and about 50% of leaf area destroyed; field appears green-flecked with brown.
75	About 75% of leaf area destroyed by blight; field looks neither predominantly brown nor green.
95	Only a few leaves left green, but stems are green.
100	All leaves dead, stem dead or dying.

[a] Anon Trans Brit Mycological Soc 1947

and precise enough for field use. Its development was assisted by a panel which helped to ensure a high degree of simplicity.

5.3 Use of Incidence Values to Estimate Severity Values

It is easier and less error prone to assess incidence by counting sampling units as diseased or healthy. Therefore, it would be advantageous to estimate severity through incidence values. However, a relationship between both sets of values may change in the course of disease progress. Seem (1984) has reviewed this possibility in great detail and has identified where and when intensity-severity relationships are applicable, which conditions are required and how to establish and formulate these relationships in mathematical terms. These relationships have to be established experimentally for each pathosystem (over a number of years and conditions) before ratios, factors or other constants can be derived for each stage of the crop and disease development. Usually good correlations can be expected if disease progress curves for both incidence and severity are linear. Severity then can be estimated by the *b*-coefficient of the linear regression equations. In the case of a non-linear relationship appropriate linearization or some other methods have to be applied (see Seem 1984). It was found that with incidences up to 30% *(Septoria nodorum)* or 60% *(Erysiphe graminis)* severity levels in low ranges between 3 to 6% can reliably be estimated. This offers an application of incidence values in early stages of epidemics. Hence, one may assess intensities early in the epidemic by incidence and switch to severities later.

Assessment of incidence, severity and, if deemed necessary, destruction may also be combined by using other methods such as those present in some field keys (Table 3.1). It may also be done in multi-stage sampling, which counts the incidence of diseased plants, and determines the severity in these plants.

5.4 Automatic Measurement and Remote Sensing

Automatic measurement of disease intensities involves techniques like image analysis, mainly done with false colours (near IR with λ of 700–950 nm), spectral radiometers or microdensitometers. The measurements may be done on detached sampling units or by means of electronic leaf area meters, or, combined with microcomputers, with film and videocamera focussed on single plants (Lindow 1983) or a larger field area (Manzer and Cooper 1982; Sharp et al. 1985). Incidence can be measured with these techniques from aircrafts or satellites, depending on the sampling unit, and the objective and the scale of the study, provided the ground truth is sufficiently well established (see Nagarajan and Ajai, Chap. 17). False colour imageries may be particularly valuable and useful in the early detection of root diseases and the assessment of their intensity from changing colours in the aerial parts of plants. Their application may be unsatisfactory for the monitoring of low intensities of disease hidden inside the canopy, particularly in early stages of leaf diseases. We believe that the use of this technology will increase with future improvement of the equipment. One feature of this technique today is its resolution, which is superior to visual assessment with medium and high disease intensities, but inferior with the more frequently encountered low levels of severity.

5.5 Measurement of Disease Intensities Through Spore Catches

Spore catches have been used in some cases for semi-automatic monitoring of disease progress in fields. Another approach is counting pycnidia, or spores washed off from defined areas of diseased tissue (see Rotem, Chap. 5). These methods might be useful for some epidemiological objectives and forecasting. Berger (1981) sees little scope for the monitoring of disease intensity through cumulative spore catches because of the large variation in catches, and cumbersome work. It is obvious that only the infectious lesions could be monitored in this way. Hence, age stratification of lesions may affect spore catches more than actual disease intensity; e.g. when old, no longer sporulating lesions dominate.

Problems must also be resolved with regard to the type of spore traps and sites of their placement in the field. For virus diseases ELISA analyses from representative samples in the field may be used for a quantitative measurement of disease (see Raccah and Irwin, Chap. 14).

6 Specific Applications of Disease Measurement

6.1 Crop Measurement

Disease measurement should always be related to defined growth stages of the crop by means of growth keys, or according to the number of days after sowing, transplanting or emergence, or other botanical criteria for plant development (see Chap. 4). For crop loss assessment disease severity has to be estimated on plant parts of the same age group or leaf insertion (see sampling units). Even for some epidemiological studies such a "canopy partitioning" (Berger 1981) may apply.

Another aspect is LAI (Chap. 4, Sect. 3.3) or measured leaf area by means of non-destructive electronic equipment, standard area diagrams, length by width measurement, counting grids, etc.

In addition to the description of growth stages and crop density, a quantifiable rating of growth vigour would be useful for loss assessment (see Teng and Johnson, Chap. 12).

Colour of foliage lends itself to quantification and can be useful for the description of the nutritional and moisture status of a crop. Finally, yield has to be measured, using as many criteria as possible, e.g. number of ears m^{-2}, number of grains per ear and thousand kernel weight (TKW). It is essential that yield is measured individually for the same plants or plots on which disease severity was assessed.

6.2 Assessment of Several Diseases in the Same Sampling Unit

Assessment of several diseases in the same sampling unit obviously requires reliable diagnosis of these diseases first. For epidemiological purposes the individual severity of each disease at each sampling date should be known for an assessment of interactions between these diseases. The relationship between incidence and severity may also be influenced by the competition of these diseases in colonizing the host tissue (see Sect. 5.3). Discriminative assessment of diseases should, if possible, be made under standard conditions, particularly with similar leaf spot diseases. Bright sunlight, an overcast sky or shadow, or leaves being dry or wet, may change the appearance of these symptoms.

For loss assessment a distinction between various diseases is only necessary, if the pathogens involved interact differently with the host metabolism, or are connected with a different loss equivalent damage coefficients (i.e. percent yield loss/percent disease intensities). However, in most cases detailed information on the action of various pathogens is not available. We then estimate the total disease intensity per sampling unit, or we estimate its complement, i.e. the remaining healthy leaf area.

6.3 Disease Assessment for Models and Forecasts

Models, forecasts and loss assessments (see also Chaps. 12, 13, 18 and 19) are often based on disease intensities or threshold values, and are often used for decision making in disease control. Accurate monitoring of disease intensities in the field is therefore essential for operation of these models. For instance, if disease severity, particularly near 0, is overestimated, the infection rates will appear lower than they actually are. Also, the impact of disease severity on yield loss may be underrated. Particularly important are the instructions worked out for sampling units, sampling sizes and methods with which disease intensities are measured. Hence, when developing models for in-field application and forecasts, the degree of accuracy and precision needed has to be predetermined as well as the sampling method which ensures inputs that are representative for fields, regions, etc. For most simulators and forecasts this monitoring requirement is neglected. During the validation of models and forecasts, the measurement of disease, if needed, has to be scrutinized accordingly.

6.4 Differences Between Disease Assessment on Shoot and Root Diseases

There is no essential difference between disease assessment of shoot and root diseases. In order to assess incidence or severity on roots, the host plants have to be lifted and adequate methods for root and soil extraction have to be employed. Roots then become a sampling unit as are leaves or stems of an aerial disease. For realistic disease progress curves of soilborne pathogens (Huisman 1982), the dynamics and pattern of root growth needs to be observed as described for shoot diseases. Assessments at regular intervals is appropriate for shoot as well as for root diseases, and only one reading at the end of the season does not suffice.

There are two principle ways of measuring disease intensities of root diseases. One method uses "additional" plants in the experimental plot which, at given sampling dates, are carefully uprooted and examined. The second method assesses intensities of root diseases according to features of the above-ground plant parts such as growth, leaf mass and colour. Relationships between root disease and the features of shoots have to be established by pilot experiments. MacNish and Lewis (1985) working with *Rhizoctonia* in wheat, compared various methods of measurements, viz. mapping on grid paper (too time-consuming), areal photography (too expensive) and scoring, which provided good correlations between patch areas, and proved to be the quickest. For patch shape, orientation and position, mapping was the method of choice (Mac Nish 1985). Apart from mapping methods (e. g. frequency distributions, indices of dispersion, like the k of the negative binomial equation, but also quadrat variance techniques), spectral analysis and lag correlations may be employed in the assessment of soilborne disease intensities and distributions (Campbell and Noe 1985).

According to Campbell and Pennypacker (1980), one consequence of a clustered versus a random distribution of lesions induced by soilborne pathogens may be the reduced accuracy of the inoculum density in clustered diseases. When inoculum density is assessed by the number of lesions in random sampling, a high variance-to-mean ratio (for both the number of lesions and the inoculum density in the soil) appears and a larger sample size may be required than the sample size needed when lesions are randomly dispersed. For a comparison of statistical methods for studying spatial patterns of soilborne pathogens in the field, see Nicot et al. (1984), and Gilligan as well as Barker and Noe (Chaps. 6 and 15).

6.5 Some Implications in Disease Assessment for Crop Losses

For crop loss assessment, i.e. loss magnitudes, loss equivalents (damage coefficients) and loss profiles (Hau et al. 1980), no distinction of the age of lesions is required. Instead, it is essential to know the intensity of a disease, and destruction or defoliation, at a given stage of host growth and its loss equivalents, or damage coefficients. Loss equivalents, in turn, are basic components of economic damage thresholds.

The following problems can arise in crop loss assessment. One is the use of exaggerated visual estimates of disease severity as independent variables in models for loss prediction (see Sect. 6.3). Another problem derives from the fact that a particular disease intensity does not necessarily have consistent loss equivalents over a season. Even at periods critical for yield formation the resulting loss may differ for the same disease intensity. At least for cassava (Hau et al. 1980) and potatoes (Rotem et al. 1983) the absolute amount of active leaf mass at such periods is more relevant. In addition to disease intensity, this mass can be affected by leaf shedding and the refoliation or new root growth. Hence, disease severity percentages assessed at a certain sampling date may not reflect the cumulative past effects of the disease. For instance, when old, diseased leaves disappear due to shedding and no new growth occurs, the absolute amount of dead leaf area appears very low. This may affect our conclusions because the highest estimates in the percentage of disease severity can be obtained just at such a stage.

A particular problem appears when several diseases and plants affect the same plants (see Sect. 6.2). If multivariate loss assessment equations, or simulators, are available and symptoms of various agents can be satisfactorily distinguished, the separate assessment of severities can provide the input for these models. For the development of such models very intensive field or phytotron studies with a large sample of 400 – 1500 individual plants is required for many years and sites. Their entire life history has to be recorded, at least for the organs that contribute most to yield formation. In these cases very complex statistical analyses are needed to unravel both individual and combined effects of various diseases and pests. These aspects and the development of crop loss models are dealt with by Teng and Johnson (Chap. 12).

6.6 Measuring Disease in Surveys

Surveys are done for various objectives such as the occurrence and economic importance of diseases in a certain area, the epidemic patterns of losses, etc. The timing of a meaningful survey should be carefully chosen in accordance with the assigned objective. An appropriately designed questionaire or field record form is essential to obtain the required information and to sustain their concordance. Pilot surveys may be needed for proper planning; and disease progress curves should be at least roughly known.

Disease measurement in surveys (see also Weltzien, Chap. 16) is usually not too scrupulous and only one or a few values of disease intensity are assessed in each unit area of crop and season. Large (1966) suggests that for wilts the 0, 0,1 – 5, 6 – 25 and 25% ratings are sufficent. For other surveys we use 0, <1, 1 – 25, 26 – 75, >75% ratings for incidence and 0, <1, 1 – 5, 6 – 25, >25% ratings for severity. Consequently, we arrive at a double figure per field e.g. 2 – 4, being equivalent for an incidence ranging from 1 – 25% and a mean severity of >25% per diseased plant.

7 References

Adams MJ, Hide GA, Lapwood DH (1985) Sampling potatoes for the incidence of tuber diseases and levels of inoculum. Ann Appl Biol 107: 189 – 203

Amanat P (1977) Modellversuche zur Ermittlung individueller und objektabhängiger Schätzfehler bei Planzenkrankheiten. Diss Univ Gießen

Analytis S, Kranz J (1972) Bestimmung des optimalen Stichprobenumfanges für phytopathologische Untersuchungen. Phytopathol Z 74: 349 – 357

Berger RD (1981) Measuring disease intensity. In: Teng PS, Krupa SV (eds) Assessment of losses which constrain production and crop improvement in agriculture and forestry. Proc of EC Stakman Commemorative Symposium, Univ Minnesota, Minneapolis, Min, pp 28 – 31

Boivin G, Sauriol P (1984) Dispersion statistics and sequential sampling plan for leaf blight caused by *Botrytis squamosa* in onions. Phytopathology 74: 1385 – 1387

British Mycological Society (1947) The measurement of potato blight. Trans Br Mycol Soc 36: 267

Campbell CL, Noe JP (1985) The spatial analysis of soilborne pathogens and root diseases. Annu Rev Phytopath 23: 129 – 148

Campbell CL, Pennypacker SP (1980) Distribution of hypocotyl rot caused in snap bean by *Rhizoctonia solani*. Phytopatholgy 70: 521 – 525

Cochran NG (1963) Sampling techniques, 2nd edn Wiley, New York
Forbes GA, Jeger MJ (1987) Factors affecting the estimation of disease intensity in simulated plant structures. Z Pflanzenkr Pflanzenschutz 94: 113–120
Hau B, Kranz J (1979) Disease assessment: individual and object dependent errors. Talk at the workshop "The Current Status of Disease Management", July 29th to August 3rd, 1973, Dept of Plant Pathology, Penn State Univ
Hau B, Kranz J, Dengel HJ, Hamelink J (1980) On the development of crop loss assessment methods in the tropics. Teng PS, Krupa SV (eds) In: Crop Loss Assessment. Misc Publ 7 (1980). Agric Exp Stn, State Univ Minnesota, pp 254–261 (see Berger)
Hebert TT (1982) The rationale for the Horsfall-Barratt plan disease assessment scale. Phytopathology 72: 1269
Horsfall JG, Barratt RW (1945) An improved grading system for measuring plant disease. Phytopathology 35: 655
Huisman OC (1982) Interrelations of root growth dynamics to epidemiology of root-invading fungi. Annu Rev Phytopath 20: 303–327
James WC (1971) A manual of assessment keys for plant diseases. Can Dep Agric Pub, No 1458
Koch H, Hau B (1980) Ein psychologischer Aspekt beim Schätzen von Planzenkrankheiten. Z Pflanzenkr Pflanzenschutz 87: 553–587
Kranz J (1970) Schätzklassen für Krankheitsbefall. Phytopathol Z 69: 131–139
Kranz J (1977) A study on maximum severity in plant diseases. Travaux dedies á G. Viennot-Bourgin, pp 169–173
Large EC (1966) Measuring plant disease. Annu Rev Phytopathol 4: 9–28
Lindow SE (1983) Estimating disease severity of single plants. Phytopathology 73: 1576–1581
MacNish GC (1985) Mapping *Rhizoctonia* patch in consecutive cereal crop in Western Australia. Plant Pathol (Lond) 34: 165–174
Mac Nish GC, Lewis S (1985) Methods of measuring *Rhizoctonia* patch of cereals in Western Australia. Plant Pathol (Lond) 34: 155–164
Manzer FE, Cooper GR (1982) Use of portable video-taping for aerial infrared detection of potato diseases. Plant Dis 66: 665–667
Nicot PC, Rouse DI, Yandell BS (1984) Comparison of statistical methods for studying spatial patterns of soilborne plant pathogens in the field. Phytopathology 74: 1399–1402
Poushinsky C, Basu PK (1984) A study of distribution and sampling of soybean plants naturally infected with *Pseudomonas syringae* p.v. *glycinea*. Phytopathology 74: 319–326
Rotem J, Bashi E, Kranz J (1983) Studies of crop loss in potato blight caused by *Phytophthora infestans*. Plant Pathol (Lond) 32: 117–122
Seem RC (1984) Disease incidence and severity relationships. Annu Rev Phytopathol 22: 133–150
Seem RC, Magarey PA, McCloud PIM, Wachtel MF (1985) A sampling procedure to detect grapevine downy mildew. Phytopathology 75: 1252–1257
Sharp EL, Perry CR, Scharen AL, Boatwright GO, Sands DC, Lautenschläger LF, Yakyaoui CM, Ravet FW (1985) Monitoring cereal rust development with a special radiometer. Phytopathology 75: 936–939
Sherwood RT, Berg CC, Hoover MR, Zeiders KE (1983) Illusions in visual assessment of Stagonospora leaf spot of orchard grass. Phytopathology 73: 173–177

Chapter 4

The Measurement and Analysis of the Effects of Crop Development on Epidemics

Robert C. Seem[1]

> *"Epidemiologists have plainly been busier with the pathogen than with the diseased plant."*
>
> *C. Populer (1978)*

1 Introduction

The above quotation is a plain indictment of plant epidemiologists who confine their interests to the dynamics of pathogen populations while failing to consider the equally important dynamics of host populations. Many scientists, who study epidemics of plant diseases, are first trained as plant pathologists with primary emphasis on the pathogens causing diseases. Consequently, epidemiological studies by plant pathologists often consider the dynamics of the pathogen under static crop conditions. Yet the host is as dynamic as each of the other two members of the plant disease triumvirate, the pathogen and the environment.

Populer (1978) along with Yarwood (1959) have eloquently described how the host and pathogen interact with respect to changes in host susceptibility over time. Greater appreciation for the consequences of host development on disease development would improve all aspects of the measurement and analysis of an epidemic. In fact, some previously unsolved epidemiological problems may be resolved by accounting for host growth. Plant growth and development is not an area of study independent from general epidemiological studies and thus should be considered by all who measure and analyze plant disease epidemics.

This chapter describes some of the more important influences of crop development on disease epidemics, and some of the special considerations and techniques used to measure and analyze the epidemic as affected by dynamic crop development. The interaction between pathogen and crop development will be considered in terms of how the host affects the pathogen as well as how the pathogen affects development of the host. We will consider both the techniques of measurement and analysis of this interaction. Finally, we will look at some special considerations of this interaction, particularly as it is affected by perennial crops.

It is also important to mention that this chapter does not deal with the static aspects of host influence on disease dynamics. Static aspects are those components of the crop that do not change over time, such as genetic resistance and stand density. Next, because of the experience of the author, topics and examples will emphasize foliar diseases caused by fungal and bacterial pathogens. Soilborne pathogens and systemic diseases are no less important, but I must leave these topics to others (Gilligan, Chap. 6).

---

[1] Department of Plant Pathology, New York State Agricultural Experiment Station, P.O. Box 462, Geneva, New York 14456, USA

2 The Dynamic Interaction of Host and Pathogen

2.1 The Pathogen Affects the Host

Disease alters plant development in many ways. However, there are general categories of interaction ranging from simple destruction of leaf tissue (area) to complex effects on host physiology. The host is usually in a state of continuous change and as a result, the impact of such factors as removal of leaf area or physiological alteration is always changing.

Charles-Edwards (1982) described five ways in which the development of a plant can be constrained. The constraints are: (1) reduction in the efficiency with which light energy is used in the production of new dry matter; (2) reduction in the amount of light energy intercepted by the plant; (3) alterations that affect the proportion of new dry matter that is partitioned to different plant parts; (4) alteration of the rate of loss of dry matter; and (5) the duration of production of particular plant parts. Diseases can alter plant growth in ways that can be attributed to all five constraints.

Plant diseases contrain plant growth by reducing the photosynthetic efficiency in complex ways. Diseases that destroy roots and plug vascular tissue will reduce the plant's photosynthetic ability by altering movement of water and assimilates in the plant. Toxins and other by-products of the pathogen can also alter the physiology of the host in ways that reduce photosynthetic efficiency.

The constraint most commonly attributed to plant disease is the reduction of the amount of light energy intercepted by the plant. Not only do lesions reduce the amount of healthy tissue intercepting solar radiation but the diseased tissue can also intercept and render useless solar radiation that otherwise could be utilized by healthy leaves lower in the canopy.

Although production of callus or galls might first come to mind as examples of the altered partitioning of new dry matter, there are other, more important examples of this constraint in diseased plants. The allocation or partitioning of new dry matter by the plant is a highly complex process and usually changes with time or condition of the plant. For example, a growing plant with a foliar disease may allocate less new dry matter to the roots since the impaired leaf tissue can only support the current growth of leaves and has little or no surplus to export to roots. Similarly, because of the disease-weakened condition of a plant, less dry matter may be allocated to the fruiting portion of the plant. Finally, a plant that has lost tissue to disease may attempt to replace that tissue by initiating additional new growth and thus partition new or existing dry matter to the new growth effort.

Plant disease may accelerate the natural loss of dry matter in plants due to senescence. Kranz (1976) has made the interesting observation that the loss of diseased leaves is related more to the duration of disease on the leaf rather than the severity of disease on the leaf. Indeed, disease on a leaf may reach an asymptotic level that may only occupy a portion of the total leaf area, yet in time, that leaf may, due to senescence, drop long before healthy leaves. Dry matter losses can occur in all plant parts, and depending on the stage of growth of the plant, may affect current growth by re-allocation of new dry matter to replace what is lost, or by altered future growth when stored dry matter reserves are lost.

The final constraint, the effect on the duration of production of a particular plant part, can best be described by an example. Rees and Platz (1983) have shown that yellow spot *(Pyrenophora tritici repentis)* on wheat delayed flowering when the disease occurred early, and hastened crop maturity when it occurred late in the season. This constraint can be most profound when the plant part affected is that part from which commercial yield is derived.

2.1.1 Plant Growth Dynamics and Loss of Photosynthetic Tissue

Because of the plant pathologists' preoccupation with the pathogen and the amount of host damaged by the pathogen, we usually are interested in quantifying the amount of diseased host tissue and then relating that factor to other measures of host such as yield. However, it is often ignored that the growing plant has the capability to compensate for a certain portion of lost capacity. As a result, a low level of diseased tissue may result in no outward reduction in the capacity of the plant to grow and yield. The effects of disease would only be apparent when the amount of diseased tissue exceeds the excess capacity of the plant. In an analogous way, Gutierrez et al. (1983) showed that a cotton crop could compensate for the loss of individual plants due to verticillium wilt because the plants adjacent to the diseased plant took advantage of the resources (sunlight, water, canopy volume) that were underutilized by the diseased plant. This is equivalent to individual plants because the resources not used by diseased tissues can be reallocated to healthy tissue, thus compensating for the affects of the disease.

Some researchers have found that the growth and production capacity of a diseased plant is more closely related to the amount of healthy tissue rather than to the amount of diseased tissue (Hooker 1979; Rotem et al. 1983). The amount of diseased tissue is important for the measurement of factors such as inoculum production; however, the measurement of the amount of healthy tissue may better reflect the influence of disease on the host plant.

2.1.2 Growth Stimulation Caused by Disease

The internal regulatory process of growth in diseased plants occasionally causes alterations of typical growth patterns. Dramatic alterations such as galls and witch's brooms are of less epidemiological significance than simple flushes of new growth. Plants which have sufficient dry matter reserves such as woody perennials, can produce new leaves if the old leaves are lost due to disease. For example, Populer (1972) has discussed production of new leaf flushes on rubber which have lost their young leaves to powdery mildew *(Oidium heveae)*. In fact, the rubber tree will produce successive flushes of new leaves until sufficient leaves have grown beyond the stage where they are susceptible to infection and defoliation by the mildew fungus.

Flushes of new leaves cause changes in the amount of susceptible tissue. Indeed, in the case of rubber powdery mildew, the proportion of susceptible leaf tissue can fluctuate depending on the stage of the refoliation. Yet the extent and duration of refoliation is a direct function of the amount of disease present.

2.2 The Host Affects the Pathogen

2.2.1 Host Susceptibility

Changing susceptibility of the host to disease is the greatest effect the host can have on a disease. Changes in susceptibility occur both in either a time or space dimension and can be termed time-dependent and space-dependent susceptibility. From an epidemiological point of view, altered susceptibility can have a great influence on pathogen development and consequently epidemic development.

In the case of rubber, the susceptibility changed in time; it was time-dependent susceptibility. Yarwood (1959) classified time-dependent susceptibility (Yarwood called it ontogenetic predisposition) into four groups: (1) juvenile susceptibility with mature resistance; (2) juvenile resistance mature susceptibility; (3) juvenile and mature plant susceptibility with a middle period of resistance; and (4) juvenile and mature plant resistance with a middle period of susceptibility. He also provided examples of each category and suggested that, in general, susceptibility to facultative saprophytes increases with age, while susceptibility to obligate parasites decreases with age. Time-dependent susceptibility can also be attributed to the emergence of different plant organs with differential levels of susceptibility (Kranz 1978).

For epidemiological studies the form of time-dependent susceptibility must be determined. Populer (1978) suggested the use of uniform inoculation at different times during the growth of the plant. The method is complex because, unless the inoculation is carried out under standard environmental conditions, inoculation at different times of the host life can be subject to very different environmental conditions and thus alter the likelihood of infection and disease development (see Sect. 4.1). Once the category of time-dependent susceptibility has been identified, the proportion of susceptible tissue must be partitioned from the resistant tissue in order to properly assess the rate of epidemic development (see Sect. 5.2).

At any one time, similar tissues of the same physiological age may differ in susceptibility due to the position on the plant. For example, leaves of the same age may differ in susceptibility due to the position within the plant canopy; leaves shaded inside the canopy may be more susceptible than exposed leaves on the outside of the canopy. Although these effects can also be attributed to indirect influence of the environment on the host, the difference in susceptibility does exist and should be considered when making quantitative measures of susceptible plant tissue, and disease.

The influence of host susceptibility on pathogen development can affect major components of epidemic development. These influences, adapted from Berger's (1977) list of specific actions of disease resistance, include: (1) reduction in the total number of infections; (2) reduction in the rate of lesion expansion; (3) reduction in the reproductive capacity of the pathogen; (4) lengthening the latent or incubation period; and (5) shortening of the infectious period (Berger 1977).

One additional influence of host plant susceptibility is on asymptotic, or maximum, disease development. Observations have been made that in different pathosystems the maximum amount of disease did not reach 100% (Kranz 1975; also Chap. 3). There are many factors that can influence this phenomenon, and one of the primary ones is the changing susceptibility of the host. If an epidemic starts during a period of host susceptibility and does not infect all susceptible tissue before the host acquires mature plant resistance, then the disease will not advance much beyond the level achieved before the change to resistance.

The analysis of epidemics could become exceedingly complex if one was to permute the above influences on epidemic development with the different possible categories of susceptible tissue on a growing host. The researcher needs to be aware of all these possible situations so that there may be a reasonable explanation when the expected results do not appear.

2.2.2 *New Tissue*

The host is rarely in a static stage of growth and is usually in the process of growing new tissue and/or losing old tissue due to natural or stress-induced senescence. The host can have a great effect on epidemic development via changes in the amount of plant organs susceptible to infection. For example, when the rate of leaf expansion by the host equals or exceeds the rate of lesion expansion of the pathogen, a zero or negative disease progress rate will occur when disease is measured as a percent (Berger and Jones 1985). However, a zero or negative infection rate should not be automatically attributed to host growth because, as Kranz (1975) pointed out, other factors such as reduced inoculum production combined with pathogen-induced leaf loss could result in similar effects.

Van der Plank (1963) tried to correct his apparent infection rates for growth of the host. The correction was an instantaneous rate of change of susceptible plant tissue and was proportional to the current amount of susceptible tissue. This simple correction has been used with mixed results; Kranz (1975) found it useful in the analysis of five pathosystems, while more recently, failure of the correction to predict apparent infection rate (Berger and Jones 1985; Kushalappa and Ludwig 1982; Lalancette and Hickey 1969) has resulted in further modifications in epidemic analysis (see Sect. 5).

2.2.3 *Plant-Altered Environment*

A final way in which a growing host plant can affect the pathogen concerns the altered environment around the plant. As the plant grows and a typical canopy fills in with new tissue, the environment that affects pathogen development changes. Light level, temperature, and wind turbulence usually decrease, while moisture (relative humidity) usually increases (Rotem 1978; Zadoks and Schein 1979). Although it is the pathogen-environment interaction that is altered, the alteration is indirectly due to the host development.

While some work has been done on the epidemiological significance of factors such as shade (Kranz and Aust 1979), very little research has been conducted to determine the effect of the plant canopy on disease development, let alone how the growing plant affects the canopy environment (but see Rotem 1982).

3 Measurement and Analysis of Host Development

Charles-Edwards (1982) pointed out that there are three distinct stages of plant development: establishment, vegetative growth, and reproductive growth. The type of measurement best suited for each stage can vary and may interact very differently with disease development.

Establishment is the stage encompassing initial growth of the host when even low levels of disease can have a profound effect on future development. Damping-off diseases typically occur in this stage and their effect is usually qualitative; either the plant survives or it dies. As a result, quantitative growth of plants in this stage is not usually measured. The resulting impact is measured in the subsequent stages of growth and is expressed as a reduced number of individuals in the host population.

The vegetative stage of growth is the time when the host plant is building the structure and reserves for reproductive growth. It is during this stage that most of the interaction between host and pathogen can take place since leaf and root expansion is greatest during this stage. Also, because it is a precursor to the reproductive stage, the interaction between pathogen and host in the vegetative stage will have an impact on transition to the reproductive stage.

The reproductive stage includes the fruiting phase of the plant. Major phenological and physiological changes take place in the plant when it shifts from vegetative to reproductive growth. Interaction between reproductive growth and disease can be either direct or indirect. For example, indirect interaction occurs when disease has altered the vegetative growth and in turn alters assimilate production and allocation to the fruiting structure of the plant. Direct interaction occurs when infection takes place on or in the fruiting structure. For example, cereal head smuts will completely replace the grain head, or infection of young apple fruit will cause severe deformation of the fruit as it grows.

For the remainder of this chapter, we will direct our attention almost entirely to the interaction of a pathogen with the vegetative growth of the host.

3.1 Physiological Time

An important concept in the measurement and analysis of plant growth is physiological time. Although plants grow and develop over time, the time scales are rarely the same between different locations and years. This is because of the environmental influences on plant growth. A time scale where growth is repeatable in one that incorporates the effects of environment. This time scale is called physiological time. The most typical form of physiological time is "degree day" in which the effect of temperature, usually greater than some minimum value, is accumulated over time. However, physiological time can take many different forms including the solar-thermal unit (Caprio 1974; see also Friesland and Schrödter, Chap. 8).

Physiological time standardizes the time component of plant growth. Therefore, it has been recommended that observations of plant growth and disease development be made (or at least analyzed) on a physiological, rather than a chronological time scale (e.g. DAS, days after sowing as 0-time). Likewise, development of pathogens is also measured better in terms of physiological time rather than chronological time. However, the measure of physiological time for the pathogen may not necessarily be the same measure of physiological time for the host.

3.2 Host Phenology

The simplest measure of host growth is by observation of its phenological growth stage, that is, rating the current stage of growth with the aid of some pictorial or descriptive key. This is the earliest form of plant growth measurement, dating back to the first half of the 18th Century (Caprio 1974). Keys describing the phenological growth of plants are now available for many agricultural crops (Chiarappa 1971; Zadoks et al. 1974) and cover the seasonal development, from establishment to reproductive growth. The stages represent easily observed and indentified phases of plant development and are used as descriptors of change in major plant structure or as counts of increasing numbers of plant units. Because they were selected primarily for observation only, the stages usually have no relationship to one another other than they follow in some defined sequence. As a result, the quantitative analysis of phenological data is not usually conducted.

Phenological growth stage is the minimal measure of plant growth required where interaction between pathogen and host growth is suspected. In a very empirical way, the phenological stage provides an estimate of the amount and type of each major plant organ. Although it is not a quantitative measure, it does provide a means of determining such things as the approximate number of leaves, when vegetative growth stopped, when flowering was initiated, and when fruit have matured. In many instances this type of information cannot be derived from more quantitative measures of plant growth.

Assuming a phenological key for the crop of interest is available, the stage of growth of the plant (or crop) should be noted every time disease observations are made. In fact, phenological observations made at frequent time intervals permit the recording of initiation of all major stages of plant growth. In this way, age of tissue types can be adequately estimated. If no key is available one can easily be made by describing or drawing the major stages of plant growth similar to published stages for other plants. Careful consideration must be given to the definition of stages that describe indeterminant growth of plant units such as the number of leaves on a shoot (Populer 1978). Leaf counts starting at the growing tip assign the same number to leaves of similar size and developmental stage, while counting leaves starting at the base of a shoot assigns the same number to a given leaf position throughout the growing season. The former method would be preferred when young leaves are most (or more) susceptible to disease (i.e., the youngest four leaves of an apple shoot are susceptible to *Venturia*), while the latter method would be useful when mature leaves are susceptible to disease (old cucumber leaves are more susceptible to *Pseudoperonospora*).

Phenological observation consists of either noting the date of first occurrence (or 50% occurrence) of a stage, or recording the stage (number) for a sample of plants on specific sampling dates. In some crops separate phenology keys may be used for vegetative and reproductive phenology (Seem and Szkolnik 1978), especially when both types of phenological stages can occur at the same time.

Analysis of phenological data can consist of either casual association of the growth stage to the time of season or quantitative association of the growth stage to physiological time. In the latter case the amount of physiological time needed to reach a particular stage is calculated. Various forms of physiological time can be used such as average temperature, degree days, and solar-thermal units (degree days and a linear index of daily solar radiation). Examples of these and other measurements of physiological time used to estimate phenology can be found in the work of White (1979).

3.3 Measurements for Growth Analysis

The more typical approach to growth analysis involves the measurement and analysis of two basic factors; the plant material present (W), and the magnitude of the assimilatory system of that plant material (A). In practice, the most common measures of W and A are total dry weight and total leaf area, respectively. If the crop under study has a continuous canopy, then the relevant measures would be the total dry weight of plant material per unit area of ground (W), and the total leaf area of the plant material per unit area of ground (A). This form of A is better known as the leaf area index (LAI). Radford (1967) lists other measures of W and A used in growth analysis and emphasizes that whatever type of measure is used, the units of measure should be clearly stated in order to compare results of different experiments.

The actual measurement of W can be dry weight or fresh weight. W may also be partitioned into different components of the plant such as leaves, stems, roots, fruit, etc. For analysis of the interaction of host and pathogen, W might be partitioned into susceptible and nonsusceptible components.

The assimilatory system of the plant (A) is typically measured as leaf area, or occasionally, leaf weight. Modern leaf area digitizers ease this process, but traditional methods of length-width ratios, planimetry, and dot-counting grids are useful means of measuring leaf area. Photographic techniques have also been used to estimate LAI of large canopies (Lakso 1980).

The plant growth measures can be incorporated into different types of analyses including: growth rate (GR); relative growth rate (RGR), net assimilation rate (NAR), and leaf area ratio (LAR) (Radford 1967). Growth rate is defined as the increase of plant material per unit time:

$$GR = \mathrm{d}W/\mathrm{d}t\,, \tag{4.1}$$

and the mean GR can be expressed in its integrated form between t_1 and t_2 as:

$$\overline{GR} = (W_2 - W_1)/(t_2 - t_1)\,. \tag{4.2}$$

The relative growth rate is the increase of plant material per unit of material present per unit of time:

$$RGR = 1/W\,\mathrm{d}W/\mathrm{d}t, \quad \text{or} \quad d/\mathrm{d}t(\ln W), \tag{4.3}$$

where ln is the natural logarithm.
This can be expressed as the mean relative growth rate between t_1 and t_2:

$$\overline{RGR} = (\ln W_2 - \ln W_1)/(t_2 - t_1)\,. \tag{4.4}$$

The net assimilation rate is defined as the increase of plant material per unit of assimilatory material per unit of time:

$$NAR = 1/A\,\mathrm{d}W/\mathrm{d}t\,. \tag{4.5}$$

The mean NAR over a period from t_1 to t_2 cannot be integrated unless the relationship between A and W (or the relationship between A and t, and W and t) is known. If it is assumed to be linear, then:

$$\overline{NAR} = (W_2 - W_1)/(A_2 - A_1)\,(\ln A_2 - \ln A_1)/(t_2 - t_1)\,. \tag{4.6}$$

Radford (1967) provides other integrated forms of $\overline{NAR}$ when A and W are not linearly related. Mean leaf ratio (LAR) is defined as the ratio of assimilatory material per unit of plant material present:

$$LAR = A/W\,. \tag{4.7}$$

The mean LAR betwenn t_1 and t_2 cannot be integrated unless either the relationship between A/W and t is known or the relationship between both A and t, and W and t are known. If we assume A/W is linearly related to time, then the integrated form of mean LAR is:

$$\overline{LAR} = 1/2\,(A_1/W_1 + A_2/W_2)\,. \tag{4.8}$$

Those seeking more information on growth analysis are referred to the works of Radford (1967) or Hunt (1982) and Zadoks and Schein (1979) with special reference to plant disease.

At this point we should note the great amount of work dealing with models of plant growth. These models are a major avenue for the study of the interaction between plant growth and pathogen development. Such modeling studies have been recently published (Berger and Jones 1985; Bloomberg 1979; Gutierrez et al. 1983; Loomis and Adams 1980) and along with general resources such as Charles-Edwards (1982); de Wit (1976); France and Thornley (1986) and Rose and Charles-Edwards (1981) should provide the reader with sufficient background to explore this important area of modeling.

4 Measuring the Dynamic Interaction

It is now necessary to consider specific techniques of measuring the interaction between pathogen development and host growth. The reader will find that the techniques are largely those used in standard epidemiological studies, as described in the other chapters of this book (see Chaps. 3, 9 and 12). However, because these techniques are now combined with measurements of development of the host, modifications are necessary to account for such factors as time- or space-dependent host susceptibility and changing amounts of host tissue.

4.1 Measuring Time- and Space-Dependent Susceptibility

If a researcher has reason to believe that time- or space-dependent susceptibility of the host to the pathogen is occurring, then the only option is to test the assumption. The process is not simple and is fraught with problems. Populer (1978) has described the difficulties and only brief mention of them will be made here.

The primary objective of this type of experimentation is to determine if the susceptibility of a plant or plant organs changes over time or location within the plant or plant canopy. The major problem is to assure that at each inoculation time the conditions for infection and disease development are equivalent, or in the case of space-dependent susceptibility, that the inoculation is done under standard, but natural conditions (see Rotem, Chap. 2).

Ensuring uniform host material is a more difficult problem because it is often difficult to grow plants to the desired stage of development under artificial conditions. The question is whether plants grown under artificial, albeit uniform, conditions possess the same time- or space-dependent susceptibility as their counterparts in the field. This can only be answered by often laborious pretesting that is still subject to the variability of the natural environment.

Another way of testing time-dependent susceptibility during a period of indeterminant growth by the host is to inoculate plant parts of different physiological age. For example, by noting the times of emergence of individual leaves on a growing shoot, the physiological age of each leaf can be determined. Thus, a single inoculation could be used to test the susceptibility of leaves with a range of physiological age.

4.2 Measuring Disease in the Presence of Changing Host Tissue

Diseased plant tissue is not usually measured as an absolute amount or area (Chap. 3), but as a proportion of plant units infected (incidence) or a proportional area of plant units infected (severity). Because the proportions are based on the amount or area of the host, if the host is changing, then the measures of disease will also change regardless of any change in the absolute amount of disease. Similarly, proportional assessments of disease on different leaves can vary even though the absolute amount of disease is constant. This is due to the morphological variation of leaves. For example, the oldest and youngest leaves on an apple terminal shoot are usually smaller than midterminal leaves. Thus, the proportional assessment of the lesion area on midterminal leaves would be less than the assessment of the same lesion area on leaves at the extremities of the terminals.

There are two basic methods of accounting for changes in measures of disease due to host growth. The first is to measure the absolute change in the amount of tissue and to adjust the pool of existing host tissue (such as the LAI, see also Sect. 5.1) correspondingly. This method assumes that all tissues are equally susceptible to disease. As a result, proportional estimates of disease can be adjusted according to the change in host tissue. The second method consists of partitioning host tissue into classes, the simplest of which would be susceptible and nonsusceptible tissue. Each class of tissue would be rated for disease separately. In the case of rapid increase of susceptible host tissue, the susceptible tissue could be further partitioned into classes of uniform physiological age. These methods can be used for both increasing and decreasing susceptible host tissue.

For those situations where the preferred measure of disease is area under the disease progress curve (AUDPC), the appropriate measure of host leaf tissue would be leaf area duration (LAD), the integral of LAI over time (Thorne 1966). Leaf area duration should not be confused with leaf area density (also abbreviated LAD) as discussed by Zadoks and Schein (1979). The difference between LAD and AUDPC, assuming a common unit of measurement, represents healthy leaf area duration (HLAD) which can serve as a very useful measure of plant production or yield potential (Waggoner and Berger 1987) as long as the disease does not affect the production of the assimilate by asymptomatic host tissue.

4.2.1 Increases in the Amount of Susceptible Tissue

In the case of increasing host tissue that is susceptible to infection, assessment of total susceptible tissue should be made at intervals no less frequent than a single multiplication cycle of the pathogen (Populer 1978). In this way infections of new tissue can be associated with assessable levels of inoculum. In reality, however, the multiplication cycles of the pathogen are usually too frequent and often overlap each other and, therefore, levels of inoculum cannot be assessed with accuracy.

A practical solution is to make observations of both disease and host growth either at predefined increments of physiological time or, in the case of event-driven diseases such as those caused by pathogens that require free water for infection, at the time of major disease events. The assessment of new host tissue can be simplified if preselected sampling sites are labeled and at each assessment time a new label is attached to the growing point. Thus, at the next time of assessment only the new growth ahead of the label needs to be measured. Sufficient numbers of sampling units must be selected, particularly if the measurements are to be made over an extended period of time since some labels may be lost and some growing points may be damaged or die. Further, care must be taken if the growth data are to be statistically analyzed because resampling the same initially random sample is different from selecting a random sample of unlabeled growing points at each observation date.

In order to accurately assess disease dynamics, the host tissue area susceptible to infection must be differentiated from nonsusceptible host tissue. Because host susceptibility can be a dynamic process as well, tissue that was susceptible at the time of infection may not be susceptible by the time symptoms develop. Consequently, the measurement of susceptible tissue cannot be based on the presence of disease. Knowledge about the host and the susceptibility of its tissue must also be known. For example, only the four youngest leaves on an actively growing apple terminal shoot are highly susceptible to infection by the apple scab pathogen, though symptom expression generally occurs on older leaves. In addition, even though infected tissue may be resistant to further infections, the lesions may continue to expand into healthy tissue.

Finally, new leaves on shoots or stems unfold or expand to reach maturity, thus leaves at the growing tip are of a different size. If new tissue is assessed by counting the number of leaves on a shoot or stem and the number starts at the growing tip, then the mean area of a sample of leaves representing each of the numbered positions can be determined. These areas can be assigned to each leaf position and the total leaf area can be estimated.

4.2.2 Decreases in the Amount of Susceptible Tissue

The situation of decreasing susceptible tissue is much less frequent than increasing tissue. There are two conditions under which susceptible tissue will decrease. First, time-dependent susceptibility of tissue can result in a decrease of susceptible tissue due to transition from susceptible to resistant stages. Measuring disease under this condition was addressed previously (see Sect. 4.1). Second, susceptible tissue can be physically removed through the processes of harvest, pruning (as a result of management), pruning as a consequence of the disease, or damage by other diseases and insects. The latter two instances can present problems in measuring disease and host because tissues pruned by the disease (i.e., girdling of stems or petioles, or disfunction of leaf veins by local lesions) can remove healthy, susceptible tissue. The decision must then be made whether to class this tissue as diseased since it no longer contributes to the growth or yield of the host, or nondiseased since it is not likely to become infectious tissue. If the lost tissue is classed as diseased, then it would be helpful to separate into infectious and noninfectious subclasses. In the case of competing pests, these tissues are also removed from the pool of total tissue, but the measurement requires much more careful discrimination between the diseased tissue of interest and damage caused by other diseases or insects (see also Chaps. 3 and 12).

5 Analysis of the Dynamics Interaction

The analysis (and interpretation) of measurements that include the dynamic interaction between the growing host and pathogen are similar to the standard methods of epidemic analysis with several notable exceptions. These exceptions include time-dependent changes in the amount of susceptible tissue and the disease asymptote, as well as the formation of host tissue classes each of which can have different infection efficiencies, latent periods, lesion expansion rates, and sporulation efficiencies.

5.1 Analysis of Changing Susceptible Host Tissue

Van der Plank (1963) was the first to propose a correction to the equation for determining apparent rates of infection. He considered the host growth to be constantly increasing in proportion to the amount of current tissue (i.e., exponential growth) and that all the new growth was susceptible to disease. Kushalappa and Ludwig (1982) adjusted the equation by permitting host growth to reach a maximum or asymptote. In both cases the amount of measured disease was adjusted by the proportional increase of host tissue. Kushalappa and Ludwig (1982) also adjusted the maximum level of disease according to the amount of host growth and made similar modifications for the exponential, monomolecular, and Gompertz models of disease progress. Rouse (1983) also proposed that the effect of host growth could be incorporated into a term of the logistic disease progress curve he called "carrying capacity". The equation has the form:

$$\mathrm{d}X/\mathrm{d}t = rX(1 - X/K)\,, \tag{4.9}$$

where X is the intrinsic measure of disease, r is the rate of disease increase, and K is the carrying capacity. K represents the maximum amount of potential disease and, according to Rouse (1983), can be a function of plant growth such as the logistic change of the carrying capacity as proposed by Turner et al. (1969). The combined term X/K is a proportional expression of disease that changes with changes in either X or K. The equation further assumes that all host tissue is susceptible. Kushalappa and Ludwig (1982) provided a more general form of the previous equation where an asymptote of disease development is included:

$$\mathrm{d}X/\mathrm{d}t = rX/K\,(A/K - X/K)\,. \tag{4.10}$$

More recently Lalancette and Hickey (1986a) have taken a different approach to analyzing the effects of host growth. They suggested that for the particular case where plants are susceptible to disease only during the host growth period, disease progress can be expressed as a function of plant growth. In essence they suggest that the time component of disease progress can be mathematically represented by the growth of the host, or as they say, "biological time". The relationship does imply that disease progress is dependent on host plant growth. They described four examples of host and pathogen interaction in which the host and pathogen have similar or dissimilar growth patterns (e.g., exponential, logistic, monomolecular, or Gompertz growth functions) and the growth of the host and pathogen initiated together or at different times. In all cases disease is computed as a function of host growth, the carrying capacities of the pathogen and host, and the combined intrinsic rates of growth for pathogen and host.

An example of this approach has been published for the apple and powdery mildew pathosystem (Lalancette and Hickey 1986b). Unfortunately, the method does not permit feedback of the pathogen on host growth, and the carrying capacities of host and pathogen are fixed.

5.2 Analysis of Changing Asymptotes

An asymptote of a disease progress curve implies some sort of equilibrium between the host and pathogen, usually that the potential infection sites on the host have been saturated by the pathogen. An implicit assumption of many methods of analysis is that the asymptote of disease is equivalent to the maximum host tissue. Yet the potential asymptote of disease is really a function of both the amount of susceptible host tissue available and the extent to which existing disease can expand into healthy tissue (amount of initial inoculum, amount of time for expansion into healthy tissue, and the rate of disease increase). Therefore, the true carrying capacity of the host is a function of these two factors. To confound matters, both factors can change with time.

Kato (1974) cited an example where growth of both host and pathogen is considered to be logistic and the asymptote of disease progress is based on the current amount of host tissue. A similar approach has been used by Jeger (1987) where host growth is based on a feedback of the amount of healthy host tissue. Assume the carrying capacity of Eq. (9) has the relationship:

$$\mathrm{d}K/\mathrm{d}t = r_{\mathrm{h}}(K - X)\,(1 - K/K_{\mathrm{max}})\,, \tag{4.11}$$

where r_{h} is the rate of host growth and K_{max} is the maximum attainable host growth. Under these conditions, if $r \geq r_{\mathrm{h}}$ than x, disease expressed as a proportion, will increase to an asymptote of 1. However, if $r < r_{\mathrm{h}}$ then x will reach an asymptote of something less than 1. Jeger (1987) showed that the lower asymptote is not stable and later considered the whole problem in more detail (Jeger 1986).

Unfortunately, as the complexity of the relationship between host growth and pathogen development increases, so does the difficulty of differentiating and integrating these equations. Indeed, some of the equations are impossible to solve, while in the case of those that can be solved, nonlinear regression methods must be used to determine values of the coefficients.

5.3 Analysis by Simulation

An alternative to the analytical approach of analyzing the interactions of host growth and pathogen development is by simulation techniques (see also Chaps. 18 and 19). Many examples of simulation of either disease development or plant growth exist, but rarely have they been combined for interaction studies.

Simulation models permit inclusion of those effects of the pathogen on the host as described in Section 2.1, namely, the five constraints described by Charles-Edwards (1982). Similarly, simulation models of the pathogen permit inclusion of those host factors that influence the development of the pathogen (Sect. 2.2). This author is not aware of any simulation model that addresses all of these interactions in combination. However, it only seems to be a matter of time until detailed interaction simulations are attempted. The work of Berger and Jones (1985) serves as a good example of such a start. In particular, they explored the influence of host growth in combination with variable length of the latent period and lesion expansion. Although the simulation techniques were straightforward, they did incorporate a feedback mechanism that permitted host growth to increase as a function of healthy tissue, thus implicating disease in the rate of host growth.

One simulation technique used by Berger and Jones (1985) to account for variable latency of infections can also be used to simulate time-dependent events of host development. The technique is called the distributed delay and is a function to generate a distribution of events that occur in stages (Manetsch 1976; Welch et al. 1978). The distributed delay is particularly useful in plant growth simulation where a single plant or a plant population may move through developmental stages, but not all at the same time. In the case of time-dependent events, such as tissue susceptibility to disease, the distributed delay would move a population of age-dependent leaves through successive stages of susceptibility. The rate of movement (how long a leaf remains susceptible) from one stage to the next can be dependent on external factors in addition to time and the distribution can be adjusted to match observed distributions of susceptibility. Minogue (1978) provided a good example of the use of a distributed delay for time-dependent susceptibility of apple leaves to apple scab infection.

6 Considerations and Modifications of Techniques for the Interaction of Perennial Crops

The perennation of crops results in different, and often special, considerations when studying the interaction of the host crop with disease development.

6.1 Survival Strategy of Perennials

The survial strategy of perennial plants is quite different from that of annual plants. In temperate regions perennials must accumulate reserves to carry through the dormant period and initiate new growth at the beginning of the next season. As a result, both vegetative and reproductive growth periods are often longer. Consequently, leaves and fruiting structure remain on the plant for a longer period of time compared to annuals. Due to the greater assimilate reserves compared to annuals, perennial plants have the capacity to produce multiple flushes of new growth if significant defoliation has occurred due to disease attack. The new growth may also be susceptible to attack and may be responsible for the irregular development of epidemics (Kranz 1978). During periods of severe stress, perennial plants are more likely to shed fruit rather than leaves in order to sustain life to the next growing season.

6.2 Size of Perennials

Perennial plants are often larger than annuals and size alone may require some special considerations for studying the interaction of disease and host development.

Large perennials are difficult to work with in chamber and glasshouses. Consequently, epidemiological studies are often conducted using seedlings or juvenile plants, and as a result, information on mature plant resistance or susceptibility may never be detected. In this situation, the researcher must either use detached material, and hope that tissue susceptibility is unaltered, or conduct experiments in the field where full-sized, mature perennials exist. In the latter situation structures may have to be constructed, such as polyethylene enclosures, to provide a certain amount of control in studies dealing with infection or inoculum production.

Another problem relating perennial size concerns the measurement and assessment of foliage of large perennial plants. This topic will be discussed in Sect. 6.4.

6.3 Time-Dependent Susceptibility of Perennials

In a previous section of this chapter time-dependent susceptibility was considered as a seasonal phenomenon. However, perennial plants can also exhibit annual and perennial variation in susceptibility to disease. Time-dependent susceptibility can occur at any time during the year for some perennials. Pathogens that invade through the woody tissue of plants can often do so most easily during the dormant period of the host. Diseases such as *Eutypa* sp. will enter grape stems during the winter when pruning or freeze injury creates open wounds in the wood. Infection during the active growing season is much less likely. Similarly, root-infecting fungi can attack the roots of perennials during their dormant periods of the year. Susceptibility can also change over a period of years, as perennials mature. Perennial species have distinct phases of perennial growth including establishment, mature growth (the main production period for fruiting perennials), and decline. All three phases can show marked differences to susceptibility to pathogens. This phenomenon is particularly true for perennials in the declining phase of growth when they become more susceptible to stressful situations and pathogen invasion. In the case of natural stands of perennials, the age structure of the stand can be very heterogeneous, and the population level, the stand may have a highly variable susceptibility to disease (Schmidt 1978).

Tropical and conifer species retain leaves for more than one growing season. Leaf tissues originating from different seasons may have a very different susceptibility to disease. These plants may never be free from disease since the pathogen can continuously reside on host tissue. In this situation a polycyclic disease may be able to carry on from one year to the next without seasonal interruption.

One distinct advantage of working with intensively managed perennial crops is that most are vegetatively propagated. The result is that a field, orchard, or plantation will have genetically uniform plants that respond similarly as a population.

The researcher who works with perennials must be aware that time-dependent susceptibility is not an event solely occurring during the seasonal period of active growth and it may change as the plant grows over a period of years. As a result, the plant tissues may have to be partitioned into structures of uniform age before disease measurement or analysis can be accomplished.

6.4 Sampling and Assessment of Perennials

One of the greatest frustrations of a researcher working with perennial crops is the difficulty in sampling and assessing growth and disease of these crops. As was mentioned in Sect. 6.2, the size of perennial crops causes severe limitations on sampling and assessment. Because individual plant units are large, sampling units must be smaller than whole plants. Sampling units must be shoots, branches, fronds, etc. The size of a perennial also makes some portions of the plant difficult to assess due to inaccessibility. Researchers often have to resort to indirect measures. For example, trunk diameter or girth has been used to estimate the amount of leaf tissue in a tree (Pearce 1976).

Indirect disease assessment methods can also be used for high-canopy crops. For example, the number of rubber tree leaves on the plantation floor is used as an indirect assessment of *Oidium* infection in the trees. This method can also be used to determine the level of disease on individual leaves that is sufficient to cause defoliation. In this and similar situations, standard grid or transect sampling techniques can be used to formalize this indirect method.

Finally, the assessment of infectious disease is also confounded in perennial plants. Disease structure such as cankers and galls may become ineffective sources of inoculum after one or more years on a perennial plant. It may be necessary to determine the age of these structures before they can be classed as infectious or noninfectious.

7 Summary and Conclusions

The measurement and analysis of the effects of crop development on plant disease epidemics is something both inherent and foreign to plant pathologists. Inherent in that the importance of the host to the development of disease is fundamental to the science. Foreign in that the host is often considered as a static component of the pathosystem. The purpose of this chapter was to demonstrate that the true dynamic nature of the host and pathogen interaction can and should be measured and analyzed in epidemiological studies. The measurements can be as simple as observation of host phenological development or as complex as area assessments of host tissue divided into organ and age classes. The technical analysis of host and pathogen dynamics is not as developed as other subjects in plant disease epidemiology. However, noteworthy advancements have been made recently, and it is likely that new ideas on the incorporation of host development into epidemic progress will be forthcoming. In the interim, epidemiologists must use analytical or simulation techniques currently available rather than using none at all.

Epidemiologists, as interpreters of disease dynamics in plant populations, must not only be intimately familiar with the pathogen and its behavior, but must be similarly knowledgeable about the host population. Only when the two systems are sufficiently well understood can the true dynamics of plant disease be described lucidly.

8 Acknowledgments

The author gratefully acknowledges the critical review of this work by D. M. Gadoury and K. L. Reynolds.

9 References

Berger RD (1977) Application of epidemiological principles to achieve plant disease control. Annu Rev Phytopathol 15: 165–83
Berger RD, Jones JW (1985) A general model for disease progress with functions for variable latency and lesion expansion on growing hosts. Phytopathology 75: 792–797
Bloomberg W (1979) A model of damping-off and root rot of Douglas fir seedlings caused by *Fusarium oxysporum*. Phytopathology 69: 74–81
Caprio JM (1974) The solar thermal unit concept in problems related to plant development and potential evapotranspiration. In: Lieth H (ed) Phenology and seasonality modeling. Springer, Berlin Heidelberg New York, pp 353–364
Charles-Edwards DA (1982) Physiological determinants of crop growth. Academic Press, London
Chiarappa L (1971) Crop loss assessment methods. Commonwealth Agricultural Bureaux, Slough, England
de Wit CT (1976) Simulation of assimilation, respiration, and transpiration of crops. Pudoc, Wageningen (simulation monographs)
France J, Thornley JHM (1986) Mathematical models in agriculture. Buttersworth, London
Gutierrez AP, De Vay JE, Pullman GS, Friebertshauser GE (1983) A model of verticillium wilt in relation to cotton growth and development. Phytopathology 73: 89–95
Hooker AL (1979) Estimating disease losses based on the amount of healthy leaf tissue during the plant reproduction period. Genetika 11: 181–192
Hunt R (1982) Plant growth curves: the functional approach to plant growth analysis. University Park Press, Baltimore
Jeger MJ (1986) The potential of analytic compared with simulation approaches to modeling in plant disease epidemiology. In: Leonard KJ, Fry WE (eds) Plant disease epidemiology: population dynamics and management, MacMillan, New York, pp 255–281
Jeger MJ (1987) Modelling the dynamics of pathogen populations. In: Wolfe MS, Caten CE (eds) Populations of plant pathogens: their dynamics and genetics. Blackwell, London, pp 91–107
Kato H (1974) Epidemiology of rice blast. Rev Plant Prot Res 7: 1–20
Kranz J (1975) Das Abklingen von Befallskurven. Z Pflanzenkr Pflanzenschutz 82: 655–664
Kranz J (1976) Einfluß einiger Pflanzenkrankheiten auf den Abgang der Blätter. Z Pflanzenkr Pflanzenschutz 83: 234–237
Kranz J (1978) Comparative anatomy of epidemics. In: Horsfall JG, Cowling EB (eds) Plant disease: an advanced treatise, vol 2. Academic Press, London, pp 33–62
Kranz J, Aust HJ (1978) Schatten als epidemiologische Einflußgröße beim Gerstenmehltau. Z Pflanzenkr Pflanzenschutz 86: 533–545
Kushalappa AC, Ludwig A (1982) Calculation of apparent infection rate in plant diseases: development of a method to correct for host growth. Phytopathology 72: 1373–1377
Lakso AN (1980) Correlations of fisheye photography to canopy structure, light climate, and biological responses to light in apple trees. J Am Soc Hortic Sci 105: 43–46
Lalancette N, Hickey KD (1986a) Disease progression as a function of plant growth. Phytopathology 76: 1171–1175
Lalancette N, Hickey KD (1986b) An apple powdery mildew model based on plant growth, primary inoculum, and fungicide concentration. Phytopathology 76: 1176–1182
Loomis RS, Adams SS (1980) The potential of dynamic physiological models for crop loss assessment. In: Teng PS, Krupa SV (eds) Crop loss assessment. Misc Publ no 7. Agric Exp Stn, Univ Minnesota St. Paul, pp 112–117
Manetsch TJ (1976) Time-varying distributed delays and their use in aggregative models for large systems. IEEE Trans Systems, Man and Cybern SMC-6: 547–553
Minogue KP (1978) A mathematical model for epidemics of apple scab. Thesis, MacDonald College, McGill University, Montreal
Pearce SC (1976) Field experimentation with fruit trees and other perennial plants. Commonwealth Agricultural Bureaux, Slough, England
Populer C (1972) Les épidémies de l'oidium de l'hévéa et la phénologie de son hôte dans le monde. Publications de l'I.N.E.A.C., Série Scientifique No 115

Populer C (1978) Changes in host susceptibility with time. In: Horsfall JG, Cowling EB (eds) Plant disease: an advanced treatise, vol 2. Academic Press, London, pp 239–262
Radford PJ (1967) Growth analysis formulae — their use and abuse. Crop Sci 7: 171–175
Rees RG, Platz GJ (1983) Effects of yellow spot on wheat: comparison of epidemics at different stages of crop development. Aust J Agric Res 34: 39–46
Rose DA, Charles-Edwards DA (eds) (1981) Mathematics and plant physiology. Academic Press, London
Rotem J (1978) Climate and weather influences on epidemics. In: Horsfall JG, Cowling EB (eds) Plant disease: an advanced treatise, vol 2. Academic Press, London, pp 317–337
Rotem J (1982) Modification of plant canopy and its impact on plant disease. In: Hatfield JL, Thomason IJ (eds) Biometeorology in integrated pest management. Academic Press, London, pp 327–342
Rotem J, Kranz J, Bashi E (1983) Measurement of healthy and diseased haulm area for assessing late blight of epidemics in potatoes. Plant Pathol (Lond) 32: 109–115
Rouse DI (1983) Plant growth models and plant disease epidemiology. In: Kommedahl T, Williams PH (eds) Challenging problems in plant health. Am Phytopathol Soc, St. Paul, pp 387–398
Schmidt RA (1978) Diseases in forest ecosystems: the importance of functional diversity. In: Horsfall JG, Cowling EB (eds) Plant disease: an advanced treatise, vol 2. Academic Press, London, pp 287–315
Seem RC, Szkolnik M (1978) Phenological development of apple trees. In: Hopp RJ (ed) Phenology: an aid to agricultural technology. Agric Exp Stn, Bull 684, Univ Vermont, Burlington, pp 16–20
Thorne GN (1966) Physiological aspects of grain yield in cereals. In: Milthorpe FL, Ivins JD (eds) Growth of cereals and grasses. Buttersworth, London, pp 88–105
Turner ME, Blumenstein BA, Sebaugh JL (1969) A generalization of the logistic law of growth. Biometrics 25: 577–580
Van der Plank JE (1963) Plant disease: epidemics and control. Academic Press, London
Waggoner PE, Berger RD (1987) Defoliation, disease and growth. Phytopathology 77: 393–398
Welch SM, Croft BA, Brunner JF, Michels MF (1978) PETE: an extension phenology modeling system for management of multi-species pest complex. Environ Entomol 7: 482–494
White LM (1979) Relationship between meteorological measurements and flowering of index species to flowering of 53 plant species. Agric Meteorol 20: 189–204
Yarwood CE (1959) Predisposition. In: Horsfall JG, Diamond AE (eds) Plant pathology, an advanced treatise, vol 1. Academic Press, London, pp 521–562
Zadoks JC, Schein RD (1979) Epidemiology and plant disease management. Oxford Univ Press, New York
Zadoks JC, Chang TT, Konzak CF (1974) A decimal code for the growth stages of cereals. Weed Res 14: 415–421

Chapter 5

Quantitative Assessment of Inoculum Production, Dispersal, Deposition, Survival and Infectiousness in Airborne Diseases

Joseph Rotem[1]

1 Introduction

The integrated effects of production, dispersal, deposition, survival and infectiousness of inoculum are manifest in the disease severity it is able to cause under the specific conditions of environmental and host factors. Evaluation of these components and of their integrated effects is often so difficult that they constitute the "black boxes" in many epidemiological analyses. There are no research techniques that will eliminate completely these black boxes. However, there are some direct and indirect approaches that may improve quantitative assessment of the individual components and their integrated action.

Buildup of inoculum, its dispersal, survival, etc. should be studied mainly in the field because (1) in growth chambers, some environmental conditions such as wind and radiation are difficult or impossible to simulate; (2) some observations require an experimental period too long for normal host development in growth rooms; and (3) the effect of micrometeorological phenomena in rather small plants in growth chambers is less pronounced than in the field (see Rotem, Chap. 2). Nevertheless, the study of some topics may, and in many cases should, be assisted by controlled-conditions experiments.

This chapter emphasizes aspects of airborne diseases, except those of long-distance dispersal (see Nagarajan and Ajai, Chap. 17) and virus diseases (see Raccah and Irwin, Chap. 14). For the relevant aspects of soilborne diseases, see Chap. 6 by Gilligan and Chap. 15 by Barker and Noe, the latter on nematodes.

2 Production, Dispersal and Buildup of Inoculum

The amount of inoculum present in the field is often estimated from the concentration of spores caught by spore-trapping instruments. The number of spores caught by these instruments is correlated with the development of specific epidemics, such as the rust epidemics in the USA (e.g. Burleigh et al. 1969). However, as discussed in Sect. 2.1, the number of spores trapped in the air indicates trends, but not amounts of spores produced on plants. Assessment of inoculum in the field is possible also by periodic collection of spores from spore-bearing leaves. The two methods may give different results and require careful interpretation.

[1] Department of Plant Pathology, Agricultural Research Organization, The Volcani Center, Bet Dagan 50250, Israel

2.1 Trapping Spores in the Air

Spores may be trapped by sedimentation or impaction on media in petri dishes, on sticky slides, rods or cylinders, by means of conical funnels, sieving or impaction filters later dissolved in solvents, liquid impingers, centrifugal samplers, electrostatic precipitants, cyclones, electrically powered suction traps and whirling arm impactors. May (1967) and Gregory (1973) discussed these methods and their values. Most of them are rarely used and are not mentioned further in this chapter.

Trapping spores on media in petri dishes exposed in the field for a given time, may demonstrate the presence of spores, according to the number of colonies that develop in the dishes. However, some pathogens fail to develop on media, and others are inhibited by the action of other organisms that land on the same dishes. Therefore, the number of colonies that develop is not necessarily correlated with the number of spores deposited. In addition, measuring spore dispersal with petri dishes fails to quantify the number of spores per unit volume of air. Nevertheless, trapping with petri dishes incorporating selective media is suitable for some purposes. In one study, 24 such dishes were installed in a sampler and moved (by batteries) at hourly intervals beneath an opening in the sampler cover (Edmonds et al. 1984).

Trapping spores with sticky cylinders (rods) or slides indicates either impaction or deposition of spores but not their concentration in the air. Because these instruments are exposed for relatively long periods, usually 24 h, these methods do not record dispersal at discrete hours of the day. According to Hirst (1959), the freely exposed surface are almost valueless for trapping small spores. Nevertheless, these methods, or at least cylinders or rods, enable a study of dispersal when the more efficient mechanical traps are not available, when no electricity is available to operate these traps in the field and when trapping has to be done at a number of sites that exceeds the number of the more expensive instruments. The correlation found by Burleigh et al. (1969) between the number of uredia counted on plants and that of uredospores impacted on glass rods, proves the value of this method.

The trapping efficiency of the suction traps is higher than that of cylinders (rods) and slides. Working with an *Alternaria* blotch on onions, Husain (1968) caught 288 times more spores in a suction trap than on a vertical slide. For further information the reader is referred to Bainbridge and Stedman (1979), who trapped *Erysiphe graminis* spores on cylinders and slides in a barley field, and *Lycopodium clavatum* spores artificially released in this field; and to Chamberlain (1975), who determined the efficiency of collection by vertical surfaces in a number of particles and targets.

Volumetric suction traps provide the easiest way of indicating diurnal cycles of spore dispersal, relative changes in spore concentration in the air and the relative effects of environmental factors on the take-off of spores. Several types of suction traps have been described (e.g. Bartlett and Bainbridge 1978; Gadoury and MacHardy 1983; Husain 1963; Kramer et al. 1976; Morris 1982; Sutton and Jones 1976). Most of them are modifications of the Hirst (1952) trap, in which air is drawn through a narrow slit at a given rate (e.g. $10 l/\mathrm{min}^{-1}$) and spores in the air are impacted onto a greased slide or band moving at a given speed (e.g. $2\,\mathrm{mmh}^{-1}$) behind the split. The Hirst-type traps record hourly changes in the number of trapped spores. Some of them operate for as long as 1 week without the need to change the sticky band. Powered by vacuum pumps, these traps usually work on alternate current which limits the sites of operation. They may also work on direct current, but require frequent charging or changing of batteries.

In specific cases a suction trap with its orifice kept close to the sporulating plant organ is used to record spore discharge from the organ. In most cases suction traps are placed at the upper level of or above the plant canopy and catch spores that originate from all the upwind parts of the field, rather than from only the close vicinity. To assess the number of spores produced in specific parts of the field the researcher may collect them from the sporulating organs as described in Sect. 2.2. Placed at varying distances from or at various heights above the field, these and other traps record dispersal gradients at these sites.

The whirling arm impactors [rotorod-type traps (Perkins 1957)] in which spores are collected by impaction on greasy, *U*-shaped rods moving rapidly through the air, are being put to increasing use. The rotorod-type traps are more convenient, simpler and cheaper than the Hirst-type traps, and can be operated on batteries for relatively long periods. However, recording hourly changes in the concentration of spores in the air involves hourly changing of sticky rods, or successive operation of several traps. Storing the arms with spores is inconvenient. I used to cover the trapping side of the arms with Scotch tape, smear it with silicon on the outside, remove it after use in the field and store it on slides.

For specific purposes one may use a portable sampler (Casselman and Berger 1970), a sampler designed for use on aerial remotely piloted vehicles (Gottwald and Tedders 1985), a sampler designed for use in a growth chamber (Aust and Kranz 1974) and samplers that catch splash-dispersed spores (Faulkner and Colhoun 1977; Fitt 1983; Zuck and Caruso 1984; and Welz, Chap. 11).

The results of trapping of airborne spores are usually expressed as the number caught per 1 m^3 air h^{-1}. This is a simple calculation in suction traps in which a known volume of air passes the instrument. The reader is referred to manufacturers' formulae for calculation of these expressions in the rotorod-type traps. In neither case do the results of trapping indicate spore deposition on plants. In specific cases results are expressed as area dose, viz. the number of spores passing through a unit field area normal to wind direction in unit time. Thus, area dose is the product of spore concentration in the air and wind speed (Landahl and Herrmann 1949).

The following are limitations which should be considered when interpretating results of mechanical spore trapping. (1) Traps may underestimate dispersal at night, when spores can be transmitted by rubbing of leaves or by falling drops of heavy dew. (2) Large masses of air hit the plant canopy but only small amounts of air pass through the traps. As a result, mechanical traps fail to record low levels of inoculum and foci of infection are generally established in the field before the first spore is caught. Similarly, mechanical traps usually fail to record the small number of spores that arrive from a distant location. Each brand of trap has its specific detection threshold, e.g. 10 spores m^{-3} air in the Hirst trap. Such a high threshold explains why spore traps are of little value in forecasting epidemics that originate from a minute local source (e.g. potato late blight), but may forecast epidemics (e.g. black rust) that arrive in a spore-laden air mass (Hirst 1959). In the case of rotorod samplers, sampling efficiency depends on the dimensions of the arms used and improves with an increase in the dimensions of the trapped spores (Edmonds 1972). The recently introduced "jet" spore sampler processes large volumes of air and catches spores alive (Schwarzbach 1979). However, we have little experimental data to evaluate its achievements. (3) The number of spores trapped indicates, but does not quantify, the number of spores produced. In addition to spores that fall on the ground and those which are trapped by neighboring leaves, this can be attributed to differences in the factors which influence sporulation and take-off of spores (Bashi et al. 1982). (4) The number of trapped spores is not always a reliable indicator of disease intensity, which is usually assessed by the proportion of live to dead leaves. Biotrophic pathogens (e.g. downy mildews) sporulate little, if at all, on dead leaves. In the final stage of such diseases the number or spores produced decreases and disease intensity may be underestimated when assessed according to the number of trapped spores. The number of trapped spores is a better indicator of disease caused by necrotrophic pathogens that sporulate on dead leaves. (5) Mechanical trapping does not distinguish between live and dead spores with the exception of the jet spore sampler. In addition to intrinsic factors which determine the rate of survival of any given species, viability depends on weather factors different from those that facilitate dispersal. A sunny, hot, dry day following a dewy night will result in extensive dispersal but may limit survival of spores of sensitive pathogens. Under overcast, cooler and more humid conditions, fewer spores are expected to disperse, but a higher proportion of them will remain alive. (6) Trapping is usually performed with a single spore trap, so the number of caught spores may reflect the position of the trap. It is not only influenced by the amount of inoculum present nearly, but also by the number of spores released upwind of the field. Therefore, with the same number of spores produced and released in a unit area of the field, the number of trapped spores will increase with enlargement of the field size. (7) Most traps fail to catch spores during periods of rainfall or overhead irrigation. During these periods spores are dispersed by the splash mechanism and can be caught by traps designed specifically for this purpose.

The efficiency of various traps has been discussed in several papers (Edmonds 1972; Eversmeyer et al. 1976; May 1967; Sutton and Jones 1976; Wili 1985). I do not question that a specific trap may be more efficient than another. However, errors deriving from environmental and field characteristics, and from using only one trap in the field, may outweigh errors deriving from the trap's specific efficiency. Considering this situation, a major factor in choosing a spore trap is the ease of its operation, its sturdiness and cost.

Wind is usually the strongest factor affecting the take-off of spores. The influence of wind is the easiest to establish, because different wind velocities can occur under conditions of similar temperature, humidity and radiation. It is more difficult to distinguish between the effects of the latter factors, because changes in their intensity often occur simultaneously. To elucidate the specific effects of any of these three factors, one must wait for the unusual combination of environmental factors as they sometimes occur: hot but overcast days, spells of unusually high or low humidity, etc. I prefer this approach to experiments in growth chambers, which are unable to control wind speed, and in wind tunnels, which are not able to control temperature and humidity. For a description and evaluation of the physical forces that affect liberation, transport and deposition of spores, see Aylor (1978 and Chap. 17).

2.2 Collecting Spores from Leaves

Samples of spore-bearing leaves are taken in the morning after the dew evaporates and before the spores start to disperse intensively. When leaves are relatively small, such as potato leaflets, they are the units that constitute a sample. In hosts with large leaves, such as tobacco, individual lesions cut from the leaves constitute a sample. Leaves or lesions constituting individual samples are put into containers with 300 to 500 ml fixative solution (usually FAA: formaldehyde, acetic acid, alcohol). Containers are then shaken vigorously for 30 to 60 min (longer for species with strongly attached spores) and filtered through plastic sieves or cheesecloth to remove leaf and dirt particles.

The number of spores per lesion varies greatly, even in apparently uniformly infected plants. Differences result from micrometeorological variations in the plant canopy, from size and age of the lesions and from shading or direct exposure to sunlight. In the case of *Phytophthora infestans* on potatoes, the number of sporangia counted on individual lesions of similar size varied by up to 500 times. Similar variations were observed with other pathogens, under controlled and field conditions. The higher the number of lesioned leaves or lesions included in one sample, the more closely spore counts reflect the actual amount of inoculum in the field. However, the actual size of the sample may be limited by other considerations such as plot size, level of disease intensity and the need to avoid destruction of spore-bearing leaves in repetitive sampling, when the number of lesioned leaves is low. Samples collected in our studies usually consisted of 50 leaves or 100 lesions, detached or cut at random drom various sites within a plot and various parts of a plant. For specific purposes, we occasionally collected separate samples from old and young leaves, or from leaves at the periphery of the plant canopy (exposed directly to wind and radiation) and from leaves inside the plant canopy, protected by foliage.

Distinguishing between leaves with lesions at different stages of development is more important in some cases than in others (see also Seem, Chap. 4). In the case of *P. infestans* in potatoes, there is usually a relatively short period of several days between the appearance of a lesion, its expansion over the whole leaf and death of the leaf. In studying this system the collected leaf samples have not been sorted according to stages of lesion development, but the completely blighted and dead leaves that produce few, if any, sporangia were not included in samples (Bashi et al. 1982). Tobacco blue mold lesions remain infectious for many days. Spore counts from lesions collected at random reflect the number of spores per "average" lesion. Spores from only chlorotic or only necrotic lesions provide information regarding their specific potential to sporulate (Rotem and Aylor 1984). Sorting the leaf samples according to stages of lesion development is important with pathogens like *Alternaria macrospora* on cotton, which produces the most spores on dead leaves that have shed, and far fewer spores on leaves covered by lesions but not yet wilted. In such a case, leaves in various stages of disease development should be collected in separate samples. A special survey establishes the frequency of categories of variously affected leaves in the field, and the number of spores produced on these leaves is calculated.

Transforming the number of spores per lesion into the number of spores in the plot, follows the calculation of the number of lesioned leaves or lesions present in the plot. This number is determined from a sample of plants (e.g. ten plants) either directly in the field or after removal to the laboratory. The total number of spores in the whole field in the morning is the product of the number of spores per lesion (or lesioned leaf) and the number of such lesions or lesioned leaves in the field.

The method for assessing the amount of inoculum present on plants in the morning can be adapted for assessing dispersed inoculum. In addition to collection in the morning, samples are also collected before sunset, and the spores are removed and counted. Since daytime conditions are usually not conductive for sporulation, the number of spores missing represents those dispersed during the day. We can assess patterns of diurnal dispersal indirectly by collecting lesions several times during the day and removing and counting spores present on these lesions. In a study with *P. infestans* in potatoes, these patterns differed from those concurrently recorded by mechanical spore traps. The peak of dispersal occurred at approximately the same time, late morning to noon, but many more spores disappeared from lesions in the late afternoon and evening than indicated by mechanical traps (Bashi et al. 1982). Consequently, counting spores retained on lesions is another indicator, but not an absolute measure, of the process of dispersal.

3 Survival and Infectiousness

As stated succinctly by Waggoner (1983), "all the words about the aerial dispersal of plant pathogens are wasted if the propagules are dead on arrival". However, data on the survival of inoculum are scarce and, for many pathogens, nonexistent. In epidemiological analyses, the number of viable spores is often based upon assumption.

Among techniques described to study survival are some which are not useful for epidemiology: survival in a water suspension of spores, and survival in petri dish cultures exposed to various temperatures. For epidemiological analyses, we need data based on inocula exposed to more natural conditions.

It is relatively easy to test survival under controlled conditions, but these conditions lack the levels of radiation that are major factors in survival in the field. Indeed, spores of species which, under controlled conditions in darkness, survive for months under harsh combinations of temperature and humidity (e.g. *Alternaria solani*), were affected by a relatively brief exposure to sunlight (Rotem et al. 1985). Despite these limitations, the relatively simple tests in the laboratory should not be abandoned.

Should we measure survival by the results of infection or may we base our conclusions on the percentage of germinating spores? Should we restrict ourselves to testing the spores or should we also measure the survival of mycelium? The answer to both questions depends upon the purpose of a given study, and the pathogen in question. The ability of spores to germinate does not always reflect their ability to infect. Some pathogens (e.g. some strains of *A. solani*) lose their pathogenicity after growing for several generations on artificial media, despite normal germinability of spores. For some biotrophic pathogens, the low percentage of spores germinating in vitro does not necessarily indicate a low ability to germinate and infect host tissue. This points to the need for measuring survival by the ability of spores to infect. On the other hand, one lesion can result from a mixture of live and dead spores, and in such a case the percentage of spores germinating in vitro provides more quantitative information. Whether to test only survival of spores, or also of mycelium (and obviously other propagules, such as sclerotia or chlamydospores), depends on the character of the system and the purpose of the research.

Some researchers assessed the viability of spores by staining them with vital stains (e.g. Caesar and Pearson 1983). I have tried several of these stains and failed to obtain uniform results. It is possible that the action fo some of these materials differs with the species of spores, and possibly also with spores of the same species at different stages of maturity. Progress in this techniques would make the study of survival easier.

3.1 Survival of Overseasoning Inoculum

Some pathogens overseason by means of spores, while others pass unfavorable seasons by means of other structures, such as sclerotia or mycelium. Even with airborne pathogens, mycelium may be the overseasoning source, as with *A. solani* in debris of potato foliage and with *P. infestans* in potato tubers. In such cases, overseasoning structures may survive differently when separated from the dead plant tissues in which they are normally present. In addition to the environment, also the tissue-invading saprophytic organisms affect pathogens overseasoning in plant debris. In some systems the saprophytes may affect survival more than abiotic factors. To distinguish between biotic and abiotic effects, the best we can do is also to test survival of the pathogen separated from the host.

We may test survival of a pathogen that overseasons in debris by assessing germinability or infectivity of the propagules present in debris, or by establishing the ability of overseasoning mycelium to produce spores after various storage periods. In all cases, samples of debris must be found easily at predetermined dates of testing. The debris is enclosed in plastic net bags, and placed on the soil surface and at various depths in the soil in the field and/or in pots. Net bags with debris are then removed periodically and examined for the number of propagules present, and for their ability to germinate and/or infect healthy test plants. When no such propagules are found, the debris is exposed in moist chambers at a temperature which promotes production of spores by the overseasoning mycelium. When visual examination of inoculum is too difficult for routine work, we often prefer to assess overseasoning inoculum by development of disease in healthy plants seeded or planted in the infected soil. We may also use the dilution plate technique, employed in the study of soilborne diseases. In this and other cases, results should be measured quantitatively: germination, as the percent of germinating spores with or without measuring the length of the germ tubes; infection, as the number of infection sites or the area of lesions produced per standard amount of spores used for inoculation and after a standard period of incubation under standard environmental conditions. For more details on infection and incubation procedures, see Rotem, Chap. 2.

Testing of overseasoning under different regimes of temperature and moisture, and with various types of soil in growth chambers, is possible only in pots. The use of pots enables the incorporation of a large number of treatments not possible when the study is carried out only in the field. However, results from pot trials may differ from those from the field and may indicate what could happen rather than what actually does happen in the field. Variations in these procedures are made according to the requirements of the particular system. For example, in studying overseasoning of *Mycosphaerella pinodes* in peas, samples of pea debris were placed in plastic net bags and, after various periods of storage, mixed with water in a blender and filtered through cheesecloth. The number of easily distinguished pycnidia in the water suspension was counted under a microscope and the suspension was sprayed onto test plants. The inoculated plants were then incubated and evaluated for symptoms (R. Neubauer, unpublished). In the early blight system, *A. solani* usually overseasons mainly as mycelium in debris. To examine overseasoning of this pathogen, containers with soil mixed with debris were kept under various regimes of temperature and moisture. The level of successful overseasoning was assessed according to the amount of collar rot that developed on tomato seedlings planted in these containers after certain storage periods. In another trial with this pathogen, debris was collected in the field before the beginning of the new growing season. It was sorted into debris collected from below or on the soil surface, and according to the cultural and soil characteristics of various fields. After incubation in moist chambers, debris was examined microscopically for the presence of spores (Rotem 1968). This easy procedure allowed inspection of many samples from various fields in a relatively short time.

There are cases in which sexual structures overwinter and release zoospores (e.g. *Plasmopara viticola*) or ascospores (e.g. *Venturia inaequalis*) at the beginning of the vegetative season. We can evaluate the success of overseasoning by collecting these spores and testing their viability. In many cases, infectivity is easier to determine than germination, especially when zoospores are the inoculum source. After storage under various environmental conditions, debris with the sexual structures can be placed under sprinklers and surrounded by healthy plants in pots. Sprinkling induces release of zoospores or ascosporess, and splash disperses them to the foliage of surrounding test plants. Viability of these spores is assessed in accordance with the frequency of infections. As mentioned previously, germinability or infectivity of spores should be evaluated quantitatively.

3.2 Survival of Inoculum in the Season

Within the vegetative season, survival of spores is the main object of investigation. The viability of spores should be measured at least from the time in the day in which sporulation in the field is normally terminated (usually morning) until the time in the day inducive for germination (i.e. the next dewy night, or rain). Attached spores survive longer than detached spores, but it is not always clear whether this is a biotic or micrometeorological effect. Whatever the cause, survival times of both have to be defined, since spores in the field remain attached for given periods before being detached and dispersed.

Survival of attached spores in the field is assessed by detaching leaves with spores at progressively later hours of the day. The leaves are shaken in water to obtain a suspension of spores which may be tested for their ability to germinate or infect.

To test survival of dispersed spores, healthy plants in pots are exposed to naturally infected plants with spores in the field. With *P. infestans* in potatoes, test plants were exposed for a 2-h period in a heavily infected field. Control plants were transferred to moist chambers after the exposure period was terminated. Other plants were transferred to another healthy plot, upwind to, and with a microclimate similar to that in the field. After various periods of exposure these plants were transferred to moist chambers, and then incubated in growth chambers until the appearance of symptoms (Bashi et al. 1982). The results were evaluated as the number of lesions per leaf area or per number of leaves on the trap plants. They show the relative effects of various treatments on survival but do not indicate the number of spores deposited and that which survived the exposure period in the field. It should be mentioned that deposition of dispersed spores on test plants is seldom uniform, and thus large numbers of replicates must be used. For preparation of trap plants used in such a study, see Sect. 4.

Testing spore survival on host plants (in vivo) simulates nature, but restricts the number of treatments. More factors can be examined in experiments under controlled conditions than by field tests, but the former does not enable estimation of the effect of radiation.

3.3 Survival of Spores on Plants Under Controlled Conditions

Survival of inoculum exposed in radiation-deficient growth rooms will differ from that in the field, and the results should be viewed as indications only. Survival of spores attached to host plants is studied by exposing plants with spores to various conditions of temperature and relative humidity (RH) in growth chambers (see Chap. 2). This technique may be used only with spores that survive for relatively short exposure times. Pathogens whose spores can survive in growth chambers for months are, in practice, not suited for this growth-chamber technique.

Air movement in some growth chambers is often too slow to remove substantial numbers of firmly attached spores and thus to affect the result. With easily detached spores of pathogens like powdery mildews, such an experiment can be performed only under nearly windless conditions. Results are evaluated in terms of germinability or infectivity of spores. When germinability is the measure, spores from leaves are removed onto membrane filters and placed on appropriate media, or shifted directly from the leaf to the media. Easily detached spores of rusts or powdery mildew can be dropped gently from leaves onto membrane filters. Transfer of species with more firmly attached spores may be done by (1) a fine brush, (2) touching the membrane filter to the leaf surface, and (3) sucking the spores into a mounted filter attached to a vacuum pump. Methods 1 and 2 are simple and suit most purposes. However, they may also remove spores that would normally not be dispersed by wind, viz. young, not yet mature, and firmly attached spores that may have lower than normal germinability. The vacuum method, while more complicated, removes mainly mature spores, normally dispersed by wind.

The following method is used when infectivity is the measure of survival. With spores of rusts, smuts and powdery mildews which are produced in quantity and easily detached, samples of spores are used for inoculation of test plants is settling towers. With spores that are strongly attached and/or produced in low quantities (but also with rusts, smuts and powdery mildews), inoculation with a water suspension of spores is the preferred technique (see Chap. 2). The number of spores in aqueous suspension is adjusted to a standard, preferably low, concentration. A high concentration of spores in the suspension may result in a high level of infection, even if some of these spores are dead.

Survival of dispersed and deposited spores is determined by exposing healthy plants to an environment rich in dispersing spores for the shortest period sufficient for spore deposition. This may be done in any large box, with healthy plants placed between heavily sporulating plants. In most of our tests, equal numbers of healthy and sporulating plants were exposed for a period of 1 h. Dispersal of spores from sporulating plants was induced by a number of strong, rotating fans. Immediately after the period of exposure, a number of test plants with spores deposited on their foliage, were placed in moist chambers to induce infection. These were then incubated in growth chambers until symptoms appeared, and were evaluated as controls. Other test plants were placed in growth chambers to examine the influence of certain temperature and humidity regimes on spore viability, or in the open to examine the effect of radiation. After the treatments, plants were moved into moist chambers and then into growth chambers until symptoms appeared and results could be evaluated.

3.4 Survival of Detached Spores Under Controlled Conditions

The factors tested are temperature and humidity. Membrane filters with spores are fixed inside hermetically closed small containers (e.g. 200 ml) above materials that maintain the desired RH. The containers are kept in incubators. Saturated salt solutions or different concentrations of sulfuric acid or glycerol with specific vapor pressures maintain the designated RH (Anon 1983; Winston and Bates 1960). Use of saturated salts is convenient, since any change in concentration of hygroscopic materials is recognized by the disappearance of excess solid salt at the bottom of the given solution. In contrast, similar changes in sulfuric acid or glycerol concentrations may pass unnoticed.

Whatever the RH-maintaining material used, a period of serveral hours may be necessary to establish the proper RH in the container. RH levels produced by various materials have been established mostly for a specific temperature (usually 20 °C) and not for others. In these cases, RH must be measured with electronic sensors left in the closed containers for a sufficient time to establish the desired RH.

Fluctuations in temperature occur in most incubators and result in some fluctuations in the level of RH. These fluctuations have a rather insignificant biological effect when samples are kept at low RH. However, at RH over 90%, fluctuations in temperature result in temporary conditions of saturation that might induce germination of spores (Schein 1964). In these cases, the containers with spores should be immersed in precise temperature baths that limit fluctuations in temperature. Effects of humidity can be tested also in systems in which spores are exposed to airflow maintained at a given RH (e.g. Hartmann et al. 1982).

3.5 Techniques for Testing the Effects of Radiation

To date, only a few quantitative studies have dealt with the effect of radiation on inoculum survival. These studies show that the effect of radiation may overweigh the effects of temperature and humidity; the ultraviolet (UV) rays of the sun's spectrum are the dominant agents (Rotem et al. 1985). The test of UV effects should be done in the field. Plants with attached spores, or detached spores on a membrane filter, are exposed to the sun. Radiation, especially ultraviolet, and other environmental factors have to be measured to obtain results applicable elsewhere. Temperature is measured with miniature sensors that touch the membrane or the leaf. Measurement of RH close to the exposed spores is almost impossible. It may be calculated from measurements made at some distance from the spores, or the distant measurement is used as indicator of the conditions under which the spores are present.

The ability of spores to germinate or infect is assessed after given periods of exposure, and always before nightfall, when dew may induce germination. Qualitative and quantitative differences in radiation can be obtained on the same day. Shading of samples by a number of layers of cheesecloth can reduce light intensity without changing its spectrum. Filters can be used to define the effects of particular parts of the spectrum. Suitable ones are the UV filters used in photography (e.g. Kodak A2). Plastic lenses cut from protective UV spectacles are studier and less expensive than photographic filters. Also in use are plastic filters that limit the intensity of UV radiation by 10% or more (Caesar and Pearson 1983). To protect large areas one may use the UV-absorbing plastic films (e.g. Honda and Yunoki 1977). These plastic films are produced by various manufacturers, have different limits of UV transmission, and should be measured in a spectrophotometer before being used in experiments.

Due to a greenhouse effect, the temperatures under the filters may be higher than ambient temperatures. As a result, spore viability is affected not only by radiation, but also by temperature and RH, which is conditioned by temperature. The net effect of radiation is determined by placing some samples in dark incubators maintained at high temperatures (Rotem et al. 1985). Another method to distinguish between the effects of high temperature and radiation is to place spore samples on top of black and white solids. All samples will receive the same flux of radiation but the temperature will differ (Bashi and Aylor 1983).

4 Testing Inoculum Potential with Potted Trap Plants

Inoculum potential refers to the combined action of all factors reflected in a naturally occurring infection. We may study inoculum potential with potted plants that catch spores during exposure in the field (see also Aust and Kranz, Chap. 1; Raccah and Irwin, Chap. 14). Part of this procedure was described for the evaluation of survival of dispersed spores. In studying inoculum potential, however, we are interested not only in survival, but in the integrated effects of all, or at least several, of the factors mentioned. The trap plants are placed among other plants in the field. They are exposed not only to inoculum lifted into the air, but also to spores transmitted by rubbing of leaves or by other means, such as splash dispersal during periods of rain or overhead irrigation. These plants are part of the natural field habitat.

The success of such a study is preconditioned by an amount of inoculum large enough to inoculate the trap plants during their relatively short exposure in the field. Attempts to employ this technique when the amounts of inoculum present are not sufficient, will probably not succeed, because the number of lesions produced on trap plants will be too small and sporadic to characterize the studied phenomena. This shortcoming can, however, be compensated by a longer exposure of the plants. A successful study is preconditioned by adequate facilities for growing the trap plants before, and incubating them after exposure in the field. The site at which the trap plants are grown should be far enough from the field to preclude their premature infestation. Preferably, the size and age of the exposed trap plants should approximate those in the field, but this is not always possible.

Deposition of dispersed spores on the trap plants is seldom uniform and thus a large number of replicates must be used: 10 to 20 trap plants per treatment. With five treatments the number of trap plants exposed in the field each day is thus 50 to 100. The number of such daily exposures must be large enough to represent the varying environmental conditions. After removal from the field, the trap plants are incubated under standard conditions, preferably in growth chambers, until the appearance of symptoms and evaluation of the results. Since each exposure in the field is followed by several days incubation in a growth chamber, the space available in the latter soon becomes the bottleneck of the study. For this reason the researcher may be forced to expose the trap plants in the field less frequently or to reduce either the number of treatments or replicates.

The results are evaluated according to the number and/or the size of lesions per unit area of the foliage (e.g. 100 cm^2) or per number of leaves on the trap plants. The results indicate the relative value of deposition and survival but not the number of spores that have actually been deposited and that survived.

Depending on the purpose of a particular test, trap plants are placed in the field at different times of day and are removed after different periods of exposure. Various questions can be addressed by manipulating the time of exposure and the post-exposure conditions. Below are examples of questions that can be studied by manipulating exposure time and post-exposure conditions.

Periodic Dispersal and Deposition of Inoculum. Test plants are exposed in the field for consecutive 2-h periods (e.g. 0800 to 1000, 1000 to 1200 h, etc.). Inoculum deposited on test plants during their periods of exposure cannot cause infection, because of the lack of moisture. In order to assess the action of the deposited inoculum, the plants, after exposure, are placed in moist chambers and then in growth chambers until symptoms appear. This procedure in not useful for studies of survival of the deposited spores, because the short periods of exposure in the field minimize the effect of factors involved in survival.

Effectiveness of Inoculum Dispersed at Various Times of the Day. The techniques described for studying "periodic dispersal and deposition of inoculum" may be modified to study the effectiveness of inoculum dispersed at various times during the day. This is done by introduction of trap plants into the field at progressively later hours. For example, the first group is placed in the field at 0800 h, the second group at 1000 h, the third group at noon, and so on until the evening. All plants are then removed at the same time in the evening. Obviously, less inoculum will settle on plants placed in the field at later hours. However, spores that remain attached to the host plant for longer periods survive better than spores that are detached, dispersed and deposited on the trap plants early in the day. Therefore, the spores dispersed late in the day may be more infectious than those dispersed early. If all the spores dispersed during the day survived until conditions that induce germination existed (e.g. dew at night), then the number of infections on plants exposed at progressively later hours should decrease in direct proportion to the amount of inoculum dispersed during the progressively shorter periods of field exposure. Such patterns are expected to occur more often with pathogens not sensitive to environmental hazards (e.g. *A. solani*) than with sensitive pathogens (e.g. *P. infestans*); and under fairly mild conditions (overcast and relatively cool days), under which also the sensitive pathogens survive, rather than under the harsh conditions (hot days with strong radiation) under which they are strongly affected (Bashi and Aylor 1983; Bashi et al. 1982). Therefore, the number of infections produced on plants placed in the field at progressively later hours may indicate not only the effectiveness of inoculum dispersed at various times of the day, but also the relative severity of the prevailing conditions.

Day and Night Dispersal, Deposition and Survival. The exposure period in the field is arbitrarily divided into day and night. Such a division involves manipulation of trap plants at inconvenient hours (sunrise and sunset) and the researcher may prefer to expose the plants at more convenient periods of morning to late afternoon, and late afternoon to morning. Results of these long exposures reflect also the environmental effects on survival of the spores. Due to such effects, the effectiveness (i.e. ultimate infectivity of spores dispersed at different times of the day may be different. In order to compare dispersal and deposition of spores on plants exposed in the dry day to that on plants during the wet night, both groups of plants must be kept in moist chambers after removal from the field.

Integrated Effects of all Components on Inoculum Potential. The integrated effects of all components affecting inoculum potential are tested with a procedure that mimics the natural situation in the field. Trap plants are exposed in the field one morning and removed from the field the next morning. These plants are placed directly into dry growth chambers without moist chamber incubation. Germination of and infection by spores deposited during field exposure are induced only by sources of moisture normally occurring in the field during the 24-h exposure period (rain, overhead irrigation or dew) and at temperatures prevalent during the moist period. However, sometimes many spores disperse, deposit on the trap plants, survive the conditions of the day, but are not able to infect the trap plant during a night with low temperature or short wetting period. The following technique aims at indicating some possible shortcomings in the conditions of such a night. Trap plants are exposed in the field one morning and removed from the field the next morning, as described previously. Half of these plants are put directly into dry growth chambers where development of infection is dependent on the humidity and temperature conditions encountered during the previous night in the field. The other plants are put overnight into moist chambers, where optimum conditions of humidity and temperature allow for more infection than is possible under conditions in the field alone. Following this wet treatment the plants are kept in dry growth chambers. The numbers of infections on plants under both post-field conditions are then compared. The closer the number of infections on plants moved from the field directly into dry growth chambers is to the number of infections on plants moved into moist chambers, the better are the natural conditions for infection encountered during the first night in the field.

5 Epilogue

There is no single or direct technique to assess production, dispersal, deposition and survival of inoculum. Use of several techniques is advised to study the phenomena, but some of these techniques provide different results. For instance, different patterns of dispersal are indicated by mechanical spore traps and by collection from leaves of the non-dispersed spores. We do not know which of these two methods is more accurate and the results of both should be subjected to critical interpretation. Similarly, a critical interpretation of the results is advised in other studies dealing with inoculum potential. This phenomenon can, in most cases, be approximated rather than quantified. All techniques may, and should, be modified in relation to local considerations and problems. The need for new techniques to study spore survival is especially urgent.

6 References

Anon (1983) In: Johnston A, Booth C (eds) Plant pathologist's pocketbook, 2nd edn. Commonwealth Agriculture Bureaux, Slough, England, pp 408–409

Aust HJ, Kranz J (1974) Eine automatische Sporenfalle für den Gebrauch in Klimaschränken. Angew Bot 48: 267–272

Aylor DE (1979) Dispersal in time and space: aerial pathogens. In: Horsfall JG, Cowling EB (eds) Plant disease, advanced treatise, vol 2. Academic Press, London, pp 159–180

Bainbridge A, Stedman OJ (1979) Dispersal of *Erysiphe graminis* and *Lycopodium clavatum* spores near to the source is a barley crop. Ann Appl Biol 91: 187–198

Bartlett JT, Bainbridge A (1978) Volumetric sampling of microorganisms in the atmosphere. In: Scott PR, Bainbridge A (eds) Plant disease epidemiology. Blackwell Scientific Publ, Oxford, England, pp 23–30

Bashi E, Aylor DE (1983) Survival of detached sporangia of *Peronospora destructor* and *Peronospora tabacina*. Phytopathology 73: 1135–1139

Bashi E, Ben-Joseph Y, Rotem J (1982) Inoculum of *Phytophthora infestans* and the development of potato late blight epidemics. Phytopathology 72: 1043–1047

Burleigh JR, Romig RW, Roelfs AP (1969) Characterization of wheat rust epidemics by numbers of uredia and numbers of urediospores. Phytopathology 59: 1229–1237

Caesar AJ, Pearson RC (1983) Environmental factors affecting survival of ascospores of *Sclerotinia sclerotiorum*. Phytopathology 73: 1024–1030

Casselman TW, Berger RD (1970) An improved portable automatic sampling spore trap. Proc Fla State Hortic Soc 83: 191–195

Chamberlain AC (1975) The movement of particles in plant communities. In: Monteith JC (ed) Vegetation and the atmosphere: 1. principles. Academic Press, London, pp 155–203

Edmonds RL (1972) Collection efficiency of rotorod samplers for sampling fungus spores in the atmosphere. Plant Dis Rep 56: 704–708

Edmonds RL, Hinshaw RW, Leslie KB (1984) A 24-hour deposition sampler for spores of *Heterobasidion annosum*. Phytopathology 74: 1032–1034

Eversmeyer MG, Kramer CL, Collins TI (1976) Three suction-type spore samplers compared. Phytopathology 66: 62–64

Faulkner MJ, Colhoun J (1977) An automatic spore trap for collecting pycnidiospores of *Leptosphaeria nodorum* and other fungi from the air during rain and maintaining them in a viable condition. Phytopathol Z 89: 50–59

Fitt BDL (1983) Evaluation of samplers for splash-dispersed fungus spores. EPPO Bull 13: 57–61

Gadoury DM, MacHardy WE (1983) A recording volumetric spore trap. Phytopathology 73: 1526–1531

Gottwald TR, Tedders WL (1985) A spore and pollen trap for use on aerial remotely piloted vehicles. Phytopathology 75: 801–807

Gregory PH (1973) The microbiology of the atmosphere, 2nd edn. Leonard Hill, London

Hartmann H, Sutton JC, Thurtell GW (1982) An apparatus for accurate control of atmospheric water potentials in studies of foliar plant pathogens. Phytopathology 72: 914–916

Hirst JM (1952) An automatic volumetric spore trap. Ann Appl Biol 39: 257–265

Hirst JM (1959) Spore liberation and dispersal. In: Holton CS (ed) Plant pathology – problems and progress, 1908–1958. Univ of Wisconsin Press, Madison, WI, pp 529–538

Honda Y, Yunoki T (1977) Control of *Sclerotinia* disease of greenhouse eggplant and cucumber by inhibition of development of apothecia. Plant Dis Rep 61: 1036–1040

Husain SM (1963) An automatic suction-impaction type spore trap and its use with onion blotch *Alternaria*. Phytopathology 53: 382–387

Kramer CL, Eversmeyer MG, Collins TI (1976) A new 7-day spore sampler. Phytopathology 66: 60–61
Landahl HD, Herrmann RG (1949) Sampling of liquid aerosols by wires, cylinders and slides, and efficiency of impaction of the droplet. J Colloid Sci 4: 103–136
May KR (1967) Physical aspects of sampling airborne microbes. In: Gregory PH, Monteith JL (eds) Airborne microbes. Cambridge Univ Press, Cambridge, England, pp 60–88
Morris JCT (1982) A simple automatic volumetric spore trap. Bull Br Mycol Soc 16: 151–154
Perkins WA (1957) The rotorod sampler. 2nd Semiannual Rep Aerosol Lab, Dept of Chemistry and Chemical Engng, Stanford Univ CA 186, pp 1–66
Rotem J (1968) Thermoxerophytic properties of *Alternaria porri* f. sp. *solani*. Phytopathology 58: 1284–1287
Rotem J, Aylor DE (1984) Development and inoculum potential of *Peronospora tabacina* in the fall season. Phytopathology 74: 309–313
Rotem J, Wooding B, Aylor DE (1985) The role of solar radiation, especially ultraviolet, in the mortality of fungal spores. Phytopathology 75: 510–514
Schein RD (1964) Comments on the moisture requirements of fungus germination. Phytopathology 54: 1427
Schwarzbach E (1979) A high throughput jet trap for collecting mildew spores on living leaves. Phytopathol Z 94: 165–171
Sutton TB, Jones AL (1976) Evaluation of four spores traps for monitoring discharge of ascospores of *Venturia inaequalis*. Phytopathology 66: 453–456
Waggoner PE (1983) The aerial dispersal of the plant pathogens of plant disease. Philos Trans R Soc Lond B Biol Sci 302: 451–462
Wili GM (1985) Comparison of the designs of two volumetric spore traps. Phytopathology 75: 380
Winston PW, Bates DH (1960) Saturated solutions for the control of humidity in biological research. Ecology 41: 232–237
Zuck, MG, Caruso FL (1984) A volumetric spore trap designed for monitoring *Venturia inaequalis* spore release in apple scab management programs. Phytopathology 74: 796 (abstr)

Chapter 6

Analysis of the Spatial Pattern of Soilborne Pathogens

C. A. Gilligan[1]

1 Introduction

Expediency in the design and conduct of experiments, and in the interpretation of field samples of soilborne plant pathogens, has induced a confusingly equivocal view of spatial pattern of these pathogens. Data are routinely interpreted as though inoculum, infection and disease occurred at random. The assumption of randomness implies that the probability of occurrence of a propagule of inoculum, an infection or a lesion at a given site within a population, is constant over all possible sites and independent of the status of neighbouring sites. This hypothesis is seldom tested, yet few pathologists would deny that it is unlikely to be true in many natural situations. Interest, however, in the description and analysis of spatial pattern of soilborne plant pathogens is increasing. A review of the subject has recently been published (Campbell and Noe 1985).

In keeping with the objectives of this book, I do not propose to attempt to review the topic extensively. Instead, I have selected three topics for description and discussion. These are: (1) definition of the objectives and requirements for the analysis of spatial pattern of soilborne pathogens; (2) the analysis of unmapped data, comprising counts, using goodness-of-fit tests and indices of aggregation; (3) the analysis of mapped data with particular emphasis on presence or absence data for row and lattice crops. Brief mention is made of the extension of methods for continuous variables such as disease severity. The statistical considerations associated with these topics are stressed throughout. Some of the techniques have already been used by pathologists but are capable of further exploitation. Others offer considerable value but have yet to be used. Two significant omissions are the analysis of locational data using nearest neighbour methods (Diggle 1983) and of methods such as the correlogram derived from time series analysis (Upton and Fingleton 1985). Another problem, testing for randomness in inoculum density experiments is not discussed here but see Gilligan (1983). Several texts dealing wholly or partly with the analysis of spatial pattern have been published, of which particular attention is drawn to Pielou (1977) and Upton and Fingleton (1985) as sources for further reference for the material discussed in this chapter. For aspects of modelling epidemics of soilborne diseases see Campbell et al. (Chap. 18), and for additional information on experimentation with nematodes the reader is referred to Barker and Noe (Chap. 15).

[1] Department of Applied Biology, University of Cambridge, Pembroke Street, Cambridge, CB2 3DX, UK

2 Objectives and Requirements for the Analysis of Spatial Pattern of Soilborne Pathogens

The spatial pattern of disease may be defined as the arrangement of disease entities relative to each other and to the architecture of the host crop (Gilligan 1982). A disease "entity" may be a lesion, a diseased root, a diseased plant, a patch or focus of disease or even a diseased field. The spatial pattern of soilborne plant pathogens, however, relates not only to disease but also to inoculum and to infections (Table 6.1). Thorough analysis of spatial patterns should involve all three components as well as consideration of the spatial pattern of host growth.

In common with other statistical procedures, two complementary objectives in the analysis of spatial pattern can be discerned, inference and modelling. Inference involves the processing of sample data to make inductions, involving hypothesis tests or estimation, about the population from which the data were taken. Once a model for spatial pattern has been validated, it can be used to predict the behaviour of potential data. Such predictions may be used to model population development spatially as well as temporally, see for example Gilligan (1985). Lin et al. (1979) and Hau et al. (1982) used simulation to test the effects of clustering of disease on the relative efficiency of sampling methods for the estimation of disease incidence within fields.

The nature, description and analysis of spatial pattern of soilborne pathogens are complex and depend upon a number of factors. Principal amongst these are: the fundamental metameter used to define spatial pattern; the entity of inoculum, the infection or disease under consideration; the sampling unit; the sampling method and the random variable used for analysis (Table 6.1).

2.1 Metameters for Measurement of Spatial Pattern

The fundamental condition against which spatial pattern is measured is randomness. By this is meant that the probability of location of an entity within a potential site in the population is constant for all sites and independent of the location of all other entities. Deviation from randomness may occur in two ways; towards clustering when, for example, there is significant secondary infection (see also Campbell et al. Chap. 18; for usage of this term) about points of primary infection; or towards regularity, for example when there is mutual inhibition amongst lesions on a single organ. Considerable attention has been given to the classification of spatial pattern of plant pathogens on the spectrum ranging from regularity through randomness to clustering (Sect. 3.1). Sometimes the use of coefficients of spatial autocorrelation are preferred as a measure of spatial pattern. These measure the degree of association e. g., of inoculum density per unit volume between contiguous areas.

Complications arise in the consideration of mosaics of disease which occur among soilborne pathogens such as *Rhizoctonia solani* (Mac Nish 1985) and *Gaeumannomyces graminis*. While the Poisson model may be used to describe the random occurrence of entities that can be represented as points (Upton and Fingleton 1985), no such unequivocal model exists for mosaics (Pielou 1977). The analysis of the spatial pattern of mosaics therefore requires definition of the particular form of "randomness" that is being used as a measure for comparison. Two models have been proposed. In the random lines model, an area is divided into polygonal cells by the intersection of randomly orientated lines. The cells are then assigned at random to one of two phases. In the random sets model, polygonal cells are obtained by constructing cells about random points such that all the sites within a cell are nearer to the random point in that cell than any other. Further complication arises in that the spatial pattern of patches of disease must be considered with respect to the pattern of the gaps and vice versa (Pielou 1977). Since both patches and gaps may be regarded as having a regular, random or clustered pattern, this gives rise to nine possible classifications overall, with respect to one criterion for randomness.

2.2 Definable Entities for Inoculum, Infection and Disease

The definable entity within the analysis of the spatial pattern of soilborne plant pathogens may be inoculum, infection or disease (Table 6.1). Direct enumeration of inoculum is possible only for relatively large propagules such as sclerotia, oospores or nematode cysts (Menzies 1963). Indirect methods of estimation, involving subsampling and dilution-plate counts or bioassay are usually necessary. Often the "units" so enumerated may be variable, representing mycelial aggregates. The technique of the most probable number is frequently used to estimate density. The Poisson distribution is routinely used as a basis for the estimation of most probable numbers but this is admissible only when there are no a priori reasons to expect clumping of inoculum. The negative binomial distribution may be used when clumping is expected.

Certain diseases, e.g. hypocotyl rot of beans caused by *Rhizoctonia solani*, produce clearly identifiable lesions within hosts that can be counted (Campbell and Pennypacker 1980). The number of infection units, *sensu* Cox and Sanders (1974), may also be counted for certain fungi and nematodes. Usually, however, the minimum resolvable entity is the diseased or infected organ (Table 6.1).

2.3 Sampling Units and Sampling Method

A sampling unit is an identifiable part of the population, the measurement or attributes of which are to be recorded. It may be a natural unit such as a root, hypocotyl, whole plant or even a rhizosphere (Stanghellini et al. 1983). The size of the sampling unit will therefore change with time. By contrast, artificial sampling units are fixed, comprising a length of row of crop or an area or volume of soil. The hierarchical nature of spatial pattern can be described by the use of a range of sizes of sampling units. Thus lesions may be clustered within roots, while the distribution of diseased roots amongst plants may be random: diseased plants may in turn be clustered with respect to small quadrats but appear random with respect to larger quadrats.

Sampling method refers to the way in which sampling units are selected. Emphasis is given to random sampling in conventional sampling theory and practice in which the principal objective is to obtain estimates of mean density with minimum bias and maximum precision. Greater precision is often achieved by division of the population into strata and then sampling within strata (see e.g. Cochran 1977). These conventional methods can be used in the analysis of spatial pattern (Sect. 3). Usually, however, in the analysis of spatial pattern, sampling units are collected systematically and the potential disadvantages of bias and inflated variance are offset by the advantage of obtaining locational information (Sect. 4).

Table 6.1. Classification of entities, sampling units and random variables used in the quantification of inoculum, infection and disease associated with soilborne plant pathogens

	Entity	Sampling units Natural	Artificial	Random variables
Inoculum	Propagule	Rhizosphere	Quadrat	Direct count
	Colony-forming unit (estimated by dilution plating)	Spermosphere	Cube	Estimated count
	Bioassay-response unit (estimated by number of lesions or infections in dilution series)			
Infection	Infection site	Organ (e.g. root, seed, hypocotyl)	Quadrat	Presence/absence
	Infected organ		Cube	Count
	Infected plant	Whole plant	Root length	Ratio
Disease	Lesion	Organ	Quadrat	Presence/absence
	Diseased organ	Whole plant	Cube	Count
	Diseased plant	Field	Root length	Ratio (e.g. proportion of total roots diseased; mean length of root diseased)
	Disease focus or patch			
	Diseased field			

Random variables comprise measures on the entity within sampling units (Table 6.1). Most emphasis is given in spatial pattern analysis to discrete (count) variables (Sect. 3) but dichotomous and continuous variables are also important.

3 Analysis of Unmapped Data

Unmapped data consist of variates obtained from sampling units without explicit consideration of the spatial arrangement of the units. The analysis of unmapped data is a simple starting point from which to make inferences about spatial pattern. There are, however, two severe and related limitations. First, because of the hierarchical nature of spatial pattern, inferences depend upon the scale of the sampling unit: thus, for example, small-scale clustering of inoculum will not be detected in sampling units that are large relative to the modal size of the clusters. Second, by not considering the relative locations of sampling units, it is impossible to test whether or not clusters of disease or inoculum are themselves clustered. Thus, identical inferences about spatial pattern at the level of the sampling unit might be obtained from populations with conspicuously different patterns at a larger scale (Nicot et al. 1984).

There are two main approaches to the analysis of unmapped data; testing the goodness-of-fit of data to probability distributions and the use of indices of aggregation. The two are linked, for parameters of distributions may be used as indices of aggregation. They differ in objective, the former are predominantly used to test specific hypothesis, while the latter are used as a measure for comparison of populations.

3.1 Goodness-of-Fit to Probability Functions

Four of the common probability distributions used in tests for spatial pattern are listed in Table 6.2. The usual procedure is first to test the null hypothesis (H_0) that inoculum, infection or disease is distributed at random amongst the sampling units, by testing goodness-of-fit of the data to the Poisson distribution. The test is usually one-sided against an alternative hypothesis (H_1) of clustering. Rejection of the H_0 is followed by a χ^2 test of goodness-of-fit to one of the contagious distributions (Table 6.2), selected on a priori assumptions. Expected values are calculated using estimates of parameters obtained by maximum-likelihood estimation for which standard programs are available. A more powerful, approximate test for agreement with the Poisson distribution involves the use of the sample variance/mean ratio (ω) for n sampling units where $(n-1)\,\omega$ is approximately distributed as χ^2 on $(n-1)$ df. Values of ω close to 1 indicate randomness: larger values indicate clustering and low values regularity.

The contagious distributions are part of a family of two-parameter distributions. Campbell and Noe (1985) list some of the applications of these distributions in plant pathology. The distributions can be distinguished empirically by the numbers of modes; the negative binomial has one mode, the Polya-Aeppli one or two, while the Neyman Type A distribution may have up to three or more equally-spaced modes

Table 6.2. Some properties of common probability distributions used in tests for spatial pattern

unit	Poisson	Negative binomial	Polya-Aeppli	Neyman Type A
Probability no occurrence within sampling unit	$e^{-\lambda}$	$\left(1+\frac{\lambda}{k}\right)^{-k}$	$e^{-\lambda_1}$	$e^{-\lambda_1}(1-e^{-\lambda_2})$
Probability r occurrences within sampling unit	$\frac{\lambda^r e^{-\lambda}}{r!}$	$\binom{r+k-1}{k-1}\left(\frac{\lambda}{\lambda+k}\right)^r\left(1+\frac{\lambda}{k}\right)^{-k}$	$\frac{\lambda_1 p}{r}\sum_{j=0}^{r-1}(r-j)(1-p)^{r-j-1}p_j \quad r\geqq 1$	$\frac{\lambda_1\lambda_2}{r}e^{-\lambda_2}\sum_{j=0}^{r-1}\frac{\lambda_2^j}{j!}p_{r-j-1} \quad r\geqq 1$
Parameters	λ	$\lambda\ k$	$\lambda_1\ \tau=(1-p)$	$\lambda_1\ \lambda_2$
Mean (overall)	λ	λ	$\lambda/(1-\tau)$	$\lambda_1\ \lambda_2$
Variance	λ	$\lambda+\lambda^2/k$	$\lambda(1+\tau)/(1-\tau)^2$	$\lambda_1\ \lambda_2(1+\lambda_2)$

Table 6.3. Summary of component distributions in the generalized Poisson and compound Poisson forms of three common contagious distributions

	Negative binomial	Polya-Aeppli	Neyman Type A
Generalized form			
Clusters	Poisson	Poisson	Poisson
Individuals within clusters	Logarithmic	Geometric	Poisson
Compound form			
Compounding distribution	Gamma	Poisson	Poisson
Individuals within clusters	Poisson	Negative binomial	Poisson

(Anscombe 1950). It is not uncommon with sample sizes frequently used in soil pathology to obtain equally good fits to more than one contagious distribution. Further complications arise in that each distribution may be derived in a number biologically feasible mechanisms. Unfortunately, information from a single dataset is insufficient to distinguish amongst these mechanism (Pielou 1977). Consider for example two broad distinctions that can be made in the derivation of the two-parameter distributions, (Table 6.3) as either generalized or compound Poisson distributions (Pielou 1977). In the generalized Poisson distributions, it is supposed that clusters are randomly dispersed within a population, while the numbers of individuals within clusters are described by specific distributions (Table 6.3). In compound Poisson distributions, it is assumed that individuals are randomly located within sampling units, but that the means of the resulting Poisson distributions vary from site to site. Realizations of the compound and generalized Poisson distributions may appear very similar. Indeed, Feller (1943) demonstrated that for any particular generalized Poisson distribution, there exists a distribution to describe the spatial variation in mean density amongst sites that leads to an identical compound Poisson distribution. To distinguish between the two, it is necessary first to consider the biological justification of a priori hypotheses concerning the distribution of individuals within clusters and the likelihood of means varying amongst sites, and second to test these hypotheses. Thus, heterogeneity in soil conditions, such as availability of nitrogen, would support a hypothesis of a compound Poisson distribution with means varying with soil condition. Additional sampling within contrasting areas would enable the hypothesis to be tested. Assumptions about the distribution of numbers of individuals within clusters may be tested by analyzing these data separately, providing that clusters can be distinguished. In practice this requires large samples (C. A. Gilligan and G. C. MacNish, unpublished).

Consideration of the mechanisms underlying two-parameter distributions leads logically to the use of the distributions to compare the spatial pattern of two or more populations, this objective has hitherto been largely restricted in plant pathology to the use of indices for crude measures of comparison (Sect. 3.2). There are, however, precedents from entomological work. Freeman (1980) and Wheatley and Freeman (1982) argued, in the context of investigating the spatial pattern of numbers of a soil pest, *Psila rosae*, that it may be appropriate to compare data sets representing samples from different treatments by fitting a common parameter. The second parameter is then estimated for each treatment. For the generalized Poisson distribution, it may be supposed that the parameter defining the distribution of individuals within clusters is constant for each treatment, while the distribution of clusters varies. This is equivalent to stating that λ/k, τ and λ_2 in Table 6.2 are all constant with treatment. Alternatively, in the formulation of a compound Poisson distribution, the distribution of clusters may be constant (i.e. k, $\lambda/2\tau$ and λ are constant) while the distribution of individuals within clusters varies (Freeman 1980). Either hypothesis may be considered for the pattern of inoculum and disease of soilborne pathogens. Outline details for fitting of these distributions are provided by Freeman (1980) for a common distribution of individuals within clusters. Bliss and Owen (1958) described a method for fitting a common k to the negative binomial distribution, which corresponds to constant density of clusters. Refinement of these methods will render the use of the contagious distributions substantially more useful in the analysis of spatial pattern.

The discrete distributions discussed above apply only to discrete variables. They have been applied to continuous variables such as length of infected root (St. John and Hunt 1983; St. John et al. 1983) and to indices of disease (Martin et al. 1983). This practice cannot be supported, however, on theoretical grounds, other than in exceptional circumstances, and cannot therefore be defended for routine analysis.

3.2 Indices of Aggregation

Of the numerous indices of aggregation advanced in the ecological literature (Southwood 1978; Pielou 1977), five heuristically distinct indices can be proposed. These are: (1) the variance/mean ratio ($\omega = s^2/m$); (2) the aggregation parameter (k) of the negative binomial distribution; (3) Lloyd's index of mean crowding ($\mathring{m} = m + s^2/m - 1$); (4) Lloyd's index of patchiness ($\gamma = \mathring{m}/m$); (5) Taylor's index b, where b is estimated from the relationship $s^2 = am^b$. [Two further commonly used indices are those of David and Moore (1954) and Morisita (1959) which are closely related to ω and γ respectively.]

Selection of one or more of the indices to summarize data on spatial pattern depends upon the properties of the particular index and the objectives of the investigation. Pielou (1977) argues that it is desirable to obtain a measure of aggregation that is independent of population density. Without this independence it is impossible to compare two populations without confounding density and pattern since it is unlikely that two or more populations will have similar densities. Of the five common indices of aggregation, only k, γ and b are independent of density. Since k is ordinarily estimated by maximum likelihood as a parameter of the negative binomial distribution, it can only logically be used if there is satisfactory agreement with this distribution (Taylor et al. 1978). Taylor's index b can be used without restriction. It does not, however, unlike the other indices have an intuitive spatial interpretation (Upton and Fingleton 1985), but see Perry and Taylor (1985) for a theoretical substance to b. Lloyd's index of patchiness, γ, combines the advantages of wide applicability, spatial interpretation and independence of density. Lloyd (1967) interpreted the index as a measure of how many times as "crowded" an individual is, on the average, as it would be if the population was random. Thus, the expectations for γ are <1, 1, >1 for regularity, randomness and aggregation respectively. Lloyd's index of crowding, $\mathring{m}$, by contrast, is a measure of the crowding of one individual by another. It is estimated by the mean number per individual of other individuals within each sampling unit.

The corollary that the indices, ω and $\mathring{m}$ are density-dependent can be used to test hypotheses about mortality factors. Consider inoculum: if deaths (or births) occur at random, irrespective of position, ω will decline in direct proportion to m. If, however, death of inoculum is density-dependent, ω will decrease more rapidly with m (Pielou 1977). For further details see Griffin and Tomimatsu (1983) who examined the relationships between $\mathring{m}$ *and* γ respectively with m, for data on the density of infection of *Cylindrocladium crotalariae* on roots of peanut.

Point estimates of indices of aggregation are of little use without corresponding interval estimates for the population parameter. Lloyd (1967) provided a method for his indices thus enabling comparisons to be made of sample data for two or more treatments. The method was based, however, on the assumptions of an underlying negative binomial distribution. An alternative procedure is given by Reed (1983). The method employs the Jackknife procedure, and is independent of assumptions concerning the underlying spatial distribution. In essence, it involves the computation of the mean and variance of a sample of size n of a series of weighted estimates of an index of aggregation. Each estimate is obtained by omitting one of the sample counts in turn. An approximate confidence interval is estimated using the t distribution, providing that n is reasonably large, i.e. >30. Reed (1983) also proposes some transformations of the data to satisfy the assumptions of normality in forming the confidence intervals when n is small, e.g. $\leqq 15$.

There is scope for more refined use of Lloyd's indices in the analysis of the spatial pattern of inoculum, infection and disease, particularly in the setting of confidence intervals and in the comparison of trends over time.

4 Analysis of Mapped Data

Maps for the spatial analysis of inoculum, infection or disease typically comprise one- or two-dimensional arrays of variates corresponding to the systematic arrangement of sampling units over an area of study. By taking explicit account of the relative positions of sampling units, map-based methods are potentially more informative than the methods described above. This is especially true in the detection of pattern at scales larger than the individual sampling units. A clear demonstration of this is given by Nicot et al. (1984).

Sampling units may be randomly selected within strata but more often they are systematically located either contiguously or spaced apart. Three-dimensional maps can be analyzed by the logical extension of the methods for two dimensions. In practice, however, it is usual to reduce three-dimensional samples to sequences of horizontal slabs. Inferences are then made for each slab separately as if it was two-dimensional.

The nature of the variate (Table 6.1) and, for infection and disease, the spatial pattern of the host plants influence the type of method used to test for randomness. In the following discussion, most emphasis is given to the analysis of dichotomous variables, i.e. presence or absence of disease for row and lattice crops. Brief consideration is given to a related method for the analysis of discrete and continuous variables, corresponding to a lattice of regions rather than of point sites. The analysis of mapped locations of points distributed in a continuum, such as diseased plants within a continuum, is not described, but see Diggle (1983) and Ripley (1981).

4.1 Analysis of Dichotomous Data

4.1.1 Transects

Consider a sequence of scores for diseased (D) and healthy (H) plants in a set of systematic samples of contiguous plants in a row:
DDHHHDDDDHDHDDHHHDDH. Two types of tests for deviation from randomness have been proposed, the runs test and doublet analysis. The runs test is based upon considerations of the numbers of runs of diseased and healthy plants present in the sequence (see e.g. Bradley 1968). Doublet analysis is based upon the consideration of the sequence of joins exhibited by successive overlapping pairs of plants. Amongst doublets, three types of join can be distinguished *DD*, *DH* and *HH*. Combinatorial

analysis is used to obtain empirical probability distributions and hence expectations and variances for the number of runs (U) or the number of joins (J) of a specific types, e.g. DD, under the H_0 of random mingling of H und D given the observed numbers of each. The observed values for U or J are then tested against the expected value, usually by use of the normal approximation, i.e.

$$Z = (U + 0.5 - E[U])/s_u , \quad (6.1)$$

in which $E[U]$ is the expectation and s_u the standard deviation of the statistic. The value 0.5 is included as a continuity correction because of the discrete nature of the variate. The calculated value for Z is then compared with the tabulated value, for a given level of significance. The assumption of normality requires that the number of sampling units is large (i.e. $\geqq 30$) and that the probability of a host being diseased is not close to 0 or 1. When the assumption of normality is untenable, the method of randomization can be used. In this all possible permutations of the data are considered or, if these are too large, a sample may be used (Sokal and Oden 1978). The test statistics U or J are calculated for each simulation and the values are sorted into ascending order. The H_0 is rejected at the significance level (α) of 5%, if the observed value lies within the extreme 5% of the values. Deviation from randomness may occur towards regularity or towards clustering. If, however, there is a priori reason to suppose that any deviation towards regularity is untenable, the test becomes one-sided: thus, the critical value for $Z(\alpha = 0.05, n > 30)$, above which H_0 is rejected, is 1.65 instead of 1.96.

Madden et al. (1982) compared the performance of the runs and doublet tests for maize dwarf mosaic virus on maize. They also compared the performance of the two tests for simulated data of random and clustered populations. The runs test gave fewer misclassifications of the simulated data than the doublet tests. However, further studies for different spatial patterns are necessary before an unequivocal recommendation is possible. Note that the original estimator of the variance in doublet analysis first used in plant pathology by Van der Plank (1946) was an overestimate (Freeman 1953; Converse et al. 1979). Use of the uncorrected estimator leads to a conservative bias in rejecting the null hypothesis.

The sequence of runs of several transects may be added (Pielou 1963) but this may introduce bias if, for example, the mean number of diseased plants differs significantly amongst the transects. Alternatively, information from several transects can be combined by calculating the probability value (p_i) for t, for each transect providing that the deviations from randomness are all in the same direction. For t transects the deviance, given by $-2 \sum_{i=1}^{t} \ln p_i$, is approximately distributed as χ^2_{2t}.

4.1.2 *Transects Across Diffuse Clumps of Disease*

Not all plants within a disease focus are necessarily diseased, yet these areas are qualitatively different from areas in a crop that are completely free of disease. Pielou (1963) extended the runs test to allow for diffuse clumps of disease in an analysis of *Armillaria* root rot on Douglas fir. The population was assumed to consist of a mosaic of "clean areas" within which no trees were diseased and "infected areas" within which a fixed proportion p of the trees was diseased. Thus, a transect might occur as $DDIIDIDHHHHII$ where I denotes the presence of a non-diseased tree in an infected area, but would be recorded as $DDHHDHDHHHHHH$. Pielou (1963) argued that the sequence of trees formed a discrete Markov chain in which the probability that the $i+1$th tree is diseased, is dependent upon the type of the ith tree but not on the type of any previous tree. From this, a transition matrix of probabilities can be constructed for the nine possible order joins, *DD, DI, DH, ID, II, IH, HD, HI, HH*, and ultimately empirical distributions for the run lengths of diseased and healthy trees. Pielou (1963) tested the null hypothesis by comparing the frequency distribution of the observed data with the expected values by the χ^2 test. The method is constrained by the restrictive assumptions in the derivation of the expected values, but it deserves further consideration and development for soilborne pathogens.

4.1.3 Disease Mosaics

The scale of trees in a plantation approximates to a discrete system but many soilborne pathogens of agricultural and horticultural crops occur as patches in a two-phase mosaic. In such cases the patch or focus rather than the plant itself is the entity of interest. Pielou (1964, 1977) has also proposed a method to test for deviation from randomness, for vegetatively reproducing plants, which may be applied to the present circumstances. Consider a transect for which the lengths of successive patches of disease and intervening gaps are recorded. The probability density functions for the lengths of diseased and healthy sections (l_D and l_H) under an H_0 of a random lines mosaic are given by (Pielou 1977):

$$f(l_D) = \lambda_D \, e^{-\lambda}_{D} {}^{1}_{D}, \qquad (6.2)$$

$$f(l_H) = \lambda_H \, e^{-\lambda}_{H} {}^{1}_{H}, \qquad (6.3)$$

in which the mean lengths of diseased patches is given by $1/\lambda_D$ and the mean length of gap by $1/\lambda_H$. A test of randomness may be carried out by the χ^2 goodness-of-fit, or by use of the normal approximation, if the sample size is large. Anisotropic effects within mosaics can be tested by comparing inferences derived from transects orientated at right angles. Data for this type of analysis can conveniently be obtained by aerial photography. An alternative method is given by Pielou (1964) involving the use of paired quadrats which is suitable for systems in which disease is diffuse within patches and boundaries are blurred.

4.1.4 Two Dimensional Lattice Maps

Most analyses of plant pathological data have tended to treat data from lattices as a series of transects. This ignores some of the information, for the observed pattern is a consequence of the order not only within rows but also across rows. Lattice tests are based upon the doublet method of joins between two contiguous points. The doublets may be orientated "vertically", horizontally and diagonally and tests are available which take account of some or all of these possibilities (Moran 1948; Freeman 1953). Three important combinations are the "bishop", "rook" and "queen" cases in which respectively, diagonal doublets only, horizontal and vertical doublets only and all three types of doublet are included. It is probable that the queen case is most useful for the analysis of soilborne pathogens. Additional inferences may be obtained with respect to possible directional correlations with contagion by considering a sequence of analyses for the linear (i.e. transect), rook and queen cases. Freeman (1953) provides equations for the means and variances of the empirical distributions for the queen and rook cases for lattices with vacancies, with allowance also for edge effects. Deviation from randomness is tested by normal approximation or by a randomization test. Freeman (1953) used the methods to analyze the spatial pattern of nettle-head virus on hops in which there were some missing plants. An alternative test has recently been proposed by Poushiansky and Basu (1984). The test is based upon the expected distribution of length of runs via horizontal and/or vertical paths to the nearest disease neighbour of a diseased plant. The test is conceptually simple, although the precise method of the χ^2 test employed is not clear. It is disadvantaged by testing only for deviation from randomness without distinction between clustering and regularity.

Considerably more information may be obtained on the spatial pattern of disease by the analysis of data for sequential observations. Freeman (1953) proposed a method for lattices to test whether or not disease is spread preferentially to those plants whose neighbours are already diseased. The method involves a 2×2 comparative trial in which the fate of healthy plants is considered at two successive times relative to the disease status of neighbouring plants at the earlier time. Freeman (1953) used a normal approximation to carry out the test but other tests such as χ^2 or log likelihood could be used. This procedure has considerable potential for the analysis and modelling of substantial data sets in which various criteria for neighbours may be considered.

4.2 Analysis of Discrete and Continuous Variables

Many methods have been proposed for the analysis of systematic maps of data involving counts and continuous variables (Cormack and Ord 1979). Historically much emphasis has centred on "quadrat methods" in which the effects upon sample variances of combining quadrats of different size or distances apart are used to make inferences about the size of clumps of, for example, diseased plants. A comparative account is given by Ludwig (1979). I propose not to discuss these methods but to consider an alternative method involving autocorrelation, due to Moran (1950).

Suppose that a lattice is regarded, not as a series of points, but as a set of contiguous rectangles, these may be occupied or not, or in the present consideration, each will be represented by a variate such as estimated numbers of propagules, numbers of diseased plants or proportion of diseased roots. If the variable is randomly distributed amongst the cells, then the autocorrelation amongst contiguous cells will be close to zero. One measure of spatial autocorrelation, I, is given by Moran (1950):

$$I = \frac{n \sum_i \sum_j W_{ij}(x_i - \bar{x})(x_j - \bar{x})}{S \sum_i (x_i - \bar{x})^2} \qquad i \neq j, \tag{6.4}$$

in which n is the number of cells, x_i is the variate of interest, W_{ij} is the weighting variable that takes the value 1 for contiguous cells and zero otherwise and S is a constant dependent upon the criterion for contiguity under consideration. Nicot et al. (1984) used the index to describe artificial populations of soil inoculum. Various criteria may be used for contiguity, such as lateral or diagonal touching. Boundaries need not be straight and the I statistic may be applied to populations involving fields. Expressions for the mean and variance of I under a H_0 of random allocation are given by Cliff and Ord (1981). The normal approximation may be used, providing $n \geqq 20$. Sequential use of I using different criteria for contiguity offers promise in the analysis of soilborne pathogens. Interest is accumulating in the use of Moran's I for the analysis of residual error in field experiments (Bartlett 1978).

5 References

Anscombe FJ (1949) Sampling theory of the negative binomial and logarithmic series distributions. Biometrics 5: 165–173
Bartlett MS (1978) Nearest neighbour models in the analysis of field experiments (with discussion). J R Statist Soc Ser B 36: 147–174
Bliss CI, Owen ARG (1958) Negative binomial distributions with a common k. Biometrika 45: 37–58
Bradley JV (1968) Distribution-free statistical tests. Prentice-Hall, Englewood Cliffs, New Jersey
Campbell CL, Noe JP (1985) The spatial analysis of soilborne pathogens and root diseases. Annu Rev Phytopathol 23: 129–148
Campbell CL, Pennypacker SP (1980) Distribution of hypocotyl rot caused in snapbean by *Rhizoctonia solani.* Phytopathology 70: 521–525
Cliff, AD, Ord JK (1981) Spatial processes: models and applications. Pion, London
Cochran WG (1977) Sampling techniques 3rd edn. Wiley, New York
Converse RH, Seeley J, Martin LW (1979) Evidence for random local spread of aphid-borne mild yellow edge virus in strawberries. Phytopathology 69: 142–144
Cormack RM, Ord JK (1979) Spatial and temporal analysis in ecology. International Co-operative Publishing House, Fairland, Maryland
Cox GC, Sanders FE (1974) Ultrastructure of the host-fungus interface in a vesicular-arbuscular mycorrhiza. New Phytol 73: 901–912
David FN, Moore PG (1954) Notes on contagious distributions in plant populations. Ann Bot (Lond) 18: 47–53
Diggle PJ (1983) Statistical analysis of spatial point patterns. Academic Press, London
Feller W (1943) On a general class of contagious distributions. Ann Math Statist 14: 389–400
Freeman GH (1953) Spread of diseases in a rectangular plantation with vacancies. Biometrika 40: 287–296
Freeman GH (1980) Fitting two-parameter discrete distributions to many data sets with one common parameter. Appl Statist 29: 259–267
Gilligan CA (1982) Statistical analysis of the spatial pattern of *Botrytis fabae* on *Vicia faba:* a methodological study. Trans Br Mycol Soc 79: 193–200
Gilligan CA (1983) A test for randomness of infection by soilborne pathogens. Phytopathology 73: 300–303
Gilligan CA (1985) Construction of temporal models: III disease progress of soilborne pathogens. In: Gilligan CA (ed) Mathematical modelling of crop disease. Academic Press, London, pp 67–105 (Advances in plant pathology, vol 3)
Griffin GJ, Tomimatsu GS (1983) Root infection pattern, infection efficiency, and infection density — disease incidence relationships of *Cylindrocladium crotalariae* on peanut in field soil. Can J Plant Pathol 5: 81–88
Hau FC, Campbell CL, Beute MK (1982) Inoculum distribution and sampling methods for *Cylindrocladium crotalariae* in a peanut field. Plant Dis 66: 568–571
Lin CS, Poushiansky GP, Mauer M (1979) An examination of five sampling methods under random and clustered disease distributions using simulations. Can J Plant Sci 59: 121–130
Lloyd M (1967) Mean crowding. J Anim Ecol 36: 1–30
Ludwig JA (1979) A test of different quadrat variance methods for the analysis of spatial pattern. In: Cormack RM, Ord JK (eds) Spatial and temporal analysis in ecology. International Co-operative Publishing House, Fairland, Maryland, pp 289–304
Mac Nish GC (1985) Methods of reducing *Rhizoctonia* patch of cereals in Western Australia. Plant Pathol (Lond) 34: 175–181

Madden LV, Louie R, Abt JJ, Knoke JK (1982) Evaluation of tests for randomness of infected plants. Phytopathology 72: 195–198
Martin SB, Campbell CL, Lucas LT (1983) Horizontal distribution and characterization of *Rhizoctonia spp.* in tall fescue turf. Phytopathology 73: 1064–1068
Menzies JD (1963) The direct assay of plant pathogen populations in soil. Annu Rev Phytopathol 1: 127–142
Moran PAP (1950) Notes on continuous stochastic phenomena. Biometrika 37: 17–23
Morisita M (1959) Measuring of the dispersion and analysis of distribution patterns. Mem Fac Sci Kyushu Univ Ser E Biol 2: 215–235
Nicot PC, Rouse DI, Yandell BS (1984) Comparison of statistical methods for studying spatial patterns of soilborne plant pathogens in the field. Phytopathology 74: 1399–1402
Perry JN, Taylor LR (1985) Ades: new ecological families of species — specific frequency distributions that describe repeated spatial samples with an intrinsic power-law variance-mean property. J Anim Ecol 54: 931–953
Pielou EC (1963) Runs of healthy and diseased trees in transects through an infected forest. Biometrics 19: 603–614
Pielou EC (1964) The spatial pattern of two-phase patchworks of vegetation. Biometrics 20: 156–167
Pielou EC (1977) Mathematical ecology, 2nd edn. Wiley, New York
Reed WJ (1983) Confidence estimation of ecological aggregation indices based on counts — robust procedure. Biometrics 39: 987–998
Ripley BD (1981) Spatial statistics. Wiley, New York
St. John TV, Coleman DC, Reid CPP (1983) Growth and spatial distribution of nutrient-absorbing organs: selective exploitation of soil heterogeneity. Plant Soil 71: 487–493
St. John TV, Hunt HW (1983) Statistical treatment of VAM infection data. Plant Soil 73: 307–313
Sokal RR, Oden NL (1978) Spatial autocorrelation in biology. I. Methodology. Biol J Linn Soc 10: 199–228
Southwood TRE (1978) Ecological methods with particular reference to the study of insect populations, 2nd edn. Chapman & Hall, London
Stanghellini ME, Stowell LJ, Kronland WC, von Bretzel P (1983) Distribution of *Pythium aphanidermatum* in rhizosphere soil and factors affecting expressions of the absolute inoculum potential. Phytopathology 73: 1463–1466
Taylor LR, Woiwood IP, Perry JN (1978) The density dependence of spatial behaviour and the rarity of randomness. J Anim Ecol 47: 383–406
Upton GJG, Fingleton B (1985) Spatial analysis by example, vol 1. Point pattern and quantitative data. Wiley, New York
Van der Plank JE (1946) A method for estimating the number of random groups of adjacent diseased plants in a homogeneous field. Trans R Soc S Afr 31: 269–278
Wheatley GA, Freeman GH (1982) A method of using proportions of undamaged carrots or parsnips to estimate the relative population densities of carrot fly *(Psila rosae)* larvae, and its practical applications. Ann Appl Biol 100: 229–244

Chapter 7

Electronic Monitoring and Use of Microprocessors in the Field

J.C. Sutton[1], T.J. Gillespie[2] and T.D.W. James[1]

1 Introduction

The requirements of field studies in plant disease epidemiology and weather-timed disease management on the farm have provided a strong impetus for developments in electronic monitoring of the crop microclimate. Recent innovations in electronic sensors have substantially improved the scope and practicality of measuring microclimatic variables while developments in microprocessor-based dataloggers have permitted efficient and continuous recording from these sensors in the field. Variables such as air, soil or plant temperature, leaf wetness duration, relative humidity and irradiance can now be monitored routinely in many kinds of crops. Today, weather monitoring is the main application of microprocessors in field epidemiology.

Weather variables are best monitored only in relation to an identified need or application (Coakley 1985). Electronic monitoring systems readily amass vast quantities of information but this may be useful only when the relationships of weather and disease are adequately known. Unnecessary monitoring should be avoided because accurate collection and processing of data requires considerable input by the investigator. Exploratory studies and careful planning make it possible to tailor the sensor and sampling frequency to the specific task at hand thus saving time and grief. (See also Chap. 8 by Friesland and Schrödter for some practical hints for the monitoring of weather factors).

2 Environmental Data Acquisition Systems

2.1 Data Recording

Reliable microprocessor-based dataloggers capable of precise recording of environmental data are widely available. Simple dataloggers measure signals from two or three sensors, while more complex systems monitor a wide range of weather variables (Table 7.1) and may allow for extensive replication of sensors in the crop environment. Sophisticated dataloggers can rapidly scan more than 100 input channels, have sensitivities exceeding 0.5 μV and are programmable for preliminary processing of the data before it is transferred to a computer.

[1] Department of Environmental Biology and
[2] Department of Land Resource Science, Ontario Agricultural College, University of Guelph, Guelph, Ontario N1G 2W1, Canada

Table 7.1. Electrical inputs and types of sensors commonly used for measuring various environmental factors with a datalogger suitable for epidemiological research

Input	Sensor type	Environmental factor measured
Analog-DC	Thermocouple or thermistor	Air or soil temperature
	Pyranometer	Irradiance
	Aspirated psychrometer	Atmospheric water vapour
Analog-AC excitation	Sulphonated polystyrene	Atmospheric water vapour
	Resistance clip	Wetness duration
	Artificial leaf or fruit	Wetness duration
	Resistance block	Soil moisture
Digital-pulse counter	Cup anemometer	Wind speed
	Tipping bucket rain gauge	Rainfall

Epidemiological research often requires frequent observations of microclimate by an array of sensors, many of which may respond quickly to abrupt changes in weather. Multichannel dataloggers with substantial memory (e.g. 16K ROM and 4K RAM) are suited to this purpose (Fig. 7.2a). In the schematic outline of such a datalogger (Fig. 7.1), sensor signals are received at A, processed and stored at C. Scan intervals, signal types and ranges, and calibration factors to convert voltages to meteorological units (e.g. °C, % relative humidity and wind speed in ms^{-1}) may be programmed at this stage. These instructions may feed back to B which controls the supply of AC or DC excitation current to sensors and the timing of other analog outputs such as power for the fan in a ventilated psychrometer. Processed input data can be stored for real-time display or passed on to output processing (D) where user-programmable summaries are generated which frequently include the time and date, sums and cardinal values. Fully processed results are retained in final storage memory (E) for transfer through external communication devices (F).

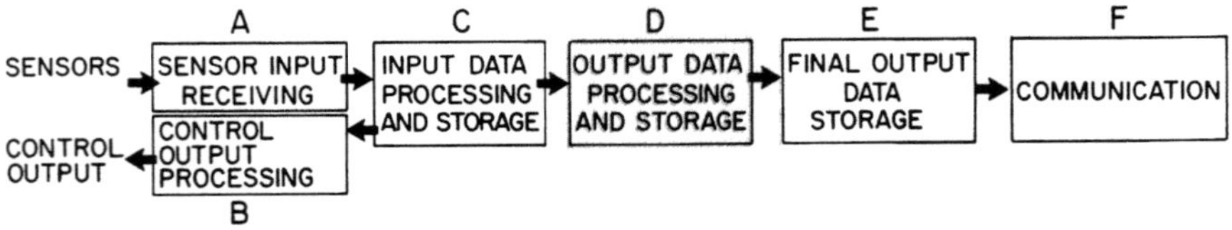

Fig. 7.1 The basic elements of a weather datalogger suitable for epidemiological research

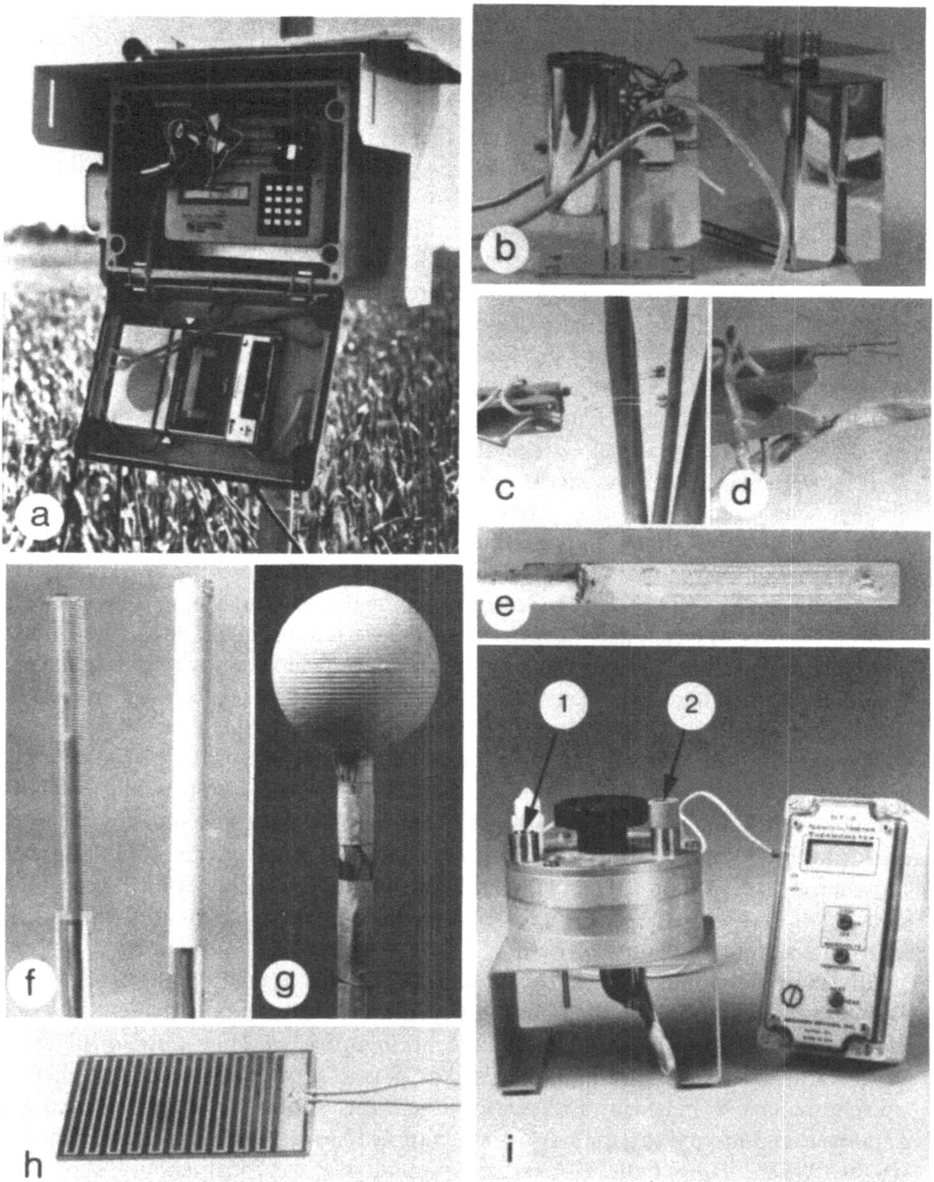

Fig. 7.2 a – i Instrumentation for electronic monitoring of weather variables in crops. **a** Multichannel datalogger with a cassette tape recorder inside a weather shelter in the field. **b** Ventilated psychrometer (Model WVU-7, Campbell Scientific Inc., Logan, UT 84321, USA) showing internal components *(left)* and shelter *(right)*. **c** Clip-type wetness sensor on an onion leaf. **d** Electrical impedance probe monitoring moisture of a dead onion leaf. **e** Plate-type leaf wetness sensor. **f** Cylindrical leaf wetness sensors, before painting *(left)* and after painting *(right)*. **g** Sensor for monitoring wetness of apple fruits. **h** Unpainted leaf wetness sensor showing electrodes bonded to printed circuit board. **i** A thermocouple psychrometer with uncapped (*1*) and capped (*2*) specimen chambers (Decagon Devices Inc., Pullman, WA 99163, USA)

All dataloggers contain elements A, C, D, E and F (Fig. 7.1). Differences among various models in sophistication and cost mainly relate to: (1) the number and variety of sensor input channels; (2) the ability to control external devices; (3) the speed of sensor scanning and sensitivity (e.g. ability to receive μV thermocouple signals without preamplification); (4) the memory size; (5) the degree of programmability by the user; (6) the capacity to process and summarize data; and (7) the degree of flexibility in communicating data.

Dataloggers typically used in on-farm weather stations may be capable of receiving signals from seven or eight different sensors, have a storage capacity of 1000 readings from each of the sensors, a user-selectable recording interval of 1 min to 10 h and software for simple data summary. Electronic weather stations can monitor environmental data pertinent to a number of different diseases and some can be programmed to give audible and/or visual signals when disease risk is high.

A simple system is useful for monitoring two or three weather variables pertinent to a specific disease. Two-channel systems that we have used to measure air temperature and leaf wetness duration have a memory capacity of about 1000 pairs of readings and average sensor readings over selected intervals. Other types record, for example, the onset of leaf wetness, the elapsed time between wetting and drying, average air temperature of the wet period, average temperature over the previous 24 h and degree days. Some of the more recent models incorporate disease forecasting software which informs the user when a fungicide spray is required (e.g. Jones et al. 1984). These models should be used *only* in studying those pathosystems for which they were developed. Use for other pathosystems can be erroneous, because the sensors (especially for wetness) may be unsuitable and the interpretive software inappropriate.

The most basic datalogging systems can provide an option when financial resources are meagre. We have built a small recorder (15 × 8 × 5 cm) from readily available components costing about US $20.00. The recorder may be attached to any electronic sensor for leaf wetness and stores the wetness data in a simple memory. Output of the recorder is displayed on a ring of 24 light-emitting diodes (LEDs) which indicate the hours of the day when the sensor was wet. A pushbutton is used to identify which LEDs light up and thus the hours of wetness.

All electronic monitoring systems require protection against operational hazards. Dataloggers, susceptible to high humidity and liquid water, should be kept in a rain-proof enclosure provided with moisture-absorbing silica gel. They are also vulnerable to power surges caused by lightning strikes. Spark gaps in sensor circuits help to prevent damage from nearby strikes and surge protectors should always be used with dataloggers connected to a main AC power supply to avoid damage from more distant strikes. When using battery-operated systems, the electrical supply should be monitored periodically since a drop in voltage can result in distortion or loss of data. Finally, it is essential that electrical shields on sensor wires are grounded only at the datalogger and that sensors in contact with soil are electrically insulated to prevent ground loops.

2.2 Data Communication

Methods for communicating weather data from the acquisition system to the user are diverse. The choice of method depends on the type and quantity of data collected, the acquisition equipment used, distance of the monitoring equipment from the user, computing equipment available, financial resources and personal preferences.

1. Liquid Crystal Display (LCD). The LCD provided on most dataloggers allows direct readout of current or stored data. It is useful for checking the operation of the datalogger and sensors, scanning past and present weather and checking battery voltage. When the LCD is the only communication device, the data must be transferred manually to a calculator or computer.

2. Printers. On-board printers provide a hard copy of weather data but are prone to malfunction in the field. Printing paper softened by damp weather tends to jam, ribbons sometimes jam or run out of ink, and batteries may drain quickly. These problems can be avoided with dataloggers equipped to communicate (at the same baud rate) with a portable printer, which can be taken periodically to the logger for data collection. Transfer of data from printed hard copies to computers must be done by hand.

3. Cassette Recorders. The cassette tape is an inexpensive medium for communicating large quantities of weather information. Data may be retrieved by cassette recorders taken to the field, or by on-line connection of the recorder and the datalogger. Commands are generated by the logger for powering the recorder and dumping of the data onto a tape. Data on the tape may be directly transferred to a computer through a standard RS-232 serial interface. High quality normal-bias tapes usually perform better than high bias chromium oxide tapes. Incorrectly adjusted or dirty recorder heads, imperfect tapes or cassettes or an incorrect volume setting on the recorder during playback may result in a "noisy" data record. While "noise" is usually removable, extensive data loss may result from stretching, tangling or breakage of the tape. Breakage is more frequent at temperatures below freezing.

4. Data Storage Modules. Erasable-programmable read-only memories (EPROMs) are integrated circuits which can be used for data storage. The stored information is rapidly transferred to a computer through on RS-232 interface. EPROMs are inexpensive, function well at below-zero temperatures and may be reused after stored information has been erased electrically or by exposure to ultraviolet light. Handling of removed EPROMs requires manual dexterity to avoid bending or breaking the pins.

5. Direct-Wire Tie to a Computer. Some dataloggers can be directly wired to a host computer for data communication under control of the logger or the host computer. This is feasible only when the datalogger is in close proximity to the host computer, as with some on-farm weather stations.

6. Telephone Modem. Protocol for telephone communication of data is contained in most sophisticated dataloggers, but simpler systems may also be modified for this purpose. A modem is required to link the datalogger to a host computer through the telephone line. This communication method allows rapid and direct transfer of weather data over long distances from individual monitoring systems or from networks of dataloggers to a host computer. The dataloggers can be remotely programmed or interrogated by the user through the computer, or interrogated automatically. Easy access of monitoring systems to telephone lines is essential, thus telephone hook-ups are most practical when the monitoring system is permanently positioned.

7. *Radio Frequency (RF) Telemetry.* RF telemetry is available with sophisticated dataloggers but is not widely used by epidemiologists. RF telemetry offers the flexibility of communicating with any datalogger operating on the same radio frequency and within the range of the RF receiving device. Operation outside this range is possible if the RF receiver is linked to a telephone line.

3 Monitoring Weather Variables

Accurate weather monitoring whether with electronic or mechanical equipment requires sensors that are in good order and properly calibrated. Most sensors should be calibrated before experiments and during monitoring. The assembled system of datalogger and sensors should be calibrated against known standards over the range of conditions expected during operation in the field.

The evaluation of weather information monitored and recorded is dealt with in Chap. 8 by Friesland and Schrödter.

3.1 Temperature

Sensors for monitoring air temperature are widely used in epidemiology and usually are positioned among the foliage. It should be recognized, however, that air temperature within the canopy often differs from that of nearby plant organs or of liquid water on the organ surfaces. Versatile thermometers for use in conjunction with dataloggers are thermistors and thermocouples.

Thermistors. The thermistor is a semiconductor with an electrical resistance which is inversely and exponentially related to temperature. Because resistance decreases about 5% when the temperature increases 1 °C, high temperature resolution (e.g. ± 0.1 °C) is possible. The semiconductor is made from metal oxides (e.g. cobalt, copper, iron, manganese and nickel) in the form of small (0.02–3.0 mm) beads, discs or cylinders (Fritschen and Gay 1979). Thermistors used for monitoring air temperatures in crops require protection from water and shielding from direct sunlight (Sutton et al. 1984). A simple shield, oriented to prevent solar radiation impinging on the sensor is effective when an accuracy of about 0.5 C suffices. More elaborate shields, in which the sensor is forcibly ventilated and well protected, allow for greater accuracy in temperature measurements.

Thermocouples. Thermocouples employ the principle that junctions of dissimilar metals generate an electromotive force which is proportional to their temperature (Seebeck effect, Fritschen and Gay 1979). Current flows when the temperatures of two interconnected junctions are different. Copper-constantan junctions are suitable for moist environments when coated with a clear lacquer. Because the thermocouple signal is small (40 μV/°C), shielded wire is required to ground currents induced by nearby electrical fields. Thermocouples are made by soldering (silver solder is stronger than soft solder) or welding the ends of the copper (acid-dipped) and constantan wires. Use of fine wire (e.g. 0.25 mm) of the same batch minimizes errors. Thermocouples used in the crop canopy require frequent checking for position and cleanliness. Thermocouples positioned on plant surfaces usually give erroneous results (Beadle et al. 1973) but accurate data may be obtained when the sensor is imbedded in fleshy tissues. Leaf temperatures may be accurately sampled by using infrared thermometry (Amiro et al. 1983).

3.2 Environmental Water

The growth and reproduction of plant pathogens are profoundly affected by the states, forms and energy of environmental water. Pathogens may be exposed to water as vapour, liquid or ice while in the atmosphere, on plant surfaces, in the intercellular spaces of living plants or in the interstices of plant residues or soil. During aerial dispersal, pathogens are exposed to a range of atmospheric water potentials (ψ atmos) and may encounter liquid water in the form of rain or fog. On foliage, pathogens are exposed to water vapour when the leaves are "dry", or to liquid water in films or discrete droplets, and to vapour among the droplets when the leaves are "wet".

Water potential is of universal signifiance in affecting the growth and development of pathogens (Duniway 1982), but other properties of water may also be critical (Sutton and Hildebrand 1985). Electronic sensors have been developed which monitor the wetness duration on plants or relative humidity (RH) from which ψ atmos also may be calculated. However, devices for monitoring other attributes of water, including the forms, ψ and rates of accumulation or loss of liquid water on living or dead foliage, have yet to be development (James et al. 1984; Sutton and Hildebrand 1985).

3.2.1 Atmospheric Humidity and Water Potentials

Ventilated Psychrometer (Fig. 7.2b). Compact psychrometers suitable for use in crop canopies are now available. These psychrometers often utilize two thermistors matched in response over a wide range of temperatures (e.g. 0–45 °C). One thermistor is kept dry (the "dry bulb") while the other (the "wet bulb") is kept continuously moistened by means of a wick. A fan which provides airflow ($>3\,\mathrm{m\,s^{-1}}$) across the thermistors is critical to assure maximum depression of the wet bulb temperature. To conserve battery power, the fan is best programmed to operate only for about 2 min before each reading. The thermistors are sheltered against rain, mud-splash, irradiance and other variables which may give rise to errors.

For a rationale to calculate RH from the wet and dry bulb reading see Sutton et al. (1984). The ψ atmos may be calculated from RH values and dry bulb readings using the following equation (Hall and MacHardy 1981):

$$\psi \text{ atmos} = 1.06(273 + {}^{\circ}\mathrm{C}) \log_{10}(\%\ \mathrm{RH}/100)\ \mathrm{MPa}\,, \qquad (7.1)$$

Because the psychrometer requires frequent attention it is better suited for use in the research plot than on the farm. The water reservoir requires replenishing with deionized water every 1–3 days. The wicks need frequent cleaning and periodic calibration against a standard instrument, such as a sling or Assman psychrometer (Fritschen and Gay 1979). Unattended operation is limited to periods of above-freezing temperatures. A well-calibrated clean psychrometer with matched and accurate (e.g. ± 0.1 °C) thermistors gives measurement errors of 2% or less over a wide range of RH.

Sulphonated Polystyrene Humidity Sensor. This sensor exploits effects of moisture adsorption and desorption on the electrical conductivity of a sulphonated polystyrene (SP) plate. Sulphonation by exposure to sulphuric acid fumes provides the structurally rigid, but porous, polystyrene with strong hygroscopic properties. Electrical conductance is sensed by a pair of separate but interlocking electrodes bonded to the plate and varies logarithmically with RH. The sensor responds within a few seconds to an RH change but may take several minutes to attain equilibrium. When the RH fluctuates rapidly, the sensor output "chases" the humidity changes without attaining equilibrium; however, sensors with an operational error not exceeding ±5% are available commercially. Unlike the psychrometer, the SP sensor functions at temperatures below freezing and is suited to year-round monitoring of RH.

When the SP sensor becomes dirty or wet it does not function properly, thus adequate shielding from dust, debris and rain is essential. Our sensors performed well for 3 month while mounted inside a simple shield suspended in an instrument shelter. The more elaborate protection of some commercial sensors may extend sensor life to about a year. Measurement of RH in the crop based on observations within shelters requires temperature measurements in both sites. Ambient RH is then calculated from:

$$RH = RH_{SP}(e_{STSP}/e_{STDB}) \,. \tag{7.2}$$

where RH_{SP} is the RH of the SP sensor, and e_{STSP} and e_{STDB} are the saturated vapour pressures at the temperatures of the SP sensor and dry bulb respectively.

3.2.2 Liquid Water on Living Foliage and Fruits

Surface moisture from dew, rain, fog, irrigation or guttation affects most foliar pathogens and is a key component of most disease forecasters. Several kinds of electronic sensors have been developed for monitoring the duration of surface wetness. In most sensors, one or more pairs of electrodes detect wetness according to electrical resistance, which is low when the surface is wet. A generalized circuit diagram for use with wetness sensors is given in Fig. 7.3.

Clip-Type Sensors (Fig. 7.2c). In these sensors, nickel wire (0.25 mm) electrodes are clipped or held directly on the plant foliage. The electrodes may be mounted on clothes pegs supported by stakes (Sutton et al. 1984). Several clips connected in a parallel circuit and positioned at various heights in the crop provide spatial sampling of wetness duration. Clip sensors are especially adaptable to plant organs which are roughly rod-shaped, fairly rigid and not prone to violent movement in gusty wind. Soft or thin tissues may be injured by the electrodes resulting in leakage of fluids and false readings. Clip sensors have been used on large leaves (Häckel 1980) and probably are feasible for large fruits like cocoa pods and bananas. Daily checking for good contact of the clip with the plant tissues is advisable.

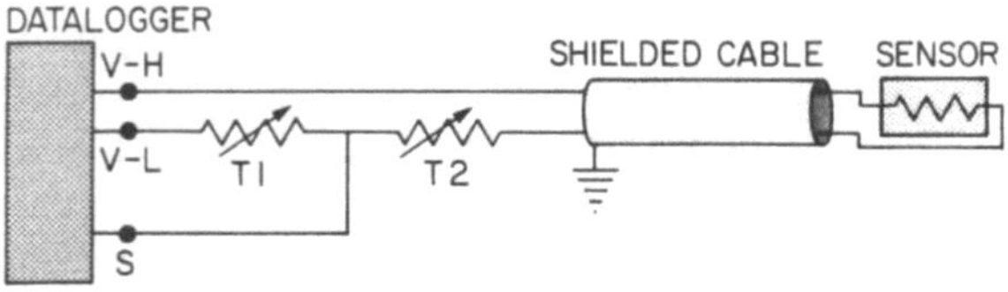

Fig. 7.3 Generalized circuit diagrams for recording signals of electronic wetness sensors. *V–H* and *V–L* are terminals of high and low supply voltage respectively. Alternating current is used to avoid polarizing water on the sensor. The wetness signal is received at *S*. T_2 and T_1 are trimming resistors (typically 0–100 kΩ) to control current through the sensor and voltage at *S*.

Clip sensors require calibration against sensible wetness because they may respond when "dry" tissues are exposed to high RH. For calibration, plants with attached sensors are exposed to increasing RH near saturation but without dew deposition. The two variable resistors (Fig. 7.3) are adjusted until only a small "dry baseline" voltage is recorded.

Wetness Sensors that Mimic Plant Organs. The electrodes of these sensors are mounted or bonded onto an artificial sensing surface which is exposed in the crop. Wetting and drying of the exposed sensor should coincide closely with that of the nearby plant organs which the sensor is intended to mimic. The size, shape, surface characteristics and thermal properties are key variables affecting the accuracy of the sensor. Accurate sensors usually resemble the plant organ in size and shape (Howard and Gillespie 1985). Surface characteristics of sensors are critical for achieving appropriate wetting and reflective properties.

Sensors usually require a suitable coating to achieve desired accuracy. The uncoated sensor is insensitive to liquid water until at least one droplet becomes large enough to bridge the space between the exposed electrodes (Davis and Hughes 1970). As further moisture is deposited, irregular "puddles" tend to form. Water films form readily as further moisture accumulates.

Silicone applied as a light spray to form a porous coat causes water to form discrete drops on the sensing surface but does not affect electrical sensitivity to deposited water. Paraffin wax applied in a striped pattern perpendicular to the electrodes achieves the same effect (Gillespie 1985, unpublished).

Sensors coated with latex-based (porous) point are highly sensitive to initial wetness which is rapidly absorbed by the paint and quickly diminishes electrical resistance. In our experience, low cost paint is best, possibly because it contains fewer additives than expensive paints. The basic white paint requires exploratory tinting to match sensor response to periods of sensible moisture on the crop of interest. For monitoring wetness on several crop species (e.g. onions, maize, strawberries) we used the following formulation: 125 parts Pittsburgh® 88–6 white latex, 20 parts water, 1 part green tint (phthalocyanine). Paint sprayed on the sensor should be dried for a day at 100 °C, immersed for 1 h in distilled water then dried for another day at 100 °C. This process removes or inactivates hygroscopic materials that cause sensor response to high RH.

Artificial Leaves. 1. Flat plates (Fig. 7.2e). Plate-type sensors usually are made from 1-mm-thick printed board, to which a pair of solder-coated copper electrodes is bonded. The electrodes may be formed as interlocking fingers spaced about 1 mm apart (Fig. 7.2h). Sensors using boards about as wide as the leaf have successfully monitored dew duration in corn (Gillespie and Kidd 1978), onions (Sutton et al. 1978) and wheat (Huband and Butler 1984).

To make a plate-type sensor, copper-clad (1 oz.) printed circuit board is cut to the required size and cleaned to remove dirt, grease and copper oxide by soap or alkaline solution, followed by 5% sulphuric acid (Howard and Gillespie 1985). The pattern of the electrodes is then drawn on the board using a resist ink pen. Copper not protected by the dried ink is removed by etching the board in a solution of ferric chloride or sodium persulphate until only the electrode pattern remains. An abrasive cleaner or steel wool may be used to remove the resist ink. The board should be dipped in 5% sulphuric acid and then tin plated to prevent oxidation of the copper when the sensor is exposed in the crop. The sensor is then coated with latex paint or silicone. Mounting of the board on a tubular handle made of nonabsorptive and nonconducting material facilitates the use of the sensor in the field. Electrical connection of sensor and datalogger is made using twisted-pair shielded cable to eliminate electrical noise.

2. Flexible type. Huband and Butler (1984) constructed a sensor using flexible, printed circuit board to simulate the disposition and motions of wheat leaves. The saturation capacity of the silicone-coated surface of the sensor declined when shaken by wind, mimicking the behaviour of real leaves. The flexible sensor was effective for monitoring wetness occurring as droplets on young cereal plants, whereas plate sensors dried long before the droplets. Flexible sensors require cleaning and treatment with silicone about every 3 weeks.

3. Cylindrical type (Fig. 7.2f). Cylindrical sensors for mimicking onion leaves were constructed using 20 cm lengths of Plexiglas® tubing (Howard and Gillespie 1985). Two parallel grooves 1.5 mm apart and 0.5 mm deep were cut in a spiral in the sensor surface which was then coated with white paint (oil-based and slightly thinned) to prevent possible absorption of water by the Plexiglas. Single strands of nickel wire (the electrodes) were wound in each groove and cemented at each end of the sensing surface. Twisted-wire shielded cable was soldered to each nickel wire at the distal end of the sensor and extended through the tubing to the datalogger. The ends of the tube were filled with electrically-insulating silicone rubber then the sensing surface was lightly sprayed with latex paint.

Artificial Fruits. Principles used for developing artificial leaves apply also to artificial fruits such as those for mimicking apples, tomatoes or cocoa pods. Fruits have a high heat capacity, on account of their mass and high water content, and exhibit slow heat exchange with the atmosphere because of their low surface-to-mass ratio. Accordingly, fruits cool down slowly in the evening, resulting in delayed dew deposition, and warm up slowly in the morning such that dew persists much later than on the leaves (Monteith and Butler 1979).

A plastic fishing float with a spherical flotation chamber about 6 cm in diameter was used to build an artificial apple fruit (M.J. Pedro and T.J. Gillespie 1980, unpublished). Parallel grooves were cut 1 mm apart in a spiral pattern from the top to the bottom of the sensor (Fig. 7.2g), nickel wire was wound in the grooves, and latex-based paint applied. For simulating the heat capacity of apples, the flotation chamber was filled with water. The sensor was usually accurate to within 30 min for dew deposition or disappearance when exposed in a tree and compared with dew on nearby apples.

3.2.3 Wetness of Dead Tissues

The water status of dead host tissues is critical to colonization, sporulation and survival of numerous pathogens, yet is rarely monitored in epidemiologic studies. Wetness sensors for green leaves are generally unsuitable for dead leaves (James et al. 1984). Only recently has attention focussed on instruments for attached dead foliage or crop residues on the soil.

James et al. (1984) used an electrical impedance probe to measure moisture of dead onion leaves in studies of *Botrytis* leaf blight (Fig. 7.2d). A pair of nickel electrodes mounted 1 mm apart on a clothes peg were inserted into the dead tissues and electrical resistance was recorded as the tissues were wetted and dried. The probe detected patterns of hydration and dehydration, but gave only a rough indication of water content, and its response was not calibrated with water potentials.

Instruments for psychrometric determination of water potentials in dead or living plant materials or soil show promise for applications in epidemiology. As an example, a thermocouple psychrometer in Fig. 7.2i accepts samples in small specimen chambers which are then sealed within the instrument. The moisture of the specimen and air in each chamber is allowed to equilibrate for 15–30 min. Thermocouples measure the temperature and wet bulb depression (nonaspirated) of the chamber atmosphere, from which ψ may be calculated using a psychrometric formula specific to the particular instrument. Salt solutions of known ψ may be used to check calibration. Because of the sensitivity of ψ to temperature change, the psychrometer should have a housing of large thermal mass and should be operated in a stable temperature environment.

3.2.4 Rain

Rain is a complex variable with a diversity of attributes of potential importance in epidemics. Among these are spatial distribution, temporal attributes (time, frequency and duration), intensity features (number, size and velocity of droplets), splash characteristics and thermal effects. Rain may be more important than dew as a source of leaf wetness in certain localities.

The tipping-bucket gauge has become a familiar workhorse for electronic monitoring of the temporal attributes and overall intensity of rain (Sutton et al. 1984). The "bucket" of the gauge is a two-chambered, V-shaped vessel positioned beneath a funnel for collecting the rain. The vessel tips back and forth in response to each 0.25 or 2.5 mm rain, sequentially dumping the water from each of the chambers. An electrical pulse is generated each time the vessel tips. Errors may result from "quenching" during heavy downpours, insufficient rain to tip the vessel during light showers and residual water on the funnel or in the vessel. Water-repellant activity of silicone sprayed periodically on the funnel and vessel helps to minimize water retention errors during collection and dumping of rain.

3.3 Wind

Wind causes movement and abrasion of the host foliage, disperses pathogens and influences temperature, environmental water and other microclimatic variables (Grace 1977, Legg and Bainbridge 1978). Air movement in and above crops may be described in terms of mean horizontal speed, direction and turbulence phenomena like gusts and eddies. Wind speed increases more or less logarithmically with height above bare soil or crops but decreases exponentially with depth into the crop canopy (Campbell 1977). Consideration of the local and momentary fluctuations in wind as well as average wind speeds is advisable when interpreting relationships of anemometric data and biological events.

The cup anemometer is still the main instrument for long-term use in measuring average wind speeds, but does not respond well to turbulence (Rosenberg et al. 1983). The instrument starts slowly because or inertia and tends to overrun because of momentum when the wind speed drops. Starting and stalling thresholds vary among different models. Recorded wind speeds usually are inaccurate below 1 to 1.5 ms^{-1}. The cup anemometer is best positioned above the canopy where the wind is stronger and less turbulent than within the canopy. Wind speeds in the upper canopy may be estimated from the wind profile theory (Campbell 1977). A propeller anemometer equipped with a vane for orientation into the wind (Rosenberg et al. 1983) may be a suitable substitute for the cup type.

Thermoelectric, sonic and ionic anemometers are sensitive to light wind speeds and respond rapidly to gusts and eddies making them potentially suitable for studies in plant canopies (Fritschen and Gay 1979; Rosenberg et al. 1983). Thermoelectric anemometers utilize heated resistance wires, films or thermocouples and measure the power needed to keep the element warm against the cooling effect of wind. These instruments are fragile, require frequent calibration and use considerable electrical power. Sonic anemometers measure the influence of wind on the speed of sound, and some resolve wind into downwind, crosswind and vertical components. Reflections of sound by plant parts may interfere with readings taken in canopies. Ionic anemometers now under study measure the deflection by wind of ions generated by a high voltage corona. They are expected to be inexpensive and reasonably robust but will require a high voltage source. Fast response by the datalogger is required when using rapid-response anemometers.

3.4 Irradiance

Solar energy impinges on crops as shortwave (0.3–3 μm) radiation (Monteith 1973; Campbell 1977). The silicon photocell pyranometer (e.g. LI-COR, Inc., Lincoln, NE 68504, USA) is a compact and practical sensor suitable for monitoring irradiance just above or beside the crop, but not inside the canopy due to its limited spectral response. The sensor responds according to the cosine of the angle of incidence of solar radiation. Thermopile pyranometers, which measure radiant energy at all solar wavelengths, may be used outside or inside the canopy (Rosenberg et al. 1983; Sutton et al. 1984).

4 Monitoring Soil Variables

The physical environment of the soil embraces temperature, water, pH and other variables affecting many diseases of roots or foliage. Nonetheless epidemiologists have only rarely attempted intensive monitoring of soil physical variables in relation to disease progress in the field. A lack of suitable sensors for routine monitoring has discouraged many investigators, but this hindrance is slowly disappearing as new instruments are developed.

4.1 Soil Temperature

A simple sensor for measuring soil temperature can be constructed by imbedding a thermistor or thermocouple in waterproof epoxy resin within a copper tube (about 15 cm long, 0.6 cm internal diameter). The sensor provides spatially-averaged temperature measurements and should be positioned horizontally at the desired depth in the soil (Sutton et al. 1984).

4.2 Soil Water

Soil water potential is of special interest in epidemiology because it directly affects the energy status of water in pathogens and host plants, and consequently their growth, development, survival and interaction. Total water potential is the algebraic sum of the matric potential (related to the adsorptive forces of the soil matrix), the solute potential (produced by salts and other solutes), the gravity potential (due to elevation above sea level) and pressure potential (caused by hydrostatic pressure) (Hall and MacHardy 1981; Hillel 1980). The matric potential usually is the main component of soil water potential and it can be monitored alone or in combination with the solute potential in the field. Alternatively, one may monitor soil water content from which water potentials can be estimated using a calibration curve for the particular soil.

Electrical Resistance Block. The two electrodes of this sensor are imbedded in a porous block of gypsum, nylon or fibreglass (Hillel 1980; Kramer 1983). The block is placed in soil where it tends to equilibrate mainly with the matric potential (Hillel 1980). Electrical conductivity of the block is a function of the absorbed water and can be scanned by a datalogger. Inert blocks (e.g. fibreglass) are highly sensitive to solutes in the absorbed water but gypsum blocks slowly dissolve such that the calcium sulphate maintains a fairly constant electrolytic concentration. This dissolution changes the porosity and the response of the gypsum block over a few months. Accuracy of resistance blocks is further limited by hysteresis, variations in the soil-block interface and disturbance of the soil when the sensor is installed.

Time-Domain Reflectometry (TDR). Topp et al. (1984) described a portable TDR instrument which measures the liquid water content of soil and has potential applications in epidemiology. A probe with a pair of stainless steel rods is driven into the ground and an electromagnetic pulse is transmitted through the soil. The rods serve as a parallel-pair transmission line and wave guide. The velocity of the pulse is a measure of the dielectric constant of the soil, which, in turn, is a function of the water content. Calibration for differing soils is unnecessary and soil variables such as texture, density, salt content and temperature do not affect the TDR measurement which Topp et al. (1984) found to be as accurate as traditional gravimetric sampling (Hillel 1980).

Thermocouple Psychrometer or Hygrometer. At equilibrium the soil water potential is equal to the potential of the water vapour in the soil air. The latter can be monitored using small soil psychrometers or hygrometers in which a thermocouple junction is suspended in a porous-walled chamber (ceramic or silver membrane) while the reference junction is imbedded in an insulating medium to ensure a constant temperature (Brunini and Thurtell 1982; Campbell et al. 1973; Hanks and Ashcroft 1980; Hillel 1980). When the sensor is buried in the soil the vapour pressure of the chamber atmosphere equilibrates with that of the soil air. A current is applied to cool the chamber junction until water condenses on it (Peltier effect). A voltage reading is then taken which is converted to wet bulb temperature (psychrometer) or dew point temperature (hygrometer), the latter being potentially more accurate. Vapour pressure can be obtained from the dew point temperature or wet bulb temperature using standard equations, graphs or tables. The soil water potential (ψ soil) is derived from the vapour pressure using the following equation:

$$\psi \text{ soil} = (RT/Vm) \ln (e/e_0) , \tag{7.3}$$

where R is the universal gas constant; T, temperature in K; Vm, molar volume of water; e, vapour pressure of the soil air; and e_0, vapour pressure of saturated air at the same temperature as the soil water. Though somewhat delicate and highly sensitive to temperature gradients between the soil and the sensor, thermocouple psychrometers and hygrometers can be used in the field and justify wider use in epidemiology.

5 Monitoring Sites

Epidemiologists and disease management specialists are inevitably faced with the decision to place weather sensors outside or within the crop.

1. Outside the Crop. Some researchers advocate an on-farm weather station, located on short grass but not close to trees or buildings (e.g. Thompson 1984). They argue that a single set of instruments should serve for the various kinds of crops on the farm. A basic assumption is that the station data can be used to estimate the microclimatic conditions in crops. Although this may be true for irradiance or rainfall, satisfactory predictive models for temperature, leaf wetness and RH are lacking. Among the potential advantages of weather stations are permanency and isolation from tillage and spray equipment.

2. Within the Crop. Direct measurement within the canopy is common and avoids the potential errors of microclimatic models. Sites should be judiciously marked to avoid damage by farm equipment but hidden from vandals (Coakley 1985). Sensors may require temporary protection during spraying. Microclimatic diversity in crops may demand several monitoring sites but prompts the question of specific site location (see also Chap. 8).

In Ontario two options have been considered but require further study: (1) Worst-Case Sites: in this option, weather is monitored in an area of the field where the microclimate is observed or predicted to be most favourable for disease. Sprays timed by "worst-case" monitoring are applied over the entire field but may not necessarily be needed in areas of less favourable microclimate. (2) Zones of Uniform Microclimate (ZUMs): in this option, the farm is divided into zones of similar biophysical environments with respect to a given type of crop. The number of ZUMs increases with diversity of topography, cropping practices, field distribution, nearby vegetation and other variables. A ZUM may include several fields on the same or adjoining farms or only a part of one field. One monitoring site is established in each ZUM and sprays may be called for in one ZUM but not another. Establishment of worst-case sites and ZUMs requires detailed epidemiological data and practical knowledge of microclimate. This challenge beckons further collaborative efforts of the epidemiologist and agrometeorologist.

6 References

Amiro D, Thurtell GW, Gillespie TJ (1983) A small infrared thermometer for measuring leaf temperature in leaf chambers. J Exp Bot 34: 1569–1576

Beadle CL, Stevenson KR, Thurtell GW (1973) Leaf temperature measurement and control in a gas-exchange cuvette. Can J Plant Sci 53: 407–412

Brunini O, Thurtell GW (1982) An improved thermocouple hygrometer for in-situ measurements of soil water potential. Soil Sci Soc Am J 46: 900–904

Campbell GS (1977) An introduction to environmental biophysics. Springer, Berlin Heidelberg New York

Campbell EC, Campbell GS, Barlow WK (1973) A dewpoint hygrometer for water potential measurement. Agric Meteorol 12: 113–121
Coakley SM (1985) Describing and quantifying the environment. Plant Dis 69: 460–466
Davis DR, Hughes JE (1970) A new approach to recording the wetting parameter by the use of electrical resistance sensors. Plant Dis Rep 54: 474–479
Duniway JM (1982) Soil-plant-water relations and disease. In: Hatfield JL, Thomason IJ (eds) Biometeorology in integrated pest management. Academic Press, London, pp 307–325
Fritschen LJ, Gay LW (1979) Environmental instrumentation. Springer, Berlin Heidelberg New York
Gillespie TJ, Kidd GE (1978) Sensing duration of leaf moisture using electrical impedance grids. Can J Plant Sci 58: 179–187
Grace J (1977) Plant response to wind. Academic Press, London
Häckel H (1980) Neues über die elektrische Methode zur Messung der Benetzungsdauer unmittelbar an der Pflanze. Agric Meteorol 22: 113–119
Hall R, MacHardy WE (1981) Water relations. In: Mace ME, Bell AA, Beckman CH (eds) Fungal wilt diseases of plants. Academic Press, London, pp 255–298
Hanks RJ, Ashcroft GL (1980) Applied soil physics. Springer, Berlin Heidelberg New York Tokyo
Hillel D (1980) Fundamentals of soil physics. Academic Press, London
Howard AM, Gillespie TJ (1985) A field guide for surface wetness duration measurement. Tech Memo 85-1 Department of Land Resource Science, University of Guelph, Ontario, Canada
Huband NDS, Butler DR (1984) A comparison of wetness sensors for use with computer or microprocessor systems designed for disease forecasting, vol 2. Proc Br Crop Prot Conf, pp 633–638
James TDW, Sutton JC, Rowell PM (1984) Monitoring wetness of dead onion leaves in relation to *Botrytis* leaf blight, vol 2. Proc Br Crop Prot Conf, pp 627–632
Jones AL, Fisher PD, Seem RC, Kroon JC, Van DeMotter PJ (1984) Development and commercialization of an in-field microcomputer delivery system for weather-driven predictive models. Plant Dis 68: 458–463
Kramer PJ (1983) Water relations of plants. Academic Press, London
Legg BJ, Bainbridge A (1978) Air movement within a crop: spore dispersal and deposition. In: Scott PR, Bainbridge A (eds) Plant disease epidemiology. Halsted Press. Wiley New York, pp 103–110
Monteith JL (1973) Principles of environmental physics. Edward Arnold, London
Monteith JL, Butler DR (1979) Dew and thermal lag: a model for cocoa pods. Q J R Meteorol Soc 105: 207–215
Rosenberg NJ, Blad BL, Verma SB (1983) Microclimate: the biological environment, 2nd edn. Wiley New York
Sutton JC, Swanton CJ, Gillespie TJ (1978) Relation of weather variables and host factors to incidence of airborne spores of *Botrytis squamosa*. Can J Bot 56: 2460–2469
Sutton JC, Gillespie TJ, Hildebrand PD (1984) Monitoring weather factors in relation to plant disease. Plant Dis 68: 78–84
Sutton JC, Hildebrand PD (1985) Environmental water in relation to *Peronospora destructor* and related pathogens. Can J Plant Pathol 7: 323–330
Thompson N (1984) Automatic acquisition of meteorological data for crop protection, vol 2. Proc Br Crop Prot Conf, pp 647–654
Topp GC, Davis JL, Bailey WG, Zebchuk WD (1984) The measurement of soil water content using a portable TDR hand probe. Can J Soil Sci 64: 313–321

Chapter 8

The Analysis of Weather Factors in Epidemiology

H. Friesland[1] and H. Schrödter[2]

1 Introduction

This chapter deals with the analysis and use of weather information monitored by thermohygrographs, thermocouples or other instruments as presented in the preceding Chap. 7 by Sutton et al. It is concerned with the formulation of biometeorological variables and their relationship with the interactions between host (see also Seem, Chap. 4), pathogen and environment. At first, however, we shall define some fundamental meteorological terms, definitions and techniques.

Sometimes irritation exists about the meaning of weather and climate, macroclimate, mesoclimate and microclimate. Weather describes the meteorological situation of a moment or for a short period of time, whereas climate is defined by the longterm average situation of a place with respect to different meteorological parameters. This sharp distinction has been dropped in the use of the word microclimate, which in agrometeorology also means the actual situation in a plant canopy and not only the characteristic and averaged conditions within the crop. The often discussed extension of the scales may be given here as:

Scale:	Range:	Examples:
Macroclimate	1000 km to 50 km	More global meteorological situation
Mesoclimate	100 km to 100 m	Meteorological situation related to a landscape or to orographical patterns
Microclimate	300 m to 1 mm	Meteorological situation confined to a plant cover

Of course these multiranged situations are interrelated and dependent on feedback. Ford and Milne (1981) define the characteristics of a dynamic system as component processes constantly changing their rates of activity which will also fit the disease triangle.

[1] Deutscher Wetterdienst, Zentrale Agrarmeteorolog. Forschungsstelle, Bundesallee 50, D-3300 Braunschweig, FRG

[2] David-Mansfeld-Weg 7, D-3300 Braunschweig, FRG

The development of a characteristic microclimate of a crop is determined by the energy budget, which for a vegetated surface equates:

$$R_n + G + H + lE + P + C = O \, . \tag{8.1}$$

Here, R_n denotes the net radiation or radiation balance, G is the soil heat flux, H is the sensible heat flux, lE is the latent or evaporative heat flux, P is the heat flux from the plant mass and C is the heat flux released chemical reactions (metabolism). Due to their very small contribution to the energy balance, P and C are mostly ignored.

The sun supplies the earth and its atmosphere with a spectrum of radiation as energy. Visible light covers only a narrow window of 400 to 700 nm wavelength, but where the spectral emittance is highest; this spectrum corresponds broadly to the Photosynthetically Active Radiation (PAR). The shortwave radiation comprises wavelengths of 150 to 3000 nm (=3 µm) including UV and near-IR radiation. The solar radiation at the earth's surface is modified by the atmosphere due to reflection, absorption and scattering, so that our measurements of global radiation catch direct solar radiation and the diffuse sun radiation. The albedo of the surface (plants, soil, water) is defined as the ratio between the incoming global radiation and the outgoing (reflected) shortwave radiation, taking values from about 0.05 for deep water to 0.9 for freshly fallen snow (Brutsaert 1982). The longwave radiation (3 µm to 60 µm) from every part of the earth, vegetation or atmosphere characterizes their temperatures after absorption of the global radiation. The radiation balance is the sum of shortwave radiation minus reflection and longwave radiation from the atmosphere minus the outgoing longwave radiation from the earth's surface.

The soil heat flux G, i.e. the heat transfer into or out of the soil, depends on heat conduction of the latter and, *inter alia*, on soil moisture and density. The sensible heat flux H into or out of the crop or plant parts is caused by advection of differently temperatured air (horizontal transport) or by turbulent mixing due to radiation exchange. The latent heat flux lE is a flow of energy from a warmer medium to the evaporating or transpirating surface. A reverse flow results in dew formation, releasing the latent heat of the water vapour. It can be deduced from the energy balance equation that in the mean no persistent warming or cooling will occur within a certain volume on the earth's surface. Nevertheless, the balance becomes positive on a sunny day or in summer and negative at night or in winter. These relationships have some bearings on the estimation of biometeorological parameters.

2 Some Practical Hints for the Monitoring of Weather Factors

When monitoring crop microclimate, mistakes of positioning of sensors of any device should be avoided (see also Chap. 7). Generally, representative and exact measurements of well-maintained instruments yield better results than measurements from many but irrelevant positions in the crop. For measurements at more than one place in the same field, the same type of calibrated sensors is needed to avoid deviations that do not stem from microclimatic effects. The measurements should correspond to the same height above soil, to the same characteristics of the crop (e.g. shady or sunny exposition), and generally uneven crop density should be avoided. Electronic data recording shows considerable advantages, as only the sensors are placed in the canopy, while the processing unit does not affect the microclimate. Reasonable time steps for measuring and storing the meteorological variables depend upon the pathogen and the phenomenon of its disease cycle and are seldom shorter than 15 min as the minimum step to calculate hourly means.

On the other hand, it suffices to monitor global or solar radiation at one place only above or beside the crop. Reflection of the plant surface does not need to be measured in every case, since the literatur supplies sufficient data on the reflective characteristics. Attempts to obtain radiation measurements from within the crop will often give doubtful results due to disturbances of the natural crop structure (e.g. movement of leaves due to wind above the sensor). The placement of the instrument should take into account the course of the sun, so that obstacles in the surroundings cannot interfere with incident radiation.

Dry and wet bulb temperatures should be measured in the layers of the canopy where the plant disease in question commonly occurs. The position of the sensors should be rather near to the foilage in order to obtain relevant data for the analysis of disease progress. This is especially important with the position of thermometers that should not be placed in the periphery of the crop but rather in the inner part of a field, where a typical microclimate prevails. Dry and wet bulb temperatures should be measured close together at the same height to make use of the psychrometric difference for the relative humidity of the air. Suitable for automatic recording are ventilated electrical thermometers (psychrometers), which are subject to regular maintenance. When the electrical differences are directly measured, precision is even more desirable than the accuracy of both the absolute temperature values (Szeicz 1975). The sensors are to be protected from radiation. A weather screen within the canopy represents more the conditions of the hut than the crop microclimate and thus should not be used.

Beneath the top of the canopy wind velocity measurements are not recommended. As in the case of radiation, useful data can be collected from above (centre of the field) or beside the crop. The site of the anemometer depends upon its height, e.g. 2 m are acceptable, if turf surrounds the mast, but a higher position will be advantageous near or in higher crops. Wind speeds in the canopy can be calculated according to the formulae in Section 4. In-field anemometers above the canopy require a long enough upwind fetch, and no obstacles should influence the wind field from any direction.

Precipitation regulates soil moisture and supplies free leaf moisture to the plant tissue by interception. A recording rain gauge, not influenced by any obstacles and regularly controlled is placed at a height of 1 m. Recordings every 24 h will be too few for the determination of leaf wetness periods. Measurements just above the crop are impractical because of crop growth and derangement of the crop. One should bear in mind that the relatively easy measurements of precipitation are subject to errors by wind and adhesive water. Also, losses of water from the instruments may occur. The crucial factor for most fungal and all bacterial diseases is the leaf wetness duration of free leaf water from rain or dew. Positions recommended for wetness sensors are both the more sun exposed and more shaded leaves in the crop layer. The evaporation process represents the most difficult factor for determining the end of leaf wetness and condensation in the case of dewfall. Nevertheless, several devices exist to monitor wetness duration (see Chap. 7). A rather advanced sensor by Häckel (1984) may be clipped to a leaf, so that the drying and wetting process can be monitored by measuring the electrical resistance.

Also, some warnings by Coakley (1985) apply to the establishment and operation of an agrometeorological station. For the practical installation, the WMO guide (1981) is recommended.

3 Techniques in Use for the Analysis of Biometeorological Data

It is essential in epidemiology as a quantitatively ecological science to analyze and quantify the influence of meteorological factors within the disease triangle host — pathogen — environment correctly. Complications arise from factors or combinations of them, which affect the diseases at the same time. Problems also develop from interactions between meteorological conditions, and the different effects resulting from them on pathogens and hosts at every development stage. According to Rapilly (1983), a certain order for the efficacy of parameters can be set up for each stage of the pathogen development, or disease cycle. Table 8.1 illustrates the diversity and complexity of the relations between plant diseases and the atmospheric environment. A comprehensive treatise on these problems and on the biometeorological aspects of epidemiology is given by Schrödter (1987). Comments on the analysis of biometeorological relations in epidemiology can also be found, among others, in Schrödter (1965), Butt and Royle (1974) and Analytis (1973, 1977, 1980).

Spore germination and *infection* of most pathogenic fungi depend above all on leaf wetness duration and temperature. In fact, the combination of temperature and duration of high relative humidity near saturation or wetness duration plays the most important role. In contrast, the powdery mildews need suitable temperatures as the highest effective variable, but relative humidity must not be totally ignored (Aust 1981; Bashi and Aust 1980).

Incubation time and *latent period* are primarily functions of temperature, but humidity or wetness may cause additive or even multiplicative effects according to the pathogen.

Table 8.1 Weather factors influencing phases in the disease cycle of airborne fungal pathogens

Phase	Meteorological factors[a]
Infection (incl. germination)	For most fungi every source of free leaf moisture such as rain, dew, fog, irrigation or guttation; temperature during these periods
Incubation, latency and lesion growth	Duration of suitable or unsuitable air or leaf temperature
Sporulation	For most fungi free leaf moisture and/or high relative humidity; temperature; light and radiation in some cases
Spore dispersal	Wind velocity (gusts), temperature, relative humidity, free leaf moisture, rain or sprinkling in the case of splash dispersal
Spore survival and overseasoning	Temperature and relative humidity, radiation (especially UV)

[a] The relevant parameters to be measured are dealt with in Sect. 3.1.1–3.6.

Also, *lesion growth* depends mainly on temperature, with leaf wetness and the water status of the plant as additional factors. This fact can be of considerable significance in connection with the application of irrigation (Rapilly et al. 1975). Sporulation of most fungal and bacterial pathogens requires leaf wetness or high relative humidity. Next in importance is temperature, whilst light and radiation may be effective in some pathosystems.

For *spore dispersal* wind and turbulence play the overwhelming role (Gregory 1968). But other factors have to be considered, too, e.g. the influence of moisture on spore release, of different spectra of light on the discharge of ascospores and sporangia, and similar effects as described by Rapilly et al. (1970) and Rapilly (1977). Furthermore, the mechanical dispersal of the spores by rain has to be mentioned, which is not only a function of precipitation duration and intensity, but also a function of the morphology of the fructifications (Pauvert et al. 1970).

3.1 Temperature

In nearly all phases of pathogen development temperature plays the key role. The different possibilities to use this factor or temperature-deduced variable in biometeorological relations shall be dealt with below under consideration of advantages and drawbacks.

3.1.1 Mean Temperatures and Daily Course of Temperature

As a principle we should aim at using hourly values of temperature as a base for evaluations in epidemiology. This not only applies to measurements and observations of the microclimate of plant canopies, but also to the analysis of data from climatological stations of international standard. Daily means of temperature are doubtlessly suited as a base for the evaluation of relations to processes in the long range, but the drawback stems from the fact that identical mean values may result from different temperature courses. Furthermore, it should be noted that the various temperatures of a day are not normally distributed, and mean temperature is not the most frequent temperature of the day at the same time. If measurements of maximum and minimum temperature exist only, the daily course may be approximated by a sine function as:

$$T = T_m + A \cdot \sin t\,, \tag{8.2}$$

where T_m denotes the daily mean from (max + min)/2 and A is the amplitude. The dimension of time can be deducted from $t = \pi/2$ for $T = T_m + A$. This approximation certainly in only a very crude one, as the true daily course of temperature usually differs considerably from a sine function. This results in the skewness of the temperature distribution as indicated above. The calculation of mean temperature for a longer period, e.g. monthly means, are of little use in annual epidemics, but can be valuable in geophytopathological studies (see Weltzien, Chap. 16).

3.1.2 Frequency of Temperatures

The use of means becomes crucial, especially when continuous measurements of temperature are not matched by equivalent continuous observations of the biological part, i.e. if the process cannot but monitored in larger time steps like weeks, a common interval for disease readings. In such cases the experimentor has to either rely upon temperature distribution as a parameter or has to deduce new variables from the frequency distribution of temperature, which considers the course of temperature. Practical use will be shown by the technique of the so-called frequency correlation analysis of Schrödter (1965), which has also been well applied as a base for models of forecasts of epidemics (see Madden and Ellis, Chap. 13).

Schrödter's method assumes an adequately linear relation between the ratio of the frequency of favourable and unfavourable temperatures and growth rate. So, the frequency distribution of temperature in different, maybe overlapping, classes is set up for every time step from the biological observation frequency, giving rise to time series of varying frequency in each class. Thus, new independent variables are created from hourly temperature data by:

$$x_k = h_i(T_k)\,, \tag{8.3}$$

which means by the number i of different frequencies h of temperatures T in the classes k. The correlation coefficients of the x_k as independent variables and the developmental rate as the dependent variable yields (as a function of the temperature of the centre of the class) a curve similar to the true temperature function and indicate the temperature optimum by the highest positive correlation coefficient.

As the linearity is adequately related to the frequency distribution of the hourly values in the single classes, this allows one to replace it by a new independent variable $y(T)$, thus:

$$y(T) = a_0 + a_1 h_1(T_1) + \ldots + a_n h_n(T_n)\,. \tag{8.4}$$

Such derived parameters can be used rather well for further statistical analysis of the interactions within the epicemiological complex (Schrödter and Ullrich 1967; Fehrmann and Weihofen 1982). These new parameters may serve as independent variables in response functions, which can be represented as third order regression functions.

The disadvantage of this technique lies in the fact that the meteorological data are not fully utilized, as no statistical significance of the correlation or regression coefficients can be obtained in the poorly filled temperature classes near the temperature limits. But the advantage of the method is its unproblematic applicability to other meteorological parameters or combinations and its suitability for multiple regression analysis (MRA) (Butt and Royle 1974; Schrödter 1987).

3.1.3 Temperature Sums

Temperature sums as degree days or degree hours are frequently used to relate temperature to the duration of biological processes and to forecast the duration of such a process. Examples are the latent period, the interval between two phenological states or changes in incidence and severity of disease (Coakley and Line 1981). Well known is the prediction of onset of flowering in fruit trees by means of sums of degree hours for the timing of preblossom sprays (see also Seem, Chap. 4).

It is an essential requirement for the use of degree days, etc. that the duration of a process entirely or predominantly depends on temperature. Of further importance are the choice of the time to start the computation and the base temperature, to be substracted from the actual temperature. Finally, the relationship between temperature and duration of a process should express itself in a hyperbolic function. This would be an approximation to Janisch's (1925) exponential law for the temperature-driven reactions in biology. When reversing dependent and independent variables deliberately, the following function applies:

$$T_m = c_1 + c_2 \cdot \frac{1}{t}, \tag{8.5}$$

where t is the duration of the process (e.g. incubation time) and T_m is the mean temperature of this time. For $t = \infty$ the equation becomes $T_m = c_1$, which signifies c_1 as the base temperature reached by the asymptote of this hyperbolic function. A transformation of coordinates yields:

$$T_m - c_1 = T_m' = c_2 \cdot \frac{1}{t} \quad \text{or} \quad T_m' \cdot t = c_2 = \text{const}, \tag{8.6}$$

which actually is a hyperbolic function related to its asymptotes as coordinates. As T_m' represents the mean temperature minus the base temperature during the process, the following equation holds:

$$T_m' \cdot t = \frac{\Sigma T'}{t} \cdot t = \Sigma T' = c_2 = \text{const}. \tag{8.7}$$

Hence, the process in question should have passed whenever the temperature sum reached the constant value of c_2.

As such a shape of the relation will not be able to specify the optimum temperature conditions for the minimum duration of a process (e.g. latent period or incubation period) Eq. (8.7) has only a very limited validity. A further disadvantage is that only relationships between the duration of the process and mean temperature during this time are expressed. Furthermore, if hourly temperature data instead of daily means are used for summation, this only implies the calculation of T_m' from hourly or daily means, so that the dimension of the temperature sums reads degree hours or degree days. But the possible effect of the temperature course, and particularly of supraoptimum temperatures, remains largely unconsidered.

Consequently, degree days are exceptable as independent variables only, if the optimum temperature required for a minimum duration of the process is not reached. In the case of supraoptimum temperatures a correction of the summation as suggested by Allen (1976) is recommended. Another possibility proposed by Analytis (1980) are 'optimum heat units' instead of degree days. This particularly applies when hourly values of the temperature are available for temperature – development functions. Both degree days and optimum heat units, however, imply that temperatures below the minimum or beyond the maximum do not revoke the state of development reached, and ensure that the process resumes unchanged once favourable temperatures occur again.

The common disadvantage of degree days and degree hours is that sums of temperature are not terms defined physically. Nevertheless, these terms, within limits, prove to be of practical use. They, however, are not measurement terms for energy and are only of limited value for modelling. More useful appears Analytis' (1980) modification of the exponential law (Janisch 1925), which describes the effect of temperature on the duration of the process y_i as:

$$y_i = M \cdot \cosh(k|T_i - T_{opt}|)\,, \tag{8.8}$$

where M denotes the minimum duration at optimum temperature T_{opt}, T_i is the actual temperature and k is a parameter determining the slope of the curve. This parameter results from the regression coefficient between the difference $T_i - T_{opt}$ as the independent variable and the expression $\ln(y_i/M + \sqrt{(y_i^2/M^2 - 1)})$ as the dependent variable. According to the examples of Analytis (1980), the advantage of this function can be found in the possibility of recognizing the differences in temperature dependence of incubation time of the same parasite due to different races. So, Eq. (8.8) is more suitable for the development of biometeorological variables than a relation of temperature sums.

3.1.4 Temperature-Response Functions

In many cases it proves profitable to turn directly from temperature to temperature-response functions. New or transformed variables are generated for the analysis of epidemics or model parameters for largely deterministic models or simulators. If the observations of the process are at hand in short time steps, so that actual or mean temperatures of short periods can be used, temperature relations may be represented by a parabola at first approximation:

$$y = y_{max} - a \cdot (T - T_{opt})^2\,. \tag{8.9}$$

Here, y_{max} symbolizes the maximum disease intensity; y is the infection intensity etc. at the optimum temperature T_{opt}; a is the regression coefficient; and $y(T)$ equals zero for temperatures of $\sqrt{y_{max}/a}$ degrees above or below T_{opt}.

This approximation has been shown to be insufficient for infection and lesion growth, as the true function often differs considerably from the parabola although it often fits well to temperature effects on incubation and latent periods. The *TE*-function developed by Schrödter (1965) proved to be more suitable for infection and lesion growth, since further biometeorological variables can be derived from it (Analytis 1973). The *TE*-function represents the mathematical expression of a squared sine function of a third degree polynomial:

$$y = \sin^2(a_1 x + a_2 x^2 + a_3 x^3)\,. \tag{8.10}$$

Here, y denotes the relative development rate with respect to the rate at optimum temperature and x is a temperature equivalent given as:

$$x = \frac{T_i - T_{min}}{T_{max} - T_{min}}\,, \tag{8.11}$$

where T_i = actual temperature, T_{max} = maximum and T_{min} = minimum temperature for development. The constants are easily determinable from the coordinates of the point of inflection in the ascending part of the function and from the requirements of $y = 1$, when $T_i = T_{opt}$; and $x = 1$ when $T_i = T_{max}$ (Analytis 1973). The function may be used for the construction of a nomogram, in which these bell-shaped curves skewed to the right are represented as straight lines (Schrödter 1965, 1987). Their treatment as further biometeorological variables has been described in detail by Analytis (1973). As to their suitability for the development of forecasts for epidemics, see Dirks and Romig (1970), Burleigh et al. (1972) and Eversmeyer et al. (1973).

The rigid tie of the *TE*-function to the squared sine shape proves to be a drawback. A fit to deviating curves is certainly possible (Dirks and Romig 1970; Analytis 1973) by variation of the constants in Eq. 8.10, but it is only feasible in narrow boundaries. Analytis (1977, 1980) elaborated a BETE-function more appropriate for practical use in the form:

$$r(\delta) = (p \cdot \delta^n \cdot (1-\delta))^m , \tag{8.12}$$

which is able to describe more flexibly the relations between development rate and temperature with respect to parasitic fungi. The symbol $r(\delta)$ denotes the relative development rate, δ the temperature equivalent like x in the *TE*-function and p denotes a proportionality factor scaled so that the development rate equals 1 at the optimum, a fraction of 1 at other temperatures and zero for both the extreme temperatures. The skewness and the flexure of the curve are determined by the parameters n and m. Though this function has been developed mainly for the quantitative description of the mycel growth of parasitic fungi in dependence of temperature, it has been proved that due to its flexibility, the function also offers a variety of applications for the development of biometeorological variables (Analytis 1980; Hau et al. 1983).

3.2 Humidity

In epidemiology humidity plays a major role both as relative air humidity and as free water. The difference between the effect of high air humidity (near saturation) and the effect of free water is often hard to discern (Rapilly 1983). In many cases humidity in free or latent form represents the restricting factor for the germination of fungal spores. In many cases each break of the necessary minimum duration of sufficient humidity decreases the capability of the inoculum to survive. There are, however, also pathogens which require such breaks to complete infection. Thus, the duration of relative humidity or wetness rather than these conditions in themselves are the relevant biometeorological parameters. This results in some consequences for the monitoring and formulation of the biometeorological variables in question.

3.2.1 Relative Humidity

The processing of hourly values from measurements of relative humidity offers the advantage of deriving possibly different biometeorological variables for humidity. In most investigations, the duration of relative humidity exceeding certain limits will be the decisive parameter, for instance, the number of hours with a relative humidity above 90%. Analogous to the technique of Analytis (1973), different biometeorological variables can be deduced from the same recorded data.

As an example we make the assumption that the result of an inspection at time $i+1$ is the consequence of the termination of two stages in pathogen development between the time i and $i+1$. Let the stage a take 4 h and stage b no less than 10 h of uninterrupted high humidity (above 90%), then the independent biometeorological variables can be formulated with the help of hourly values as:

$$H_a = \frac{n_{F4}}{N} \quad \text{and} \quad H_b = \frac{n_{F10}}{N} . \tag{8.13}$$

Here, n_{F4} denotes the number of hourly values above 90% belonging to periods of at least 4 h above 90%, whereas n_{F10} signifies the corresponding number belonging to periods of at least 10 h, and N is the total number of hourly data.

If measurements in the screen at 2 m height (weather screen) are used, it has to be understood that these data only give rough estimates for the unfavourable or favourable humidity conditions within the canopy for the biological process in question. With relative humidity the difference between macro- and microclimate can be rather considerable. In his investigation on take-all in cereal, Siebrasse (cited by Schrödter, 1987) found that in screen measurements 50% relative humidity indicated above 80% RH within the canopy. Hence, monitoring of the microclimate should be preferred, but as far as possible the measurements should be correlated with records from the weatherhut to facilitate the use of these data, e.g. for forecasts. If, according to microclimatic data the duration of high air humidity near saturation is the governing factor, then it cannot be excluded that humidity acts as free water rather than the relative humidity. This often results from the fact that already a very small difference between air and leaf temperature at humidities near saturation suffices for reaching the dew point and initiating the condensation process at the leaf surface. Thus, it is possible that the relative humidity data are only indicators for wetness and its duration (Wallin 1963; Van Eimern 1964).

3.2.2 Duration of Wetness

The measurement of wetness duration is a problem for several reasons (see also Chap. 7). At first it is only meaningful when conducted within the microclimate of crops. If mechanical or electronic instruments are used, the problem is that the artificial collecting or sensing areas can never have the same physical properties as a leaf. Therefore, the artificial areas react differently to the cooling process leading to dew formation (see energy balance in Sect. 1). This may cause considerable differences of measured wetness duration. Also, these instruments are not able to record guttation water, which may be a source of free water necessary for fungal development (Yarwood 1952).

The wetting by rain is monitored by those instruments only at the start, but not at the end of the wetness duration. The morphological characteristics of the plants and the structure and density of the crop play an important role in the wetting and drying processes during and after rain. They also determine the wetness duration by their interception capacity and evaporation of intercepted rain. These facts are expecially illustrated for longer wetness durations following rainfall and stand in close connection with the energy balance of the plant canopy (Payen 1983).

Methods for direct measurement of wetness close to the plants exist (e.g. Schurer and Van der Wal 1972; Häckel 1984). The problem of mean wetness duration typical for the canopy can be solved when the investigator works with a larger number of sensors within the canopy which are randomly distributed within the canopy. If this cannot be realized, measuring positions have to be found, which are typical for the structure of the canopy as well as for the single plant with respect to the parasite in question.

For the development of forecasts a possibility exists for the indirect determination of wetness with the help of meteorological parameters (Choisnel and Payen 1981; Payen 1983). For interception capacity and evaporation of intercepted water, the energy and water balance of the plant, canopy and soil form the base for such a physical model.

If instruments for measuring the wetness duration are not at hand, data of relative humidity of the air and of precipitation may be used for an approximation. It can be assumed that in many cases wetness has occurred at the surface of the canopy when humidities above 90% have been measured at a height of 2 m. The same may be supposed if a lower humidity is monitored but rain of at least 0.1 mmh^{-1} must have fallen at the same time. From the combination of both requirements the probable wetness duration can be derived. But because of the different structures of the canopies it is recommended to examine which values have to be taken in each single case for using humidity or precipitation as an indicator for wetness duration (cf. Sect. 3.2.1).

The formulation of biometeorological variables for wetness or its duration may follow Analytis (1973) or Sect. 3.2.1. The problem of finding the suitable boundary conditions when using humidity data which were derived by substitution has to be solved by a meaningful variation of the biometeorological variables in an iterative way.

3.3 Temperature-Humidity Combination

In general, several meteorological factors related with each other are associated with the disease. This is especially true for the parameters temperature and humidity in their combined effect. The latter shall now be treated as an example also for other parameters.

The effect of temperature on pathogen and disease development often becomes distinct only when optimum humidity conditions prevail at the same time. The analogy holds true for the effect of humidity. Therefore, it is necessary to develop suitable biometeorological variables for the combined effects of both parameters from the data at hand.

For hourly values of temperature, only those are considered, at which relative humidity exceeds the critical value at the same time or when wetness prevails. This critical value may be 90%, and the temperature dependence of pathogen development may be described by the *TE*-function of Eq. (8.10) Then a biometeorological variable including both parameters can be formulated after Analytis (1973) from the hourly temperature and relative humidity data as:

$$W_i = \sum_{i=1}^{n} f(x_i) \bigg/ \sum_{i=1}^{N} f(x_{\text{opt}}) , \tag{8.14}$$

where W_i denotes the variable of disease observation at the time i, n is the number of hours of a relative humidity above 90%, N is the number of hours of the time from $i-1$ to i, $f(x_i)$ is the *TE*-function and $f(x_{\text{opt}}) = 1$ accordingly. Similar formulations result if wetness duration is inserted in the place of n. A similar procedure can be used, if, vice versa the dependence on relative humidity at optimum temperature is to be analyzed.

The disadvantages of this treatment result from the constant level of one of the factors. In neither the intercorrelation between temperature and humidity nor in their interaction is the factor-for-factor category of the compensation phenomenon (Rotem 1978; Aust et al. 1980) considered. However, we are able to eliminate the intercorrelation of the variables by suitable transformation with the help of the analysis of principal components, and to identify the relative significance of both the independent variables (Analytis 1973). The complex of the independent variables has to be resolved in orthogonal rates of effectiveness under determination of the eigen values of the variance-covariance matrix and of the eigen vectors of these values.

Furthermore, the so-called method of the generalized components of interactions may be used (Schrödter 1987). This is advantageous because the non-linearity of the relations to the variables and their intercorrelation are taken into account by a best-fit, squared function of the general shape:

$$y = a_0 + a_1 x_1 + \cdots + a_n x_n + a_{12} x_1 x_2 + \cdots + a_{n-1,n} x_{n-1} x_n + a_{11} x_1^2 + \cdots + a_{nn} x_n^2 , \quad (8.15)$$

and for n independent variables of x. This function can be interpreted as an n-dimensional paraboloid in the $(n+1)$ dimensional space with generalized conic sections as surfaces of equal response. Such a response surface specifies the effects and countereffects of the parameters and thus, also considers compensatory effects. The literature cited above is referred to with regard to the details of both the presented techniques for the formulation of biometeorological variables of combined effect.

3.4 Precipitation

Measurements of precipitation, determined as sums of months, decades or days, are only of very limited value in epidemiology. They furnish neither information about wetness duration nor about phenomena due to precipitation like spore release, their splash dispersal or washout for deposition. Therefore, continuous registrations are preferred in the monitoring of the rain duration and its intensity. They give suitable data when measured with standard pluviographs with a daily run of the tape. They allow for a rather precise interpretation of rain duration necessary for the calculation of wetness duration, if direct monitoring is impossible. A similar definition of biometeorological variables may be found from precipitation duration or hourly amounts above a certain intensity as shown in Sect. 3.2.1 and 3.2.2 with hourly data of relative humidity or wetness duration.

A further advantage of precipitation records in contrast to single measurements can be seen in the simultaneous recording of rain intensity. This is important not only for the efficacy of the washout effect, but also for direct spore dispersal by raindrops. The mechanism of the splash-dispersal effect and the washout effect by rain has been exhaustively studied by Gregory et al. (1959). But the effectiveness not only depends on rain intensity, but also on drop size and velocity of fall, where for each spore type a characteristic drop size exists (Gregory 1973).

The fact that this mechanism of spore dispersal has not been possible to record by the usual spore traps has been a problem for the adequate estimation of the splash-dispersal effect for a long time. Spores contained in drops of different size cause technical difficulties in monitoring them by means of sporetraps as Fitt and Bainbridge (1983) have recently shown.

Additional problems arise from using precipitation measurements as an indicator for changes in soil humidity for the investigation of soilborne parasites. The type and structure of the canopy influence the interception of rain amount, its intensity, and the intercepted portion of rain (which is intermediately stored without reaching the soil and then evaporates). But also the proportion of precipitation falling or dripping through the canopy changes soil humidity. Direct measurement of the latter, therefore, is preferred.

3.5 Wind and Turbulence

Spore dispersal by air currents is one of the most important mechanisms for the spread of plant diseases (e.g. Schrödter 1960; Gregory 1968, 1973; Pedgley 1982). Measurements of the wind speed alone above the canopy give only insufficient information about the possible distance of the spread, since they only monitor the horizontal component of air currents. According to this measurement alone, spores could hardly be transported higher than their point of departure, and very close limits would be set to their dispersal, thus contradicting our practical experience. It can be shown by a relatively simple, intelligible theory, which components, responsible for the vertical transport, have to be taken into consideration (Schrödter 1960).

If a cloud of spores has reached the ever turbulent atmosphere in a direct or indirect way, the possible trajectory of its centre of gravity in the $x-z$ plane is given by:

$$z = 0.4769 \sqrt{4At/\varrho} - ct, \tag{8.16}$$

where z is the height of a point of this trajectory at time t, A is the coefficient of turbulent mass transfer, ϱ is the density of the air and c is the velocity of fall of the spores in calm air. From this distance, height and duration of flight can be derived. For example, the possible flight distance x, i.e. the distance where the point $z=0$ is reached again by the trajectory equals:

$$x = 0.91 \frac{Au}{\varrho c^2}, \tag{8.17}$$

where u denotes the horizontal wind speed. However, x is proportional to wind speed but also proportional to the mass exchange A or to the eddy diffusivity $A/\varrho = a$. Additionally, flight distance is inversely proportional to the square of the velocity of fall c in calm air. In spite of the low magnitude of c it should not be neglected compared with turbulence. With the help of Stoke's Law it can be estimated as:

$$c = \frac{2\delta g}{9\eta} \cdot \sqrt[3]{a} \cdot b \sqrt[3]{b^2}. \tag{8.18}$$

Here, δ = specific weight, η = viscosity of the air and g = gravity which may be regarded sufficiently constant. Thus, c only depends on the size of the spores, for a and b denote the lengths of the main axes of an ellipsoid of rotation around the shape of the spore.

The eddy diffusivity shows a distinct daily course with a maximum at noon and a minimum at night. The roughness of the earth's surface determines the degree of turbulence. So, it depends on the characteristics of the canopy and of the single plants representing elements of roughness. For the estimation of the magnitude of turbulence as the base for the determination of possible spore dispersal, it is important to measure the gradient of wind speed above the canopy and to choose at least three different heights. Further details are discussed in Sect. 4.

Turbulence with its vertical and horizontal components plays a decisive role in the change of spore density per volume of air as an epidemiologically very important quantity for the spread of pathogens. Spore concentration generally may be recorded rather well by volumetric spore traps, whereas the true flight distance can hardly be determined due to the rapid decrease of concentration in wind direction starting from spore take-off. It is common practice in trials to separate the experimental plots by natural shelterbelts to avoid infection by spores. Indeed, such shelterbelts, like hedges, possess a certain filtering effect, but transmission cannot be totally prevented. Rather, it has to be taken into account that directly leeside of shelterbelts, a further increase in spore concentration occurs as a result of turbulent mixing of the air. Dimensions of a shelterbelt with a given filtering effect, at which advected spores are totally eliminated, can be theoretically assessed. But in reality, trials show that such a complete wind shield cannot exist, because the optimum extension depends on turbulence and wind speed (Schrödter 1987).

3.6 Light and Radiation

Light and radiation interfere with the biology and epidemiology of pathogens in different ways, indirectly and directly. On the one hand, they influence the induction of periodic events in the development cycle of the parasite by the change of light and darkness and by seasonal variation of day length. On the other hand, there is an influence of radiation and its intensity at different wavelengths on the parasite and host plant.

If the standard high-precision solarimeters are not at hand for registration of global radiation, sunshine duration as biometeorological variable will not be a good substitute, as it can hardly give evidence about radiation intensity. It would be more advantageous to use sunshine duration for the estimation of global radiation according to function (8.19) with the help of extraterrestrial radiation and maximum possible sunshine.

A greater problem is the treatment of the visible part of the spectrum (0.38 to 0.76 nm), i.e. the light radiation, as the term light is not clearly defined. Photometry, in contrast, is occupied with a physiologically defined measurement of light related to the human eye as receptor. Photometers, as normally used, are adjusted to the selective sensitivity of the eye with a distinct maximum at 0.55 nm. Since measurements by these instruments cannot be made according to the units of radiation, the arbitrary unit lux was introduced for illumination values. But the spectral response function of photosynthesis possesses only its minimum at the maximum sensitivity of the photometer. Therefore, the photometric measurements are only of very limited value for epidemiological research.

According to Aust (1981), the infectivity of the conidia of barley powdery mildew is influenced, amongst others, by the intensity of light, which directly affects the infectivity via the metabolism of the host plant. This may be a more common phenomenon and it is advisable to shift from the photometric measurements of light to those of Photosynthetically Active Radiation (PAR) (see Chap. 7). In contrast to photometers, PAR sensors show a spectral sensitivity corresponding to the photosynthetic response function and furnish data, clearly defined and physically comparable, of the biologically efficient radiation intensity.

As radiation measurements are not recommended for use within the crop canopy, calculations are preferred. Also due to the sparse network of radiation instruments, approaches have been made for empirical formulae of global radiation R_s. A good estimate yields the equation:

$$R_s = (a + b \cdot n/N) \cdot R_a \quad (J\,\mathrm{cm}^{-2}), \tag{8.19}$$

where $a = 0.25$, $b = 0.5$ as empirical values, n = recorded daily sunshine hours, N = maximum possibly daily sunshine hours and R_a = extraterrestrial radiation dependent on latitude and time of year. Doorenbos and Pruitt (1977) provide a table for R_a and some slightly deviating values for a and b, depending on location and author in their appendix. As R_a in this table is calculated as an evaporation equivalent in mm, the values have to be multiplied by 245 in order to obtain global radiation in $\mathrm{J\,cm^{-2}}$.

The vertical radiation distribution (profile) within the canopy can be described exponentially so that, depending on the leaf area index and the geometry of the crop, the radiation at the bottom of the stand is reduced to a small portion of the flux at the top. The profile changes according to reflection (inclination of the sun) and to the properties of the leaves, e.g. leaf angle or transmittance of diffuse radiation. The calculation of the curve for radiation extinction or interception has been shown by Goudriaan (1977). Denmead (1976) suggests the equation:

$$R_n(\mathrm{LAI}) = R_n(0) \cdot \exp(-\kappa \cdot \mathrm{LAI}), \tag{8.20}$$

where $R_n(\mathrm{LAI})$ denotes the net radiation (flux density) below the vegetation surface dependent on the leaf area index (LAI), $R_n(0)$ is the net radiation at the surface and κ the extinction coefficient which depends on leaf type.

4 Estimation of Missing Micrometeorological Data

Although epidemiologists should try to do direct measurements whenever possible, in some cases indirect estimation of biometeorological parameters may be required. This may apply for many researchers.

The *temperature distribution in a canopy* always tends towards an equilibrium between the net energy input from all-wave radiation (output for the night) and the total loss of heat fluxes. The same holds for a leaf so that for its surface temperature T_0, the equation reads according to Monteith (1981):

$$T_0 = T' + \gamma^* \cdot r_H \cdot A/(\bar{\varrho} \cdot c_p \cdot (\Delta' + \gamma^*)). \tag{8.21}$$

This physically based equation is appropriate for deterministic models, as it includes all energy fluxes which maintain the balance between air and leaf temperature (energy budget). The expressions are explained as follows:

T' = wet bulb temperature of adjacent air;
$\gamma^* = \gamma(r_v/r_H) = \gamma(1 + r_s/r_H)$, where $\gamma = c_P/L$ and
r_V, r_H, r_s are the resistances to vapor and heat exchange and of the stomata;
c_p = specific heat capacity of the air at constant pressure;
L = latent heat of vaporization.

For A in equation 8.21 the following explanation applies

$A = R_n - G = LE + H$, enthalpy flux (cf. Eq. 8.1);
$\bar{\varrho}$ = mean air density;
$\Delta' = \partial q_s(T')/\partial T'$, rate of change of saturation vapour pressure with wet bulb temperature, where
q_s = saturated specific humidity, calculated as
$q_s = 0.622 \cdot e_s/(p - 0.379 \cdot e_s)$
with the units g water vapour/kg moist air and p = air pressure,
e_s = saturation vapour pressure in hPa.

The *diffusion resistances* to water vapour and heat transfer (r_V and r_H) are often combined to an aerodynamic or boundary layer resistance r_a, which according to Gates (1980), takes minimum values of 30 to 120 s m^{-1} for smaller leaves in still air or light wind, while the minimum stomatal resistance r_s ranges from 30 to 500 s m^{-1} (mean of 150 s m^{-1}), depending on plant type. r_H can be assumed to be of the same value as r_a, thereby ignoring small differences.

For spore dispersal sometimes also *wind speeds* within the crop have to be calculated from outside measurements. Wind velocity decreases logarithmically with the height above vegetated or bare surface. For a mean wind speed u_z in the height z, the equation reads:

$$\bar{u}_z = u_*/k) \cdot \ln(z/z_0)\,, \tag{8.22}$$

where u_* = friction velocity in $m\,s^{-1}$, k = von Karman's constant (0.4), z_0 = roughness length of the underlying surface, denoting the height at which the wind speed reaches practically zero. In high, dense canopies z_0 naturally takes higher values than in short grass or bare ground. Gates (1980) furnishes some values of roughness length for different ground covers. For vegetated surface Eq. (8.22) takes the form (Goudriaan 1977):

$$u = (u_*/k) \cdot \ln(z - d/z_0)\,, \tag{8.23}$$

where d denotes the zero plane displacement. It equals $0.64 \cdot h$ for relatively dense vegetation (h = crop height) and $z_0 = 0.13 \cdot h$ as approximations.
The friction velocity under neutral conditions is defined as:

$$u_* = k \cdot z \cdot (\Delta \bar{u}/\Delta z)\,, \tag{8.24}$$

which takes more complicated forms in the case of stable or unstable conditions (see Oke 1978).

Within the plant stand the calculation of wind velocity meets other difficulties like turbulence structure as the result of crop architecture and temperature stratification. Thom (1975) showed wind profile approaches consistent with the exponential decay of eddy diffusivity (diffusion coefficient of momentum) as:

$$u(z) = u(h) \cdot \exp(-n\zeta)\,. \tag{8.25}$$

Here, $u(h)$ at the crop surface can be estimated by the above logarithmic wind profile. The coefficient n characterizes the vegetation type in the range from 2 to 4, while the so-called fractional depth is $\zeta = 1 - z/h$, so that for $z = h$ the exponential function comes to unity, and wind velocity calculations at the crop surface – air interface reach identical solutions from both directions.

The amount of intercepted water after every rain event is an often desired quantity for the estimation of free leaf moisture and subsequent calculation of *leaf wetness duration*. The relation between duration and amount of rain, number of rainfalls per day and the leaf area index (m^2 leaf area/m^2 ground) can be used in formulae. They yield the intercepted water and the throughfall which only contributes to soil moisture. It is easy to imagine many showers of low intensity producing a considerable amount of intercepted water per day and giving rise to long wetness duration and often favourable pathogen development. Plants with a high leaf area index (LAI) thus produce a high humidity microclimate. A rather elegant and physically based equation of Braden (1985) shall be presented, as all its parameters are easy to measure:

$$N_i = a \cdot \mathrm{LAI} \cdot (1 - (1 + N_0 \cdot b/(a \cdot \mathrm{LAI}))^{-1}), \tag{8.26}$$

where N_i = intercepted rain amount (mm or $1\,\mathrm{m}^{-2}$) by the crop;
a = crop-specific maximum capacity coefficient;
N_0 = precipitation amount outside (above) the canopy;
b = degree of ground cover.

This relation holds for single rain events and therefore has to be used repeatedly, when several showers have occurred during the day. Only if precipitation reaches the saturation value of interception $a \cdot$ LAI, will the soil be wetted in denser plant stands.

For wetness duration the summation hours above 90% relative humidity in the weather screen yields a rather gross estimation (Smith 1975). This varies with the climatic region and may be valid for more open plant stands like orchards, but especially in dense canopies the lower leaf layers experience longer duration.

Dew formation begins when the upper leaves cool down below the dew point of the surrounding air, especially on nights with clear skies. As a rule, a maximum of 0.4 mm dew forms at night, and an hourly rate of 0.067 mm ($\pm$10%) has been reported by Rutter (1975) as a physical upper limit. Pedro and Gillespie (1982) used an energy balance approach and calculated dew duration from hourly weather bulletins. They estimated the error to be 1 h, and for shaded leaves the wetness was about 1 h longer. This will probably compensate for a later beginning of dew formation on the lower leaves, but also the different, efficient evening and morning temperatures have to be regarded for the pathogen cycle. When investigating the development of a plant disease in relation to the microclimate of a field, the practical application of these experiments would be easier, if the meteorological parameters as measured in the field are correlated with those monitored in a nearby station of the standard weather service (in a range closer than 1 km). After the experimental period any meteorological transfer model eventually should become available for practical application, e.g. in forecasting when micrometeorological measurements are not available. In such a case advantage must normally be taken from routine weather data (Adams and Seager, 1977; Schrödter 1983), from which parameters complying with the micrometeorology within the crop can be generated.

5 Epilogue

This chapter has attempted to outline which meteorological parameters are needed in epidemology, how and where they are to be measured and what their impact is on the plant, on the pathogen, on the microclimate itself, and what their interactions are. It has been stressed that within-crop measurements are essential and that their experimental realization may be subject to errors and failures. Sometimes calculations or estimates of parameters from outside the canopy or from routine weather data are possible and successful. Some procedures and equations have been suggested for use in the analysis of plant disease progress. The very important interdisciplinary research of agrometeorologists, plant pathologists and other related scientists is emphasized.

6 References

Adams RJ, Seager JM (1977) Agrometeorological use of the synoptic data bank in plant disease warning services. Met Mag 106: 112–116

Allen JC (1976) A modified sine wave method for calculating degree days. Environ Entomol 5: 388–396

Analytis S (1973) Zur Methodik der Analyse von Epidemien, dargestellt am Apfelschorf (*Venturia inaequalis* (Cooke) Aderh). Acta Phytomed 1: 1–76

Analytis S (1977) Über die Relation zwischen biologischer Entwicklung und Temperatur bei phytopathogenen Pilzen. Phytopathol Z 90: 64–76

Analytis S (1980) Obtaining of sub-models for modeling the entire life cycle of a pathogen. Z Pflanzenkr Pflanzenschutz 87: 371–382

Aust HJ (1981) Über den Verlauf von Mehltauepidemien innerhalb des Agro-Ökosystems Gerstenfeld. Acta Phytomed 7: 1–76

Aust HJ, Bashi E, Rotem J (1980) Flexibility of plant pathogens in exploiting ecological and biotic conditions in the development of epidemics. In: Palti J, Kranz J (eds) Comparative epidemiology – a tool for better disease management. Pudoc Wageningen, pp 46–56

Bashi E, Aust HJ (1980) Quality of spores produced in cucumber powdery mildew compensates for their quantity. Z Pflanzenkr Pflanzenschutz 87: 594–599

Braden H (1985) Ein Energiehaushalts- und Verdunstungsmodell für Wasser- und Stoffhaushaltsuntersuchungen landwirtschaftlich genutzter Einzugsgebiete. Mitt Dtsch Bodenkundl Ges 42: 294–299

Brutsaert W (1982) Evaporation into the atmosphere. Theory, history, and applications. Reidel, Dordrecht

Burleigh JR, Eversmeyer MG, Roelfs AP (1972) Development of linear equations for predicting wheat leaf rust. Phytopathology 62: 947–953

Butt DJ, Royle DJ (1974) Multiple regression analysis in the epidemiology of plant diseases. In: Kranz J (ed) Epidemics of plant diseases, mathematical analysis and modeling. Ecological studies, vol. 13. Springer, Berlin Heidelberg New York, pp 78–114

Choisnel E, Payen D (1981) Étude de l'interception d'eau par le capitule du tournesol. Concept de modélisation et vérification. Inf techn CETIOM 75: 7–10

Coakley SM (1985) Describing and quantifying the environment. Plant Dis 69: 461–466

Coakley SM, Line RF (1981) Climatic variables that control development of stripe rust on winter wheat. Climat Change 3: 303–315

Denmead OT (1976) Temperature in cereals. In: Monteith JL (ed) Vegetation and the atmosphere, vol 2. Case studies. Academic Press, London

Dirks VA, Romig RW (1970) Linear models applied to variation in numbers of cereal rust spores. Phytopathology 60: 246–251

Doorenbos J, Pruitt WO (1977) Guidelines for predicting crop water requirements. FAO irrigation and drainage paper no 24, revised form, Rome

Eversmeyer MG, Burleigh JR, Roelfs AP (1973) Equations for predicting wheat stem rust development. Phytopathology 63: 348–351
Fehrmann H, Weihofen U (1982) Untersuchungen zur Epidemiologie von *Pseudocercosporella herpotrichoides* in Wintergerste. Phytopathol Z 104: 60–77
Fitt BDL, Bainbridge A (1983) Dispersal of *Pseudocercosporella herpotrichoides* spores from infected wheat straw. Phytopathol Z 106: 214–225
Ford ED, Milne R (1981) Assessing plant response to the weather. In: Grace J, Ford ED, Jarvis PG (eds) Plants and their atmospheric environment. Blackwell, Oxford, pp 333–362
Gates DM (1980) Biophysical ecology. Springer, Heidelberg Berlin New York
Goudriaan J (1977) Crop micrometeorology: a simulation study. PUDOC, Wageningen
Gregory PH (1968) Interpreting plant disease dispersal gradients. Annu Rev Phytopathol 6: 189–212
Gregory PH (1973) The microbiology of the atmosphere, 2nd edn. Leonard Hill, London
Gregory PH, Guthrie EJ, Bunce ME (1959) Experiments on splash dispersal of fungus spores. J Gen Microbiol 20: 328–354
Häckel H (1984) Zur Messung der Benetzungsdauer von Pflanzen: Verfahren und Ergebnisse. Meteorol Rdsch 37: 97–104
Hau B, Aust HJ, Kranz J (1983) Problems in modelling powdery mildew epidemics. EPPO Bull 13: 259–262
Janisch E (1925) Über die Temperaturabhängigkeit biologischer Vorgänge und ihre kurvenmäßige Analyse. Pflügers Archiv Physiol 209: 414–436
Monteith JL (1981) Evaporation and surface temperature. Q J R Meteorol Soc 107: 1–27
Oke TR (1978) Boundary layer climates. Methuen, London
Pauvert P, Fournet J, Rapilly F (1970) Dissémination d'un inoculum fongique par des gouttes d'eau; influence de la morphologie des fructifications. Ann Phytopathol 2: 43–53
Payen D (1983) Modélisation de l'épidémiologie de *Sclerotinia sclerotiorum* sur tournesol. EPPO Bull 13: 277–281
Pedgley DE (1982) Windborne pests and diseases (meteorology of airborne organisms). Ellis Horwood, Chichester
Pedro MJ, Gillespie TJ (1982) Estimating dew duration. II. Utilizing standard weather station data. Agric Meteorol 25: 297–310
Rapilly F (1977) Réflexions sur les notions de propagule et d' unité de dissémination en épidémiologie végétale: cas des champignons parasites des organes aériens des plantes. Ann Phytopathol 9: 161–176
Rapilly F (1983) Effets de quelques facteurs physiques du climat sur diverses séquences épidémiques. EPPO Bull 13: 63–68
Rapilly F, Foucault B, Bonnet A (1975) Réalisation d'un appareil permettant l'étude au champ de la dispersion des spores de champignons par des gouttes d'eau; application à *Kabatiella zeae*, *Septoria nodorum* et *Fusarium roseum*. Ann Phytopathol 7: 45–50
Rapilly F, Fournet J, Skajennikoff M (1970) Études sur l'épidémiologie et la biologie de la rouille du blé *Puccinia striformis* Westend. Ann Phytopathol 2: 5–31
Rotem J (1978) Climatic and weather influences on epidemics. In: Horsfall JG, Cowling FB (eds) Plant diseases, an advanced treatise, vol 2. Academic Press, London, pp 317–337
Rutter AJ (1975) The hydrological cycle in vegetation. In: Monteith JL (ed) Vegetation and the atmosphere, vol 1. Principles. Academic Press, London, pp 111–154
Schrödter H (1960) Dispersal by air and water — the flight and landing. In: Horsfall JG, Dimond AE (eds) Plant pathology — and advanced treatise, vol 3. Academic Press, London, 169–227
Schrödter H (1965) Methodisches zur Bearbeitung phytometeoropathologischer Untersuchungen, dargestellt am Beispiel der Temperaturrelation. Phytopathol Z 53: 154–166
Schrödter H (1983) Meteorological problems in the practical use of disease-forecasting models. EPPO Bull 13: 307–310
Schrödter H (1987) Wetter und Pflanzenkrankheiten — Biometeorologische Grundlagen der Epidemiologie. Hochschultext. Springer, Berlin Heidelberg New York
Schrödter H, Ullrich J (1967) Eine mathematisch-statistische Lösung des Problems der Prognose von Epidemien mit Hilfe meteorologischer Parameter, dargestellt am Beispiel der Kartoffelkrautfäule *(Phytophthora infestans)*. Agric Meteorol 4: 119–135

Schurer K, Van der Wal AF (1972) An electronic leaf wetness recorder. Neth J Plant Pathol 78: 29–32

Smith LP (1975) Methods in agricultural meteorology. Elsevier, Amsterdam Oxford New York (Developments in atmospheric science 3)

Szeicz G (1975) Instruments and their exposure. In: Monteith JL (ed) Vegetation and the atmosphere, vol 1. Principles. Academic Press, London, pp 229–273

Thom AS (1975) Momentum, mass and heat exchange of plant communities. In: Monteith JL (ed) Vegetation and the atmosphere, vol 1. Principles. Academic Press, London, pp 57–109

Van Eimern J (1964) Untersuchungen über das Klima in Pflanzenbeständen als Grundlage einer agrarmeteorologischen Beratung insbesondere für den Pflanzenschutz. Ber Dtsch Wetterdienst Nr. 96

Wallin JR (1963) Dew, its significance and measurement in phytopathology. Phytopathology 53: 1210–1216

World Meteorological Organization (1981) Guide to agricultural meteorological practices, 2nd edn. WMO, Geneva

Yarwood CE (1952) Guttation due to leaf pressure favors fungus infections. Phytopathology 42: 520

III Special Topics

Chapter 9

The Analysis of Effects of Control Measures on the Development of Epidemics

R. D. Berger[1]

1 Introduction

The raison d'être for the study of plant pathology is to reduce crop loss due to disease (Carlson and Main 1979). To achieve these lower crop losses, plant diseases are controlled with chemicals applied before or during the epidemic, with plant resistance to disease, and with sanitation and other crop management practices. To determine the superior treatment among the many choices that may be available for the control of disease, or to predict the disease progress after the application of a control measure, is frequently not an easy task. It is the purpose of this chapter to provide suggestions, methods, and experimental techniques to aid in the interpretation of the disease progress associated with control procedures. Hopefully, the improved interpretation of disease progress will enable the researcher to make the choice of the best control measure, to predict future disease with or without the application of the control measure, and to correlate disease progress to spore numbers, weather, and plant growth which should lead to better advice to growers on the control of diseases.

2 Epidemiological Principles of Control

The ultimate purpose of disease control is to have as little disease as possible at a critical host stage or at the end of the crop season so that yield and the quality of the harvested product are not seriously reduced. If we consider that many foliar diseases proceed logistically or nearly so, then the disease (y_t) at some later time is dependent upon the level of initial disease (y_0), the rate at which the epidemic proceeds (r), and the duration of the epidemic (t). This can be expressed in a simple equation:
can be expressed in a simple equation:

$$\mathrm{logit}(y_t) = \mathrm{logit}(y_0) + rt\,, \tag{9.1}$$

where $\mathrm{logit}(y) = \log_e(y/(1-y))$. Equation (1) is frequently used to predict the severity of future disease when the present disease is known and an average epidemic rate has been determined from historical data. Tabular values of logit (y) are provided in Table 2 in Van der Plank (1963) and Appendix 6.1 in Zadoks and Schein (1979).

Since the purpose of disease control is to obtain a lower value of y_t, then if any of the three components (y_0, r, or t) on the right side of Eq. (9.1) are reduced, disease is reduced at all time thereafter (Berger 1977; Van der Plank 1963).

[1] Plant Pathology Department, University of Florida, Gainesville, FL 32611, USA

If, e.g., y_0 is reduced from 1 to 0.1% (which is generally easy to do), then the disease 20 days later is reduced from 50 to 9% if the average epidemic rate is 0.23. For:

$$\text{logit}(y_t) = \text{logit}(0.01) + 0.23 \times 20$$
$$0 = -4.60 + 4.60; \text{ hence, } y_t = 0.5;$$

and:

$$\text{logit}(y_t) = \text{logit}(0.001) + 0.23 \times 20$$
$$-2.31 = -6.91 + 4.60; \text{ hence, } y_t = 0.09.$$

Obviously, superior control is obtained by combining control measures that reduce y_0, r, and t simultaneously (Berger 1977).

3 Analysis of the Effect of Sanitation

Sanitation measures reduce the amount of initial inoculum and subsequently also reduce the amount of initial disease. Several good examples of the various sanitation measures are provided by Berger (1977), Van der Plank (1963), and Walker (1957). The progress of foliar diseases following sanitation practices has been studied for only a few pathosystems (Berger 1977).

3.1 The Sanitation Ratio

Van der Plank (1963) derived what he called the "sanitation ratio" to calculate the benefit in time of the delay of the epidemic brought about by reducing the initial disease a specified amount. The ratio y_0/y_{0s} was for initial disease without and with sanitation. If the epidemic rate is known or estimated, the expected delay in time (dt) can be calculated (Van der Plank 1963) as:

$$dt = \log_e(y_0/y_{0s})/r\,. \tag{9.2}$$

For example, if initial disease was reduced from $y_0 = 0.01$ to $y_{0s} = 0.001$ by sanitation, and $r = 0.23$, the delay (dt) would be:

$$dt = \log_e(0.01/0.001)/0.23 = 10 \text{ days.}$$

This effect is depicted graphically in Fig. 9.1. If disease without sanitation is above $y_0 \approx 0.05$, then the delay in time cannot be calculated accurately by the sanitation ratio as Van der Plank described because the correction factor $(1-y)$ for logistic increase is not included. For cases in which y_0 exceeds ≈ 0.05, the delay should be calculated (Plaut and Berger 1981) as:

$$dt = [\text{logit}(y_0)\text{-logit}(y_{0s})]/r\,. \tag{9.3}$$

Van der Plank (1963) made it quite clear that the length of the delay was entirely dependent upon the epidemic rate (r), and that the disease following sanitation proceeds at the same rate to that of disease progress in the pathosystem without sanitation. That is, the delay (dt) would be the same for all time periods. "The effect of sanitation remains proportionately the same at all times in the logarithmic stage" (Van der Plank 1963, p 124). "The delay dt at the end of the logarithmic phase is the delay at all later levels of disease" (Van der Plank 1963, p 127).

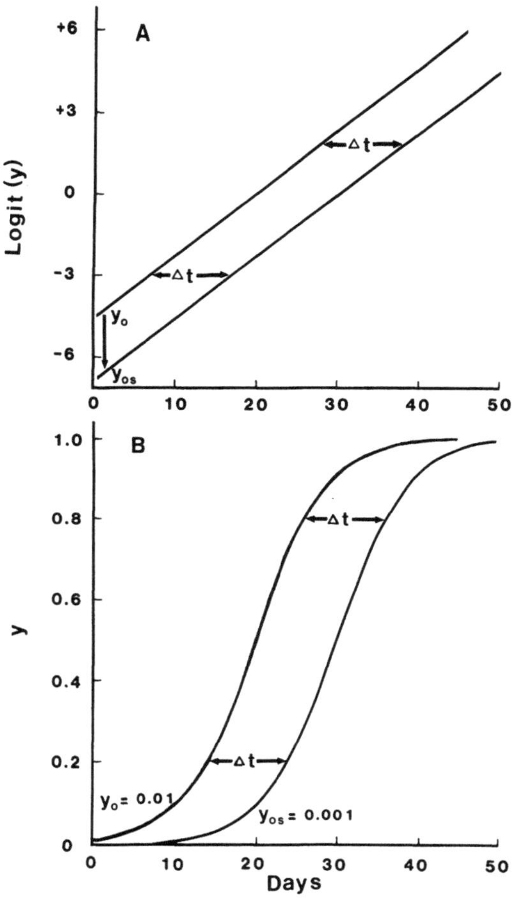

Fig. 9.1 a and b. The delay (Δt) in the development of an epidemic based on Van der Plank's sanitation ratio: $\Delta t = \log_e(y_0/y_{0s})/r$ (Zadoks and Schein, 1979). **A** Logit scale; **B** linear scale. The reduction in initial disease by sanitation from $y_0 = 0.01$ to $y_{0s} = 0.001$ yielded a theoretical Δt of 10 days for all time periods; r was 0.23

3.2 An Alternate Hypothesis for the Effect of Sanitation

Is the hypothetical effect of sanitation on disease progress as envisioned by Van der Plank (1963) characteristic of natural systems? Is r independent of y and therefore is the delay in time brought about by sanitation the same regardless of the level of y_0? How can the hypothesis of the sanitation ratio be tested? In simulated epidemics of *Cercospora apii* on celery, disease progress was initiated from different levels of initial disease (Berger, unpublished). Peculiarly, as initial disease decreased, the epidemic rate became faster. That is, the more rapid epidemic rate largely counteracted the benefits grained by the reduction in initial disease. Consequently, final disease levels for the epidemics begun at the different initial levels of disease were not much different. Was this lack of parallelism for disease progress from the different initial levels of disease an aberration of the simulation model, or was the seeming dependence of r on y a real phenomenon that occurs in natural epidemics? Plaut and Berger (1981) verified the simulated results with disease progress observed for epidemics of three pathosystems (*Cercospora arachidicola* on peanut, *Botrytis cinerea* on *Begonia,* and *Uromyces phaseoli* on *Phaseolus*) all begun with three levels of initial disease. As epidemics were begun from lower and lower levels of initial disease, disease progress was faster and faster. These results were confirmed independently for *Helminthosporium maydis* on corn (Gregory et al. 1981) and *Erysiphe graminis* on wheat (Rouse et al. 1981). Other pathosystems and a range of environments need to be examined to determine whether the phenomenon is universal.

3.3 Methods to Test the Effect of Sanitation

To test the effect of the initial level of disease on subsequent disease progress, a broad range of initial disease, e.g., $y_0 = 0.01$, 0.001, 0.0001, and 0.00001, should be tested. Plants can be inoculated with different inoculum densities, and after disease appearance, leaves with lesions can be trimmed to achieve the desired intensity of initial disease. These source plants can be used to initiate epidemics in polycyclic experimentation in special greenhouse chambers or in the field. The experimental design with appropriate buffers of space or nonhost areas must be such that interplot interference is reduced to near zero. The disease should be assessed every one-fourth to one-third latent period (p) for the first few cycles; then the interval between assessments can be lengthened to one-half to one period (Berger 1980). For more details on interplot interference and assessment procedures, see Aust and Kranz, Chap. 1, Kranz, Chap. 3, and Teng and Johnson, Chap. 12, respectively.

3.4 Other Considerations of Sanitation on the Epidemic

The significance of these more rapid epidemic rates from lower levels of initial disease is of tremendous importance in the analysis of epidemics. Equation (9.2) of Van der Plank (1963) to determine the delay in time caused by sanitation should be considered to be the maximum delay. For most pathosystems, and under usual environmental conditions, less delay is most likely. The degree of increase of the epidemic rate seemingly depends upon the pathosystem, the environment, and particularly the host area available for infection. The delay provided by sanitation practices may be particularly beneficial for those pathosystems that typically have epidemics that develop late in the crop season. The limited amount of inoculum that usually occurs in these early epidemic stages is probably more efficient than inoculum later in the epidemic (under a similar environment) because of the greater likelihood of the inoculum landing on susceptible sites. If historical information on the epidemic rates for a pathosystem is available, the epidemic rates from reduced initial disease can be approximated if a certain value of y_{max} is assumed at a given t for disease progress from the highest level of y_0 (Fig. 9.2). Since many pathosystems naturally have faster logistic rates in early epidemic stages, a closer approximation of the epidemic rates can be likely obtained if a transformation such as the Gompertz (Berger 1981) is used instead of the logistic. Small changes in y_0, or even errors in the estimation of y_0 cause great changes in the logistically transformed values; this problem is less acute for the Gompertz transformation (Luke and Berger 1982). A table of gompit (y), i.e. $Y = -\log_e(-\log_e(y))$, is provided by Berger (1981).

Even though the epidemics may go faster from lower levels of initial disease, and the final disease levels may be similar, there is still some gain from the sanitation practice.

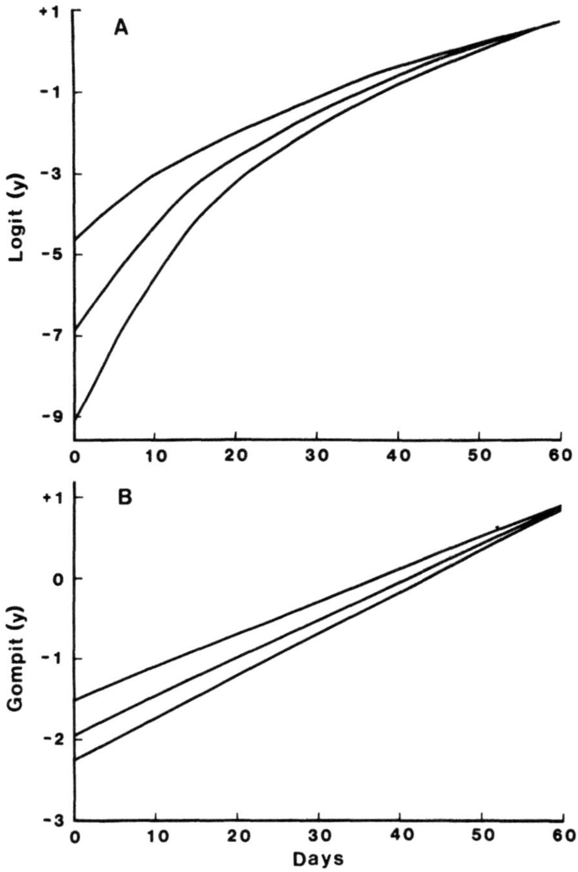

Fig. 9.2a and b. Disease progress from three levels of initial disease (after Plaut and Berger 1981). Compare Fig. 2a (natural representation) to Fig. 1a (theoretical). The level of disease intensity at $t = 60$ was arbitrarily chosen as a point of convergence. Disease progress on logit scale **(A)** would be more difficult to estimate than on gompit scale **(B)**

Unfortunately, the gain may be somewhat minor as it occurs primarily at low levels of disease when there is not a great affect on the host. This gain from sanitation is most easily measured as the area under the disease progress curve (AUDPC), see Sect. 4.2.

Sanitation practices applied during the epidemic, i.e., the removal of diseased tissue or the roguing of diseased plants, have been examined quantitatively for only a few cases (Berger 1977). This practice slows the epidemic rate, the analysis of which is handled in the next section.

4 Analysis of Effects of Control Measures on Epidemic Rate

4.1 Methods

The many fungicides that are applied to the foliage of plants to control leaf spots and blights primarily slow the epidemic rate. The repeated applications of fungicide in a season-long program should result in a disease-progress curve that has a slow epidemic rate and a low level of final disease compared to the curve for the nonsprayed crop. These curves should be plotted as y, the derivative dy/dt, or transformed y versus time (Kranz 1974). The disease progress on plants with horizontal disease resistance is nearly identical to the disease progress of the sprayed crop. "Fungicide equivalents" can be determined for the degree of host resistance (Fry 1978). The degree of suppression of the epidemic depends upon the degree of effectiveness of the spray or of the resistance. To compare the effects of various fungicides, it is important that several rates of each material be tested. Since it is statistically improper to compare rates of application among and between materials with multiple comparison procedures (Chew 1977), at least four levels of each material should be evaluated to provide sufficient values to obtain a response which can be tested with linear or nonlinear regression (Cornell and Berger 1987). There are no clear-cut guidelines for which levels to choose, but the following rates generally provide a broad family of disease-progress curve when the fungicides are applied to the crop at a relatively close interval (≈ 7 days) and under disease-favorable conditions (nx recommended rate; $n = 2, 1, 0.5, 0.1, 0.01, 0.001$, and 0). It is far better to adjust rates of the fungicide than to adjust spray interval to achieve these families of disease-progress curves. Adjustment of spray intervals frequently results in great flushes and lapses in disease appearance and this complicates the interpretation of the disease-progress curves.

4.2 Analysis of Disease Progress

The determination of the disease response to a control measure is dependent solely upon the frequency, accuracy, and precision of the disease assessments. Too few assessments or an unfortunate choice in assessment dates may lead to an erroneous interpretation of the disease-progress curve (Fig. 9.3). Since the general response for many control measures, including host resistance, is a slowing of the epidemic rate, many treatment comparisons are based on these rates. There are at least three ways to calculate the epidemic rates: (1) two-point method, (2) linear regression of the transformed disease-intensity values, and (3) nonlinear fitting of disease-intensity values to various growth functions.

In the two-point method, two values are selected along the disease-progress curve in an area of interest and the epidemic rate is calculated from them. If the curve has a logistic shape, the rate (r) is calculated as:

$$r = [\mathrm{logit}(y_2) - \mathrm{logit}(y_1)]/dt\,, \tag{9.4}$$

where logit (y) is $\log_e(y/(1-y))$ and dt is the time interval between the selected points. If the disease-progress curves are asymmetrically sigmoidal, the Gompertz transformation (Berger 1981) should be used. The Gompertz rate is calculated similarly:

$$k = [\mathrm{gompit}(y_2) - \mathrm{gompit}(y_1)]/dt\,, \tag{9.5}$$

where gompit (y) is $-\log(-\log(y))$. Unless the rate is relatively constant for much of the run of the curve, the two-point method may not provide a very close estimate of the true epidemic rate. Because of the problems with sampling errors, sampling size, variability, bias, scale, heteroscedasticity, etc. caution should be exercised when interpretations are based on the epidemic rates calculated by the two-point method.

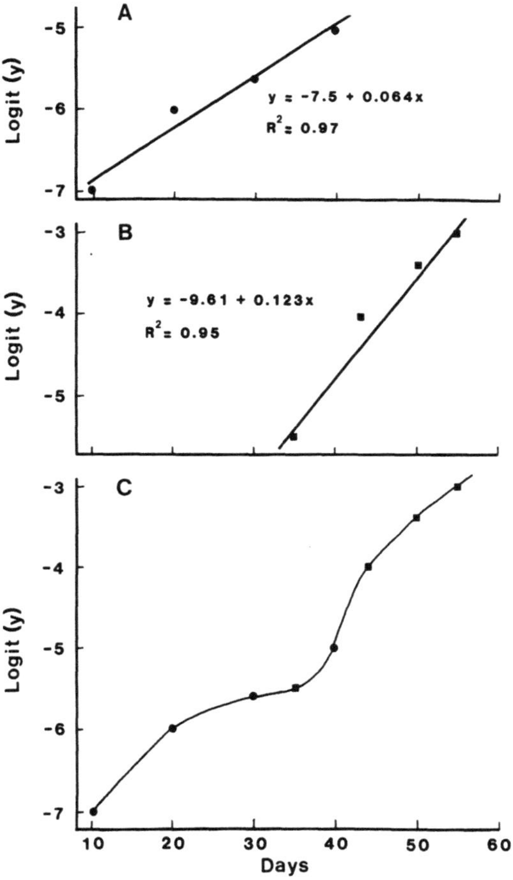

Fig. 9.3 a – c. The inaccurate estimation of true disease progress due to the choice of assessment dates. **A** Four early dates, $r = 0.064$; **B** four late dates, $r = 0.123$; **C** the complete disease progress for the period. The relative rate of the rapidly changing epidemic rate would be more descriptive of the response than an average epidemic rate

A common technique to obtain a reasonable estimate of the epidemic rate is to apply linear regression to the transformed diseased-intensity values; common models for transformation are logits and gompits (as above). The regression yields a slope parameter (b) which is an estimate of the epidemic rate. Particular attention must be paid to any pattern of residuals, as this determines whether the correct model was chosen for linearization (Cornell and Berger 1987). With the logistic transformation of many disease-progress curves, the transformed values for both low and high levels of disease intensity are frequently below the average logit line for the complete curve (see, e.g., Ayers et al. 1976; Larios and Moreno 1977). For such curves, the Gompertz transformation usually provides a better statistical fit (Berger 1981). The regression of the transformed values also yields an estimate of the intercept (a), which is the estimate of y_0 at time zero.

Nonlinear curve fitting is the third way to obtain estimates of the epidemic rate. Various growth functions (see, e.e., Table 2 in Kranz 1974) are fit to the disease-intensity values by minimum least squares. Standard computer packages such as SAS-NLIN (SAS Institute 1982) can be used for this purpose. Alternatively a program such as that provided in Appendix II can be used. The major disadvantage of these curve-fitting programs is that many values (>10) of the response to be fitted are needed to obtain a realistic estimate of the rate parameter, although many researchers frequently use fewer values. Some models are sensitive to the range of the values to be fitted (Berger 1982; Thal et al. 1984). Spurious or erroneous estimates of the parameters for these models are obtained with limited or narrow-range data sets. The rates estimated by the nonlinear programs for the various treatments can be compared by goodness-of-fit statistics.

Other epidemic parameters commonly used in treatment comparisons are the AUDPC (see also Chap. 12), maximum amount of disease (y_{max}), time to reach 50% disease (t at $y_t = 0.5$), and amount of disease at a given time or crop stage ($y_t = n$). Of these parameters, the AUDPC is perhaps the most useful for treatment comparisons because much of the total epidemic can be summarized into this measure of "disease stress". The AUDPC is also frequently used to correlate disease intensity to estimates of crop loss. The accuracy of the AUDPC value as a measure of the epidemic is a function of the closeness of the interval between disease assessments; the narrower the interval, the more accurate is the determination of the area under the curve (Fig. 9.4). The AUDPC is usually calculated by trapezoidal integration as:

$$\mathrm{AUDPC} = \sum (y_i + y_{i+1})/2 \times dt_i \,, \tag{9.6}$$

where dt_i is the time interval between each two observations of y_i and y_{i+1} (Shaner and Finney 1977). The AUDPC can be normalized by dividing the AUDPC value by the total time duration of the graph (Fry 1978). A computer program in the BASIC language for the calculation of AUDPC is given in Appendix I.

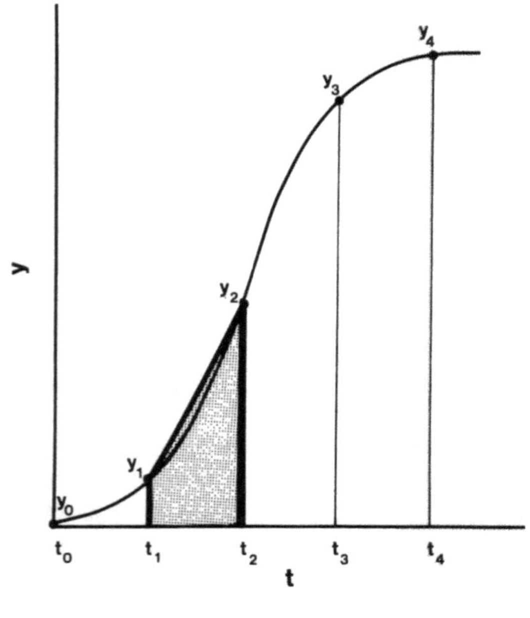

Fig. 9.4 Trapezoidal integration to determine the area under the disease progress curve (AUDPC); $\mathrm{AUDPC} = \Sigma(y_i + y_{i+1})/2 \times (t_{i+1} - t_i)$. For the *shaded* portion of the curve between y_1 and y_2, the calculated area would overestimate the actual area because of the length of the interval from t_1 to t_2

Appendix I. Computer program in Microsoft BASIC to calculate the area under disease progress curve (AUDPC) by trapezoidal integration

```
 10 DIM N(50),T(50),AU(50)
 20 CLS:PRINT "THIS PROGRAM CALCULATES THE AREA UNDER THE CURVE"
 30 PRINT :PRINT :INPUT "INPUT THE NUMBER OF DISEASE VALUES ";X
 40 PRINT :PRINT "NOW, INPUT THE ";X;" DISEASE VALUES"
 50 FOR I=1 TO X:INPUT N(I):NEXT I
 60 PRINT :PRINT "NOW, INPUT THE ";X-1;" TIME INTERVALS"
 70 FOR I=1 TO X-1:INPUT T(I):NEXT I
 80 AD=Ø
 85 PRINT :PRINT "SUMMARY"
 90 FOR I=1 TO X-1
100 AU(I)=((N(I)+N(I+1))/2)*T(I)
110 PRINT N(I);" TO ";N(I+1);" BY ";T(I);" HAS AREA OF ";AU(I)
120 AD=AU(I)+AD
130 NEXT I
140 PRINT :PRINT "TOTAL AUDPC = ";AD
150 END
```

Appendix II. Computer program in Microsoft BASIC to fit data to mathematical functions by nonlinear least squares. Examples of one-parameter models (lines 1630–1650) and a two-parameter model (line 1660) are given

```
  50 DIM D(1Ø),S2(1Ø),D3(1Ø),D2(1Ø),Y(1ØØ),X(1ØØ)
 100 CLS:LOCATE 5,35:PRINT "N L I N"
 110 LOCATE 8,16:PRINT "Non-linear least squares curve-fitting program"
 120 LOCATE 18,5:INPUT "Push return to continue ",A$
1010 CLS:PRINT "YOU MUST INPUT YOUR DATA" :PRINT :PRINT
1020 INPUT "HOW MANY PAIRS OF DATA DO YOU HAVE ? ",M
1040 PRINT :PRINT :PRINT "YOU MUST INPUT ALL DISEASE PROPORTIONS"
1050 PRINT "FIRST, THEN INPUT ALL TIME VALUES"
1060 PRINT :PRINT :PRINT "DISEASE PROPORTIONS :"
1065 SUM=Ø
1070 FOR I=1 TO M:INPUT Y(I):SUM=SUM+Y(I):NEXT I
1080 MUY=SUM/M:SSY=Ø
1090 FOR I=1 TO M:SSY=SSY+(Y(I)-MUY)^2:NEXT I
1120 PRINT :PRINT "NOW INPUT ALL TIME VALUES :"
1130 FOR I=1 TO M:INPUT X(I):NEXT I
1200 CLS:PRINT "CHOOSE A MODEL TO BE FITTED" :PRINT :PRINT
1205 X5=.1:I4=1:J1=Ø:I9=Ø:N5=5Ø:X4=.Ø1
1210 PRINT "CHOOSE A MODEL TO BE FITTED"
1220 PRINT "1. LOGISTIC"
1230 PRINT "2. GOMPERTZ"
1240 PRINT "3. MONOMOLECULAR"
1250 PRINT "4. WEIBULL"
1260 PRINT :PRINT :PRINT: INPUT "WHICH ";B%
1320 ON B% GOTO 134Ø,137Ø,139Ø,14ØØ
1340 MO$="LOGISTIC" :LN=LOG(Y(1)/(1-Y(1))) :M2=1:GOTO 15ØØ
1370 MO$="GOMPERTZ" :B=-LOG(Y(1)):M2=1:GOTO 15ØØ
1390 MO$="MONOMOLECULAR":M2=1:GOTO 15ØØ
1400 MO$="WEIBULL":A=-1:M2=2
1410 INPUT "INPUT INITIAL ESTIMATE FOR SHAPE PARAMETER C (0.9<c<5.9)
     ",D(2)
1500 CLS:INPUT "INPUT INITIAL ESTIMATE FOR RATE ",D(1)
1550 PRINT :PRINT
```

```
1600 S5=∅°:REM CALC OF SUMS OF SQUARES
1610 FOR I=1 TO X(M)
1620 ON B% GOTO 163∅,164∅,165∅,166∅
1630 Y1=1/(1+EXP(-(LN+D(1)*X(I)))) :GOTO 17∅∅
1640 Y1=EXP (-B*EXP(-D(1)*X(I))) :GOTO 17∅∅
1650 Y1=1-EXP(-D(1)*X(I)):GOTO 17∅∅
1660 Y1=1-EXP(-(((X(I)-A)*D(1))^D(2)))
1700 S5=(Y(I)-Y1)^2+S5
1710 NEXT I
1730 PRINT "SS= ";S5
1740 IF S5=∅ THEN GOTO 244∅
1750 IF I9<>∅ THEN GOTO 2∅5∅
1760 J1=J1+1:IF J1>1 GOTO 182∅
1780 PRINT "ITERATION="; I4
1790 PRINT "EACH PARAMETER IS CHANGED BY AN AMOUNT=";X5
1800 PRINT :S3=S5 :GOTO 185∅
1820 S2(J1-1)=S5:D(J1-1)=D(J1-1)-D3(J1-1)
1840 IF J1>M2 THEN GOTO 188∅
1850 D3(J1)=D(J1) *X5:D(J1)=D(J1)+D3(J1):GOTO 16∅∅
1880 D5=∅
1890 FOR J1=1 TO M2
1900 D2(J1)=S2(J1)-S3:A5=ABS(D2(J1))
1920 IF A5<=D5 THEN GOTO 195∅
1930 J4=J1:D5=A5
1950 NEXT J1
1960 D3(J4)=X5*D(J4)
1970 IF D2(J4)>∅ THEN GOTO 199∅
1980 GOTO 2∅∅∅
1990 D3(J4)=-D3(J4)
2000 FOR J1=1 TO M2
2010 IF J1=J4 THEN GOTO 2∅3∅
2020 D3(J1)=-D2(J1)*X5*D(J1)/ABS (D2(J4))
2030 NEXT J1
2040 S9=S3:IF S9<.∅∅∅∅∅∅1 THEN GOTO 244∅
2050 I9=I9+1
2060 IF I9<>1 THEN GOTO 213∅
2090 FOR J1=1 TO M2:D(J1)=D(J1)+D3(J1):NEXT J1
2120 GOTO 16∅∅
2130 IF S5>=S9 THEN GOTO 216∅
2140 S9=S5:GOTO 2∅9∅
2160 FOR J1=1 TO M2:D(J1)=D(J1)-D3(J1):NEXT J1
2190 PRINT :PRINT "BEST FOR ROUND=";I4
2210 FOR I=1 TO M2:PRINT "PARAMETER ";I;"=";D(I):NEXT I
2230 PRINT "SUM OF SQUARES (OLD)=";S9
2235 PRINT :PRINT
2240 I4=I4+1
2250 IF I4>N5 THEN GOTO 244∅
2260 FOR I=1 TO M2:D9=D(I)*X4:A9=ABS(D3(I)):Q=∅
2300 IF A9>D9 THEN GOTO 232∅
2310 GOTO 2340
2320 Q=1:I=M2+1
2340 NEXT I
2350 IF Q=1 THEN GOTO 237∅
2360 GOTO 244∅
2370 J1=0:S5=S9
2390 IF I9=2 THEN GOTO 241∅
2400 GOTO 242∅
```

```
2410 X5=.5*X5
2420 I9=Ø
2430 GOTO 175Ø
2440 REM PRINT END PARMS
2450 R2=1-S9/SSY
2460 FOR I=1 TO M2:PRINT "PARAMETER ";I;"=";D(I):NEXT I
2480 PRINT "SUM OF SQUARES =";S9;"  R2=";R2
2490 PRINT "MODEL=";MO$
2600 REM END SUB
2610 FOR P=1 TO 3ØØ
2620 IF P=1ØØ OR P=2ØØ OR P=3ØØ THEN BEEP
2630 NEXT P
2650 PRINT "END !!"
2660 PRINT :INPUT "ANOTHER RUN WITH THIS DATA SET (Y/N) ? ",Y$
2670 IF Y$="Y" OR Y$="y" GOTO 12ØØ
2680 END
```

4.3 Disease Progress After a Single Application of a Fungicide

The single application of an effective, protectant fungicide to the foliage of a crop leads to a major interruption in the progress of the fungal disease. Special analytical approaches are needed to aid in the interpretation of these rapidly changing intensities of disease.

4.3.1 The Theoretical Disease-Progress Curve

The theoretical interpretation of this interruption is as given in Zadoks (1971, Figs. 9 and 13). A generalized curve is given in Fig. 9.5. Fungicides affect the number (N) and the efficiency (e) of fungal spores on the treated host surface. Since the basic infection rate (Van der Plank 1963), R (the number of daughter lesions per mother lesion per day), can be expressed as $R = N e$, then the changes in the actual value of R determines the effect of fungicide on disease progress. A simple computer program using the basic infection rate (R) to simulate the effect of spraying on disease progress is given elsewhere (Berger 1987). In Fig. 9.5A, note that the effect of the fungicide begins one latent period after the application and lasts for a specific "effective period." The length of the effective period is a function of the fungicide efficacy, the dose, and the residual breakdown curve. The key characteristics of this theoretical response are the sharpness of the swing of disease into the effective period, the zero-disease increase during the effective period, and the parallelism of disease progress to the nontreated control (with a time delay equal to the effective period) after the effect of the interruption has passed.

4.3.2 The Actual Disease-Progress Curve

The actual disease response following epidemic interruption has not been studied in great detail for many pathosystems. For those cases where it has been studied, the disease proceeds quite unlike the theoretical response; the general response is depicted in Fig. 9.5B. The disease has more of a gradual swing into the effective period, primarily because some latent infections are variable in their appearance; the laggards cause the smoothing of the curve. When host growth is minimal during this time, the effective period usually has a slight upward slant of increasing disease because of the incomplete control of all infections by the fungicide, some infections occur on the unprotected new

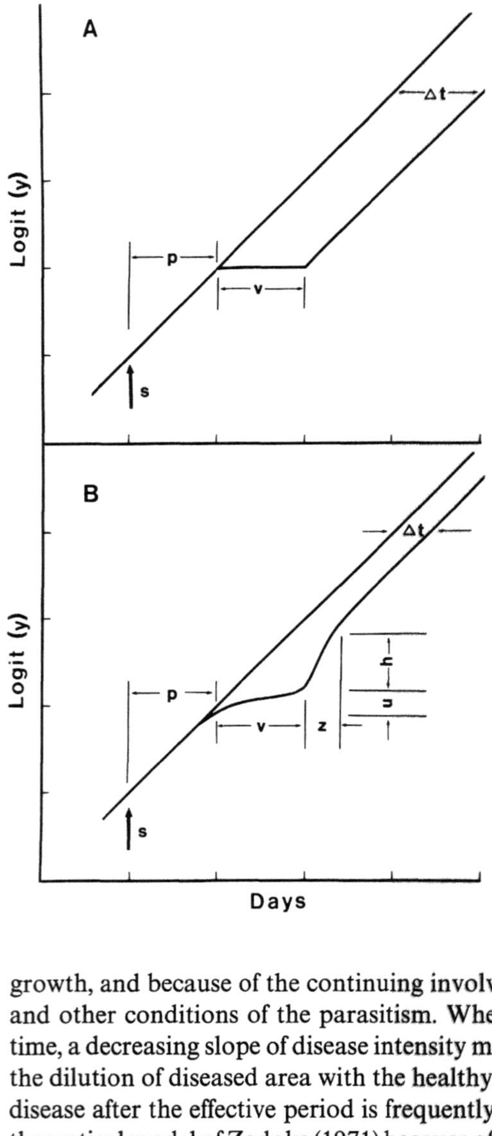

Fig. 9.5a and b Disease progress after epidermic interruption by an effective, protectant fungicide. The *straight line* represents the disease progress in the nonsprayed crop. **A** In the theoretical response (after Zadoks 1971), the effective period (v) begins abruptly one latent period (p) after the day of spraying (s). The epidemic delay (Δt) is equal to v at all subsequent times. **B** In the representation of the natural response, disease intensifies gradually into the effective period (v) at approximately one latent period (p) after spraying (s). Disease increases at amount u during the effective period. A period (z) of rapid disease progress of intensity (h) occurs after the effective period. The epidemic delay (Δt) is considerably less than the effective period. See text for variations

growth, and because of the continuing involvement of host tissue by lesion expansion and other conditions of the parasitism. When host growth is substantial during this time, a decreasing slope of disease intensity may occur in the effective period because of the dilution of diseased area with the healthy growth. The upward swing of increasing disease after the effective period is frequently much greater than that predicted in the theoretical model of Zadoks (1971) because of the rapid infection of vast amounts of the then available healthy host tissue in the few days immediately after the interruption has ceased (Berger 1977; Berger and Jones 1985). This rapid increase in disease corresponds to a high infection efficiency for the spores on the newly available host tissue.

The level of disease intensity at the time of the interruption affects the subsequent disease progress. As the level of disease is higher at the time of the interruption, the swing into the effective period is more gradual, the effective period has a steeper slope (more disease increase), and the effective period is of shorter duration (Berger 1977; Berger and Jones 1985). At high levels of disease, the infection efficiency (e) becomes lower and lower. Even though many spores (N) may be available for infection, the low efficiency still reduces R in the quantity, $R = Ne$.

4.3.3 Prediction of the Increase in Disease

The exact speed of increase and the duration of this rapid rate after the effective period are two of the most difficult variables to predict in epidemic progress. Seemingly, both variables are largely dependent upon the amount of healthy host area available for infection after the interruption rather than dependent upon the pathogen's inoculum, since ample inoculum is usually present to saturate available infection sites. When more susceptible host tissue is available, the disease progress is faster and longer in duration. After this rapid increase, the epidemic swings into a phase of generalized increase which may be more or less parallel to the disease increase in the nontreated crop or the uninterrupted epidemic. However, this later disease progress is nearly always at a considerable shorter dt than the original effective period. A general model of epidemics has been published (Berger and Jones 1985) which is suitable to estimate the disease increase after epidemic interruption, if certain particulars of the pathosystem are known.

The above description was largely for the effect observed for a protectant fungicide. With a systemic fungicide, the time after the application to the effective period is shorter than one latent period (because the fungus is killed in the most recent latent infections), the effective period is longer, and the disease increase during the effective period is less. The rapid increase of disease after the effective period is much the same for both types of fungicides.

Since the epidemic rate changes so rapidly over a period of a few days when the epidemic is interrupted by spraying or unfavorable environmental conditions, treatment comparisons based on the epidemic rate are virtually meaningless (see Fig. 9.3). The responses noted here are the main reason that comparisons of fungicides should be made with alteration of fungicide rates rather than an adjustment of spray interval. Since the epidemic rate is so variable under these conditions, the AUDPC, time of delays, time to $y_t = n$, and y_{max} are more appropriate parameters to compare treatments. If the proportions of y_{max} and $y_t = n$ cover a wide range of values, they should be transformed before analysis by arc sin $\sqrt{y}$. The best comparisons of the disease progress can be made by plotting y, the derivative dy/dt, or transformed y for the various treatments on the same graph. If the assessments were accurate and made at a close interval, then the time and duration of the delays, and magnitude of changes in the disease progress are immediately evident.

5 Epilogue

The epidemiological analysis of measures applied to control polycyclic diseases has not been studied in great detail except for a few pathosystems. The paucity of information on this important aspect in plant pathology is likely because of the frequency of assessments needed to define the rapidly changing disease intensity, and the general lack of confidence plant pathologists have had in the accuracy of the assessments. After a quiescent period of a few decades with little improvement in methods for disease assessment, instruments are now available to measure respiration and photosynthesis, to sense thermal and infrared radiation, videorecorders for leaf imaging, etc. (see Kranz, Chap. 3). With the advent of these methods to assess disease and host growth with more speed and accuracy, researchers may be encouraged to examine the intricacies of the disease response associated with the changing environment and control measures.

The epidemic rates calculated by the logistic, Gompertz, and other equations have been useful to compare epidemics in a general sense. However, these rates, as well as other parameters, such as y_{max} and $y_t = 0.5$, and others to compare epidemics critically have been rather disappointing (Fry 1978; Shaner and Finney 1980). The AUDPC has been useful, but this measure also has its limitations (Fry 1978; Shaner and Finney 1980). That is, plant pathologists frequently have been so overly concerned with the assessment and progress of disease that they have forgotten the host plant on which the epidemic is occurring. Thus, a more relevant approach might be to determine the amount and duration of the photosynthetically active tissue, particularly during the rapidly changing disease intensity. The duration of healthy leaf area may then have the added benefit of being used as an accurate and reliable predictor of yield and yield loss (Waggoner and Berger 1987).

6 References

Ayers JE, Nelson RR, Castor LL, Blanco MH (1976) Yield losses in corn caused by *Helminthosporium maydis* race T. Plant Dis Rep 60: 331–335

Berger RD (1977) Application of epidemiological principles to achieve plant disease control. Annu Rev Phytopathol 15: 165–183

Berger RD (1980) Measuring disease intensity. In: Stakman EC Proc. Commemorative symposium on crop loss assessment. Univ Minnesota Misc Publ 7, pp 28–31

Berger RD (1981) Comparison of the Gompertz and logistic equations to describe disease progress. Phytopathology 71: 716–719

Berger RD (1982) Problems encountered in fitting disease-progress curves to the Weibull function. Phytopathology 72: 1006

Berger RD (1987) Description and application so some general models for plant disease epidemics. In: Leonard KJ, Fry WE (eds) Plant disease epidemiology, vol 2 (in press)

Berger RD, Jones JW (1985) A general model for disease progress with functions for variable latency and lesion expansion on growing host plants. Phytopathology 75: 792 797

Carlson GA, Main CE (1979) Economics of disease-loss management. Annu Rev Phytopathol 14: 381–403

Chew V (1977) Comparisons among treatment means in analysis of variance. US Dep Agric Tech Bull H-6

Cornell JA, Berger RD (1987) Factors that influence the value of the coefficient of determination in simple linear and nonlinear regression models. Phytopathology 77: 63–70

Fry WE (1978) Quantification of general resistance of potato cultivars and fungicide effects for integrated control of potato late blight. Phytopathology 68: 1650–1655

Gregory LV, Ayers JE, Nelson RR (1981) Reliability of apparent infection rates in epidemiological research. Phytopathol Z 100: 135–142

Kranz J (1974) The role and scope of mathematical analysis and modeling in epidemiology. In: Kranz J (ed) Epidemics of plant diseases: mathematical analysis and modeling. Springer, Berlin Heidelberg New York, pp 7–55

Larios JF, Moreno RA (1977) Epidemiologia de algunas enfermedades foliares de la yucca en diferentes sistemas de cultivo. II. Roya y muerte descendente. Turrialba 27: 151–156

Luke HH, Berger RD (1982) Slow rusting in oats compared with the logistic and Gompertz models. Phytopathology 72: 400–402

Plaut JL, Berger RD (1981) Infection rates in three pathosystem epidemics initiated with reduced disease severities. Phytopathology 71: 917–921

Rouse DI, MacKenzie DR, Nelson RR (1981) A relationship between initial inoculum and apparent infection rate in a set of disease progress data for powdery mildew on wheat. Phytopathol Z 100: 143–149
SAS Institute (1982) SAS user's guide: statistics. SAS Institute, Cary, NC
Shaner G, Finney RE (1977) The effect of nitrogen fertilization on the expression of slow-mildewing resistance in Knox wheat. Phytopathology 67: 1051–1056
Shaner G, Finney RE (1980) New sources of slow leaf rusting resistance in wheat. Phytopathology 70: 1183–1186
Thal WM, Campbell CL, Madden LV (1984) Sensitivity of Weibull model parameter estimates to variation in simulated disease progression data. Phytopathology 74: 1425–1430
Van der Plank JE (1963) Plant diseases: epidemics and control. Academic Press, London
Waggoner PE, Berger RD (1987) Defoliation, disease, and growth. Phytopathology 77: 393–398
Walker JC (1957) Plant pathology, 2nd edn. McGraw-Hill, New York
Zadoks JC (1971) Systems analysis and the dynamics of epidemics. Phytopathology 61: 600–610
Zadoks JC, Schein RD (1979) Epidemiology and plant disease management. Oxford University Press, New York.

Chapter 10

How to Detect and Measure Fungicide Resistance

J. Dekker[1]

1 Introduction

In recent years the effectiveness of control by fungicides is threatened by the phenomenon of fungicide resistance. The latter is defined as the stable, inheritable adjustment of a fungal cell or a fungal population to a fungicide, resulting in a less than normal sensitivity to that fungicide. The term fungicide resistance is used for strains of a sensitive species which have become, usually by mutation, significantly less sensitive to a fungicide. Application of the fungicide will select for the resistant strains, and thus favour their multiplication. When the major part of the pathogen population has become resistant, failure of disease control may occur (Delp and Dekker 1985). Problems caused by fungicide resistance have increased over the last 2 decades, since selective fungicides were introduced, many of which act systemically in the plant. These fungicides act at specific sites in the fungal metabolism. In contrast to these specific-site inhibitors, there have hardly been any resistance problems with conventional fungicides, such as copper compounds and dithiocarbamates. The latter interfere with many enzyme systems in the fungal metabolism and are calles multisite inhibitors. The relation between specific-site inhibitors and mutation towards resistance is briefly discussed in Section 3.1. For more detailed information about the genetic and biochemical basis of resistance, see Georgopoulos (1977) and Dekker (1985).

Development of fungicide resistance affects the farmer, who may suffer unexpected crop losses, and the manufacturer, who may lose his profits and even may be confronted with law-suits by the growers. Fungicide resistance affects also the advisory services, the regulatory authorities, the agrochemical industry as a whole, and in some cases even the consumer and the national economy (Schwinn 1982a).

Before introduction of a new fungicide the potential of fungi to develop resistance to it should be checked in laboratory, greenhouse and field tests. If the compound appears to be resistance-prone appropriate measures should be taken to avoid or delay the appearance of resistance (Staub and Sozzi 1984; Dekker 1982b). Monitoring of fungicide sensitivity should be made for early warning that resistance is starting to build up, so that counter measures can be taken in time. In order to estimate whether the presence of resistant strains might have practical implications, it is also important to measure the level of resistance and the fitness of resistant strains in absence of the fungicide (Dekker 1982a).

[1] Laboratory of Phytopathology, Agricultural University, Wageningen 6709 PD, The Netherlands

When disease control fails, it is necessary to check whether it is due to fungicide resistance or to other causes, e.g. improper application of the compound. Proof that failure of control is due to resistance, requires first detection of the presence of resistant strains, then assessment of the proportion of resistant cells in the pathogen population and finally demonstration that fungicide-treated plants, inoculated by such strains, can become diseased.

Execution of these experiments require easy and reproducible methods to detect and measure resistance. Understanding the principles of fungicide resistance and employment of generally accepted and standard methods of measurement is required for comparison of results obtained by researchers at different places. Description of these principles and methods is the subject of this treatise.

2 Measurement of Resistance

2.1 Principles

Recognition of fungicide-resistant strains must be made by comparison with the sensitive, wild-type fungus. Therefore, it is necessary to have the baseline data of the fungicide sensitivity of the latter. The dosage-response curve of the wild-type pathogen is then compared with the curve of the investigated strains. Determination of the proportion of resistant cells in the pathogen population may be done by using one discriminating concentration of the fungicide. When the resistant strains have several levels of sensitivity, the use of two or more concentrations is advised.

The baseline data of the wild-type pathogen should preferably be collected before large-scale applications of the fungicide have started. It has happened that, after a period of use of the new fungicide, the original wild-type population has disappeared and we are forced to use baseline data described in the literature.

In some cases the difference in sensitivity between resistant and wild-type strains may be large, so that it is easy to distinguish between the two. In other cases (e.g. with triazoles and other inhibitors of sterol biosynthesis), the differences may be small, so that it is much more difficult to prove that development of resistance has occurred. This happens when the wild-type population has a wide variation in sensitivity. Since the effect of a fungicide is affected by the medium, the type and amount of inoculum and the environmental conditions, the experiments should be carried out under the same conditions and with the same techniques used for determination of the baseline data. It is recommended to include the wild-type pathogen, especially when an accurate regulation of the environmental conditions cannot be achieved.

Especially important is the correct preparation of the fungicide concentrations used for discrimination between resistant and sensitive isolates. For routine detection of fungicide resistance, we may use the formulated commercial product. When isolates are tested on whole plants (e.g. in the case of obligate parasites), use of the formulated product is recommended because it secures an adequate leaf coverage. However, when a very accurate determination of the level of resistance is required (e.g. when the differences in sensitivity between resistance and sensitivity are relatively small) and we measure it on artificial media, it is advisable to use the fungicide in a form as pure as possible. This avoids the possible complications due to the activity of adjuvants. Also, the solubility of the fungicide in water should be taken into account and concentrations beyond saturation should not be used, since these may lead to an overestimation of resistance. As solubility in water is low for most fungicides, they may be dissolved in inorganic solvents, which are mixable with water. However, these solvents may be fungitoxic and they shall be used in small amounts not exceeding 1 or 2%; the same amount of solvent should be added to the controls (Georgopoulos and Dekker 1982; Georgopoulos 1982a).

2.2 Use of Artificial Media

Radial Growth of Mycelium on Fungicide-Amended Agar Medium. When small agar blocks with mycelium, taken from the periphery of an actively growing colony, are placed and incubated in a petri dish, the diameter of the colony is a reliable measure of growth (Brancato and Golding 1953). The toxicity of the fungicide is represented by the diameter of the colony on amended agar and is compared to that on unamended agar, both after subtraction of the diameter of the agar block. A disadvantage of this method is that only linear growth is measured, and that the compactness of the colony or areal mycelium growth is not taken into consideration. This commonly used method is described by Uesugi (1982) for testing the sensitivity of *Pyricularia oryzae* to kasugamycin and Kitazin P. By testing the organism in a series of plates with different fungicide concentrations, a dosage-response curve can be made, providing information about the level of resistance. This has been described for resistance to benomyl in *Cercospora beticola* (Georgopoulos 1982b).

Spore Germination and Colony Formation on Fungicide-Amended Agar Medium

A spore suspension is seeled on agar medium containing the fungicide, and germination of these spores is measured microscopically or colony formation is observed macroscopically. Georgopoulos (1982b) used this method to investigate the benomyl sensitivity of conidia of *Cercospora beticola*. Placement of drops with conidia on premarked sites on the agar allowed the microscope estimation of the percentage of resistant cells within 24–48 h. When early determination of resistance is not required, growth of the resistant colonies can be measured in due time. In studies of resistance to benomyl or dodine Gilpatrick (1982) streaked a conidial suspension of *Venturia inaequalis* on fungicide-free and fungicide-amended agar and recorded visible growth after 4–7 days.

Spore Germination in Fungicide Solution

Fungicide sensitivity of spores can also be determined in the "slide germination" test. With this technique the germination of spores is observed in drops of a spore suspension, placed on a glass slide and kept under moist conditions to prevent drying. The fungicide may be deposited on the slide before introducing the drops, but the preferred method is to mix in test tubes small quantities of a dense spore suspension with various concentrations of fungicide. This method has been described in detail by a committee of the American Phytopathological Society on standardization of fungicidal tests (Anonymous 1947). It has been used, for instance, for the determination of dodine and benomyl sensitivity of ascospores of *Venturia inaequalis* (Gilpatrick 1982). To test fungicide activity against pathogens, which produce zoospores, such as *Phytophthora,* observation of the motility time of zoospores enables a very rapid assessment of fungicide sensitivity (Sonoda et al. 1970).

Paper Disc Bioassay

Paper discs, usually 12 mm in diameter, are dipped in a fungicide solution and placed on agar medium seeded with fungal spores. After incubation, which allows the spores to germinate and produce visible growth, the diameter of the inhibition zone is measured. This can, after subtraction of the disc diameter, serve as a measure of fungicide sensitivity. This method, widely used in medicine, has also been applied for the determination of antibiotics active against fungal plant pathogens (Dekker and Ark 1959). It can also be used for measuring fungitoxic action of other chemicals. The size of the inhibition zone is not only influenced by the concentration of the fungicide and the sensitivity of the fungus, but also by the diffusion of the fungicide in the agar. A dosage-response curve obtained after dipping the discs in a range of fungicide concentrations, allows a comparison between resistant and sensitive strains. This method has been used by Jones and Walker (1976) for the detection of resistance to dodine in conidia of *Venturia inaequalis*.

Germination and Growth in Liquid Nutrient Medium

Flasks with liquid nutrient medium are inoculated with fungal spores or with mycelium and, when necessary, shaken during incubation. After a suitable period of growth the mycelium is harvested, dried and weighed. Since the total increase of mycelium is determined, this method is more accurate than measuring radial growth on agar, but it is more laborious. For instance, it has been used for assessment of benomyl resistance in *Cercospora beticola* (Georgopoulos and Dovas 1973). Measurement of turbidity of the nutrient medium is a much easier and non-destructive method, but it can be used only with pathogens with yeastlike growth. It has been used for the determination of chloroneb resistance in *Ustilago maydis* (Tillman and Sisler 1973).

Inhibition of the Metabolic Process

When dealing with a fungicide, which interferes with a specific process in the fungal metabolism, measurement of its inhibition provides information about the fungicide sensitivity. Such experiments may be carries out with fungal cells or with cell-free extracts. This method has been used for the study of carboxamide resistance in *Ustilago maydis* (Georgopoulos et al. 1972).

2.3 Use of Living Plant Material

Although experiments in vitro quantify resistance, useful is also information obtained from experiments with living plant material. The proof that failure of disease control is due to development of fungicide resistance, is completed by the demonstration that the resistant strains can cause disease on fungicide-treated plants. The use of plant material is obviously indispensable when dealing with obligate parasites. In specific cases whole plants, plant parts, fruits or seeds are used (Zehr et al. 1978).

Plants

Most common is the use of seedlings. The fungicide is sprayed or dusted on foliage or administered to roots as in the case of systemic fungicides. After inoculation with the studied strain and a subsequent incubation, disease is assessed and compared with that caused by the wild-type pathogen. The test is used mainly in greenhouses, but sometimes also in small field plots. Examples are tests on fungicide resistance with *Phytophthora infestans* on potato (Schwinn 1982c), *Plasmopara viticola* on grapes (Schwinn 1982b), *Pyricularia oryzae* on rice (Uesugi 1982), *Venturia inaequalis* on apple (Gilpatrick 1982) and *Sphaerotheca fuliginea* on cucumber (Dekker 1982c). Application of fungicides in a nutrient solution to roots has been described for cucumber powdery mildew (Dekker 1982c), and soil application, for tests with benomyl against *Cercospora beticola,* on beets (Georgopoulos 1982b). In tests with heavily sporulating pathogens, it will often be necessary to keep the isolates separate, e.g. by covering the seedlings with plastic bags.

Detached leaves

The use of detached leaves saves space and plant material. Usually leaves are first sprayed with the fungicide, then detached, placed on moist filter paper in large petri dishes and inoculated with the pathogen. In specific experiments, inoculation may be made in an earlier phase. This method has been described for the determination of resistance to acylalanines in *Phytophthora infestans* (Schwinn 1982c).

Leaf Discs

Leaf discs, usually between 10 and 15 mm in diameter, are cut from vigorous leaves, and floated on a solution of the tested fungicide in a small petri dish. For tests with powdery mildews the discs have been inoculated by conidia dusted on the upper surface (Dekker 1982c), and for tests with downy mildews, drops of spore suspensions have been placed on discs floating upside down (Schwinn 1982b). This method can be used only with systemic fungicides. Instead of round discs also leaf segments of various shapes or sizes may be used.

Seeds

Seeds must be used to check the sensitivity of seedborne pathogens to fungicides in vivo. They are inoculated by dipping them in a spore suspension of the tested strains. The chemical, usually a dust or slurry, is applied to the surface of the seeds by rolling or shaking. To simplify the treating process, the liquid compounds are frequently adsorbed in talc or other diluents (Torgeson 1967). The seeds are then sown in soil or placed in a moist environment for germination and for evaluation of symptoms.

Seed treatment may also be used to test systemic fungicides supposed to protect seedlings from invasion of aerial pathogens. This has been described by Brent (1982) for testing ethirimol sensitivity of barley powdery mildew. Leaf segments of seedlings, grown from treated seeds, have been embedded in water agar at their one end and inoculated by dusting conidia on the leaf surface. The amount of visible mildew on each leaf piece has been recorded after 6 days of incubation at 19 °C.

Fruits

Determination in vivo of the fungicide sensitivity of pathogens, which primarily attack the fruit, requires treatment of fruit, before or after inoculation, with the pathogen. This has been done in tests with *Penicillium italicum* and *P. expansum,* causing citrus fruit rot (Eckert 1982) and with *Monilinia* species, causing brown rot of fruit and nut crops (Ogawa 1982).

3 Estimation of Resistance Risk

3.1 Emergence of Resistance on Agar Medium

The potential of fungi to mutate towards fungicide resistance can be studied in laboratory experiments. Fungal spores are plated on agar medium, containing a discriminating concentration of the fungicide. Recommended is a concentration which is slightly higher than the minimal inhibitory concentration, because it leads to a complete inhibition of normally sensitive cells. Only those cells that have become resistant will survive this treatment and form a colony. The frequency of mutation for resistance is calculated by counting the number of colonies. In view of the low mutation frequency (usually between 10^{-4} and 10^{-9}), the spores are often irradiated with mutagenic rays, or treated with a mutagenic chemical. In order to ensure a strong mutagenic effect on the surviving cells, it is recommended to use a dose of the mutagen, which kills the major part of the fungal cells (Georgopoulos 1977).

Experience shows that emergence of resistant strains in such in vitro tests occur with virtually all specific-site inhibitors. Due to the large number of fungal propagules produced in the field, the frequency of resistant mutants will rarely be the limiting factor; if resistant mutants can arise at all, they will certainly arise in the field. However, the emergence of resistant mutants in the field does not automatically imply that resistance problems will arise in the field immediately. This will not happen before the major part of the pathogen population has become resistant due to the selection pressure of the fungicide.

The described phenomenon is rare with conventional multisite inhibitors, such as dithiocarbamates or other metal toxicants. Mutations, affecting all sites of fungicide action are unlikely, and development of resistance due to decreased uptake or increased detoxification appears to be rare. When fungi exposed to these fungicides do not produce resistant mutants in agar tests combined with mutagenic treatment, no emergence of resistant mutants in the field has to be expected.

3.2 Fitness and Buildup of Resistance in the Field

Buildup of a resistant pathogen population in the field from occasional mutants and the rate of such buildup depends inter alia on the fitness of resistant strains in relation to fitness of the wild-type pathogen in the absence and presence of the fungicide. Convenient parameters for assessing fitness in vitro are spore germination, mycelium growth on nutrient media and sporulation on agar. Important parameters of fitness on the host plant are the frequency of infection, the rate of colonization of the host tissue and the degree of sporulation.

There is evidence that increased resistance, at least to some but not all fungicides, is linked to decreased fitness in absence of the acting fungicide. This phenomenon depends on the action of different fungicides, and the resistance mechanism of fungi (Dekker 1982a). On this basis, fungicides may roughly be classified as imposing high, moderate or low risk for development of resistance. For example, the benzimidazole fungicides and the acylalanines impose a high risk, the antibiotic kasugamycin, a moderate risk and some groups of sterol biosynthesis inhibitors impose a low risk. An inverse relation between resistance and fitness is also related to the level of resistance. Strains with a high level of resistance, but low fitness, will lose ground to strains with a lower level of resistance. The latter might still be controlled by normal or slightly increased doses of the fungicide.

The fitness of resistant and susceptible strains is compared in competition tests which may be carried out in the greenhouse. In such tests plants are inoculated with a mixture of resistant and susceptible spores and changes in the ratio of resistant to sensitive spores are recorded after several transfers to new batches of plants. When these experiments are carried out in absence of the fungicide they indicate the chance for survival of resistant mutants either during intervals between control treatments or when application of the fungicide is stopped.

Estimation of fitness of resistant strains helps to evaluate the risks of rise of the resistance problems in the field. Besides the parameter of fitness, which is inherent to the type of fungicide and the mechanism of resistance, the phenomenon is affected by other factors such as the type of disease, the life cycle of the pathogen, the selection pressure by the fungicide and environmental conditions (Dekker 1982a).

3.3 Use of Models

The question has been raised whether models might be helpful to estimate the risks of development of resistance problems in the field. Models have been presented by Kable and Jeffery (1980), Skylakakis (1982) and Levy et al. (1983). Although these models are not yet accurate for prediction of resistance in practice, they may help to increase our knowledge about the quantitative effects of various parameters on the buildup of resistance. Under certain conditions they may be useful for describing the population dynamics of fungicide resistant strains (Zadoks 1982).

4 Detection an Monitoring of Resistance in the Field

4.1 Sampling

Failure of disease control does not necessarily derive from the development of fungicide resistance in the pathogen, but may result from the influence of other factors. To prove that it is caused by the presence of resistant strains, it has to be demonstrated that they constitute a major proportion of the pathogen population, and that they induce disease in fungicide-treated plants. Such proof can be provided with a small number of samples collected from sites where severe infection occurs in spite of the fungicide treatment. If resistance appears to be the cause of control failure, application of the fungicide should of course be stopped. In order to monitor changes in the fungicide resistance it may then be desirable to check the proportion of the resistant population and level of its resistance at regular time intervals. It is still a matter of discussion whether the application of the fungicide can be renewed after resistance has dropped to a low level, but it is obvious that a renewed application is senseless when the proportion of resistant cells remains high.

It may also be useful to monitor fungicide sensitivity before failure of disease control occurs, especially when the fungicide applied is suspected or known to be risky. Early detection of the buildup of resistance will allow one to apply countermeasures to slow down the development of resistance and/or to avoid losses. However, for detection of resistance appearing at a very low frequency fairly large samples of diseased material are needed. Studying benomyl resistance of *Monilinia* in fruit crops, Ogawa et al. (1983) calculated that when 1% of the isolates are resistant, 458 and 298 samples respectively, are required to detect resistance at the 1 and 5% levels of certaintly. This illustrates that much labour is needed for the detection of a low frequency of resistant cells. However, when dealing with an explosive disease, such as potato late blight, the detection of only 1% of resistant strains may be too late for application of countermeasures. Therefore, the effort required for monitoring for fungicide sensitivity is usually not justified, until the first signs of reduced disease control are observed; exceptions are those cases, where buildup of resistance is expected to be gradual and slow, because of low selection pressure or use of a fungicide with moderate risk of resistance.

Less labour is required for extensive sampling, when many sites of the target disease in a region or country are examined, and one or a few bulk samples of the pathogen are collected at each site. This method is particularly useful when it is directed to follow gradual and general changes in fungicide sensitivity, instead of the detection of a small percentage of propagules with pronounced resistance. It has successfully been used in surveys of ethirimol and triadimefon resistance in barley powdery mildew (Brent 1982).

If, for certain reasons, no monitoring of fungicide resistance is carried out, it is recommended to follow the effectivity of the used materials. Early detection of reduced control caused by fungicide resistance will allow the farmer to shift to other fungicides or other control methods in his plots, and to warn other growers.

4.2 Handling and Processing of Samples

The sampled material should be as fresh as possible and free of old, senescent leaves. Care should be taken to avoid interference by residues of the fungicides, with which the plants were treated in the field. If such residues are suspected to be present, the samples should be washed, so that the spores are also removed. The samples should then be placed under conditions suitable for sporulation. Subculturing may be necessary to produce inoculum for the experiments. Repeated subculturing should be minimized, as changes in the fungal material may occur during that period. One way to avoid shifts in sensitivity during storage and processing is to expose potted and fungicide-treated test plants in the field, and then to transfer them to the glasshouse for incubation and assessment (see Rotem, Chap. 5). This method is particularly useful for obligate parasites, such as powdery mildews (Brent 1982). The labour required for surveying a large area may be reduced by mounting the potted test plants on top of a car, driven around the area to sample the population of spores (Wolfe and Minchin 1976).

More detailed information about the collection and processing of samples is provided in a number of standardized test methods published by FAO. These methods concern the detection of fungicide resistance in *Venturia inaequalis* (Gilpatrick 1982), *Monilinia* sp. in fruit and nut crops (Ogawa 1982), *Sphaerotheca fuliginea* in cucumber (Dekker 1982c), *Cercospora beticola* in beets (Georgopoulos 1982b), *Plasmopara viticola* in vines (Schwinn 1982b), *Phytophthora infestans* in potatoes (Schwinn 1982c) and *Pyricularia oryzae* in rice (Uesugi 1982).

4.3 Recording and Interpretation of Results

The parameters most often used for comparison of resistant and sensitive strains are the minimal inhibitory concentration (MIC) (i.e. the lowest concentration which provides complete inhibition), and the concentrations which inhibit 50% (ED_{50}) or 95% (ED_{95}) of the pathogen population. The ED_{50} ratio of the resistant to the wild-type strain is often used as measure for resistance. The steepness of the slope provides information about the spread in fungicide sensitivity.

When resistence builds up rapidly as happens with benzimidazol fungicides, interpretation of the results is easy. Then it is also easy to measure which part of the population has become resistant. Interpretation becomes difficult, when mutation for resistance proceeds in small steps, and the frequency distribution curves of sensitivity partly overlap in the mutant and wild-type cells. Changes in the ED_{50} values may sometime occur without mutation towards resistance. This may happen when the sensitivity distribution of the wild-type fungus is very wide, and the fungicide selects for the least sensitive cells of the fungal population. It is therefore important to describe accurately what is meant by resistance and to rate resistance and sensitivity in terms of quantitative data (Delp and Dekker 1985).

Care should be taken also in extrapolating results of greenhouse tests for prediction of the fungicide effect in the field. The dose needed for good control in the field will usually be substantially higher than in the greenhouse or in small experimental plots, where plants are sprayed very carefully.

5 Epiloque

The development of fungicide resistance has become a major problem in plant disease control. In order to devise tactics and strategies to cope with these problems, methods are necessary to detect and measure resistance, and to estimate the involved risks. Standart tests based on accepted principles are needed for improved communication between researchers, growers and manufacturers. It is important to use standard tests based on generally accepted principles.

6 References

Anonymous (1947) Test tube dilution for use with the slide germination method of evaluation protectant fungicides. Phytopathology 37: 354–356

Brancato FP, Golding NS (1953) The diameter of the mold colony as a reliable measure of growth. Mycologia 45: 848–864

Brent KJ (1982) Powdery mildews of barley and cucumber. In: Dekker J, Georgopoulos SG (eds) Fungicide Resistance in Plant Protection. Pudoc, Wageningen, pp 219–230

Dekker J (1982a) Can we estimate the fungicide resistance hazard in the field from laboratory and greenhouse tests? In: Dekker J, Georgopoulos SG (eds) Fungicide Resistance in Crop Protection. Pudoc, Wageninden, pp 128–138

Dekker J (1982b) Countermeasures for avoiding fungicide resistance In: Dekker J, Georgopoulos SG (eds) Fungicide Resistance in Crop Protection. Pudoc, Wageningen, pp 177–186

Dekker J (1982c) Method for fungicide resistance in powdery mildew of cucumber. Plant Prot Bull FAO 30: 59–61

Dekker J (1985) The development of resistance to fungicides. In: Hutson DH, Robers TR (eds) Progress in pesticide biochemistry and toxicology, vol 4. pp 165–218
Dekker J, Ark PA (1959) Protection of antibiotic pimaricin from oxidation and ultraviolet light by chlorophyllin and other compounds. Antibiot Chemother 9: 327–332
Delp CJ, Dekker J (1985) Fungicide resistance: definition and use of terms. EPPO Bull 15: 333–335
Eckert JW (1982) *Penicillium* decay of citrus friuts. In: Dekker J, Georgopoulos SG (eds) Fungicide Resistance in Crop Protection. Pudoc, Wageningen, pp 231–255
Georgopoulos SG (1977) Development of fungal resistance to fungicides. In: Siegel MR, Sisler HD (eds) Antifungal compounds, vol 2. pp 409–495
Georgopoulos SG (1982a) Detection and measurement of fungicide-resistance. In: Dekker J, Georgopoulos SG (eds) Fungicide-Resistance in Plant Protection. Pudoc, Wageningen, pp 24–31
Georgopoulos SG (1982b) Method for resistance to benzimidazoles in *Cercospora beticola*. Plant Prot Bull FAO 30: 63–65
Georgopoulos SG, Dekker J (1982) Detection and measurement of fungicide-resistance, general principles. Plant Prot Bull FAO 30: 39–42
Georgopoulos SG, Dovas C (1973) A serious outbreak of strain of *Cercospora beticola* resistant to benzimidazole fungicides in Northern Greece. Plant Dis Rep 57: 321–324
Georgopoulos SG, Alexandri E, Crysayi M (1972) Genetic evidence for the action of oxathiin and triazole derivates on the succinic dehydrogenase system of *Ustilago maydis* mitochondria. J Bacteriol 110: 809–817
Gilpatrick JD (1982) Method for fungicide resistance in *Venturia* spp. of apple and pear. Plant Prot Bull FAO 30: 51–54
Jones AL, Walker RJ (1976) Tolerance of *Venturia inaequalis* to dodine and benzimidazole fungicides in Michigan. Plant Dis Rep 60: 40–44
Kable PF, Jeffery H (1980) Selection for tolerance in organisms exposed to sprays of biocide mixtures: a theoretical model. Phytopathology 70: 8–12
Levy Y, Levi R, Cohen Y (1983) Buildup of a pathogen subpopulation resistant to a systemic fungicide under various control strategies: a flexible simulation model. Phytopathology 73: 1475–1480
Ogawa JM (1982) Method for resistance to benzimidazoles in *Monilinia* spp. on fruit and nut crops. Plant Prot Bull FAO 30: 55–57
Ogawa JM, Manji BT, Heaton CR, Petrie J, Sonoda RM (1983) Methods for detecting and monitoring the resistance of plant pathogens to chemicals. In: Georghiou GP, Saito T (eds) Pest Resistance to Pesticides. Plenum, New York p 117–162
Schwinn FJ (1982a) Socio-economic impact of fungicide resistance. In: Dekker J, Georgopoulos SG (eds) Fungicide-Resistance in Crop Protection. Pudoc, Wageningen, pp 16–23
Schwinn FJ (1982b) Method for fungicide resistance in downy mildew of grapes. Plant Prot Bull FAO 30: 67-68
Schwinn FJ (1982c) Method for fungicide resistance in late blight of potato. Plant Prot Bull FAO 30: 69–70
Skylakakis G (1982) The development and use of models describing outbreaks of fungicide resistance. Crop Prot 1: 249–262
Sonoda RM, Ogawa JM, Lyons T, Hanson JA (1970) Correlation between immobilization of zoospores by fungicides and the control of *Phytophthora* root and crown rot of transplanted tomatoes. Phytopathology 60: 783–787
Staub T, Sozzi D (1984) Fungicide resistance: a continuing challenge. Plant Dis 68: 1026–1031
Tillman RW, Sisler HD (1973) Effect of chloroneb on the growth and metabolism of *Ustilago maydis*. Phytopathology 63: 219–225
Torgeson DC (1967) Determination and measurement of fungitoxicity. In: Fungicides, an Advanced Treatise, vol 1. Academic Press, London, pp 93–123
Uesugi Y (1982) Method for fungicide resistance in *Pyricularia oryzae*. Plant Prot Bull FAO 30: 141–143
Wolfe MS, Minchin PN (1976) Quantitative assessment of variation in field populations of *Erysiphe graminis* f. sp. *hordei* using mobile nurseries. Trans Br Mycol Soc 66: 332–334
Zadoks JC (1982) Can we use models describing the population dynamics of fungicide-resistant

strains? In: Dekker J, Georgopoulos SG (eds) Fungicide-Resistance in Crop Protection. Pudoc, Wageningen, pp 149–160
Zehr I, Bird GW, Fisher KD, Kickey KD, Lewis FH, Line RF, Rickard S (eds) (1978) Methods for evaluating plant fungicides, nematocides and bacteriocides. American Phytopathological Society, St. Paul, USA

Chapter 11

Analysis of Virulence in Pathogen Populations

Günter Welz[1]

1 Introduction

Virulence analyses are performed to measure the frequency and/or distribution of specific virulence genes or factors in populations of plant pathogens. However, also insects or nematodes that interact with their hosts in gene-for-gene systems may be subject to virulence analyses. Virulence will be considered here as a qualitative or binary trait related to race-specific resistance genes or factors, while the term aggressivenes will be related to the quantitative ability of a pathogen isolate to cause disease (Van der Plank 1975). Virulence and aggressiveness are the main components of pathogen fitness. Since the latter term relates to survival and dominance over time it may also include other components, particularly when periods longer than a single epidemic are considered. For example, the ability to overwinter or to remain infectious during dissemination may be important components of fitness. Fitness is defined in relative and absolute terms. Relative fitness is usually related to the phenotype with the highest reproduction in the sample and calculated from frequency changes. Absolute fitness is usually expressed as the absolute number of progeny per individual. It can be estimated by measuring the components of aggressiveness (latent period, disease efficiency, colony growth, etc.). If the word fitness is used in this text without an adjective, relative fitness is meant.

Every race analysis is also an analysis of virulence. The opposite is not always true. This is because virulence may be analyzed with just a single resistant host line, whereas a race can only be determined on a set of differential host lines. Therefore, the term "virulence analysis" has a less strict meaning.

Virulence analyses may be performed for more practical or purely scientific reasons. Rather practical are the objectives of nationwide virulence surveys. They are meant to provide an early warning against new or dangerous virulence phenotypes (Roelfs 1984; Wolfe 1984) and to characterize the virulence or race distribution of a population. This information may be of immediate use to breeders and farmers. A more scientific aspect of virulence analyses is the evolution of pathogens (Luig and Watson 1970; Wolfe and Schwarzbach 1978). In epidemiology virulence analyses are used to study changes in virulence patterns of pathogen populations in time and space.

This chapter is to give a critical overwiew of relevant methods and procedures for use in epidemiology. It is mainly based on the work with cereal mildews or rusts.

[1] Phytopathologie und Angewandte Entomologie, Tropeninstitut der Justus-Liebig-Universität, Schottstr. 2, 6300 Gießen, FRG

2 Methods of Virulence Analysis

2.1 Sampling the Population

Techniques to sample truly at random from a pathogen population and to handle large sample sizes are still a major problem of many virulence surveys (Roelfs 1984). For sampling see also Dekker (Chap. 10) and, in general terms, Kranz (Chap. 3).

2.1.1 Sampling Techniques

Samples should be representative of a population. Since little is known about the spatial distribution of races within fields, some form of randomized sampling is recommended. In diseases where epidemics arise from foci and not from a rather uniformly distributed starter population or when the population is from a heterogeneous environment, the sampling area should be stratified and subsamples be drawn at random.

Sampling Populations from Individual Fields

With respect to the inoculation of the differential set, virulence tests can be direct or indirect. This is important for sampling. The former tests imply the direct exposure of the differential cultivars to natural infection, e.g. as "mobile nurseries" (see below). For indirect virulence tests the fungus is isolated and/or multiplied under controlled conditions prior to the inoculation of the differential set. Sampling units for indirect tests may be collected, diseased leaves or colonies from spores caught on susceptible seedlings.

The simplest technique to collect isolates is to cut leaves from diseased plants and transfer spores to fresh plant material in the laboratory (Wolfe et al. 1981; Luig 1983; Roelfs et al. 1984). It can be particularly helpful for detecting new races among plants of resistant cultivars and to give clues to the identification of resistance genes in new cultivars (Wolfe et al. 1984). For the purpose of randomness not more than a single colony should be isolated from one collected leaf. For large samples from single fields other methods seem more adequate: susceptible seedlings can be exposed to infection in the field, either static (Wolfe and Minchin 1976) or by dragging plants in small wire cages through the canopy (Wolfe et al. 1983). When disease levels are low, the potted seedlings can be left in the field for a number of days. However, this will increase the risk to catch exogenous inoculum imported from outside the field. Not more than five colonies should be isolated from the ca. 15 seedlings of one pot.

Jet spore traps (Schwarzbach 1979) are also convenient to collect airborne spores on living leaves. They may contain whole seedlings in small pots or leaf segments placed in petri dishes on water agar containing ca. 50 ppm benzimidazole. In large fields the jet trap can be fixed to some kind of support and carried through the canopy, e.g. diagonally, for disease readings. The running time of the trap has to be adjusted to the spore content of the air which may vary by a factor of 1.000 during an epidemic of *Erysiphe graminis* f. sp. *hordei* (Limpert et al. 1984). At the peak of an epidemic a few minutes may yield enough infections, while adverse conditions may require hours of operation. An electric timer can support the automatic operation of a jet spore trap. The seedlings should preferably represent more than one host genotype, as unknown resistance genes may be present in a susceptible cultivar. Direct and thus rapid estimates of virulence frequencies in mildew populations are provided by mobile nurseries of differential varieties (Eyal et al. 1973). However, in the laboratory density and uniformity of inoculations will be better controlled and produce more reliable estimates of virulence frequencies.

Sampling of the "Overall Population"

For random samples to be taken from large areas or even whole countries, sophisticated techniques have been developed. The United States race survey of *Puccinia graminis* f. sp. *tritici* is made over a ca. 20,000 km route covering several ecological regions of the pathogen (Roelfs et al. 1984). Stops are made at a commercial field each 32 km or at the first field thereafter. In addition, samples from disease nurseries are tested. Kiyosawa (1980) recommended stratified sampling of large populations. Subsamples should be weighted according to the area covered by each resistance genotype and according to their disease severity.

Somewhat easier and more representative are spore catches from the aerial spore pool (Wolfe and Knott 1982). This was proved with a static jet spore trap and trap plants placed on top of high buildings. The spore populations which were caught in the air over an agricultural area represented virulence frequencies that were highly correlated with those in samples collected with mobile spore traps on the ground (Wolfe et al. 1984; Limpert unpublished). With the latter mobile spore traps even populations in large areas can be sampled. Two types have been used in surveys of cereal powdery mildew in Europe: the "wind impaction spore trap" (WIST: Wolfe et al. 1981) and a special jet spore trap (Limpert and Schwarzbach 1981). Either of the traps was mounted on the roof of a car and moved along transects or circular routes through growing areas of barley and wheat. The resulting virulence frequencies were highly reproducible. The operation time of mobile spore traps should also relate to the spore content of the air (cf. Limpert et al. 1984). The car can be driven at speeds greater than 80 kmh^{-1}.

Plot Experiments

Several sources of error can be relevant if virulence changes within individual fields are studied. During the first generations of pathogen reproduction there is still an import of exogenous inoculum. Therefore, selection and migration may both contribute to qualitative change. This interplot interference (see Aust and Kranz, Chap. 1) is minimized by large plot sizes and avoidance of neighbouring sources of inoculum. Smaller plots can also reveal changes during the esodemic but are only recommended if they are widely spaced and isolated by border rows. Large plot sizes are particularly required when the pathogen's population size is small (resistant varieties). Wherever possible the plot size should equal that of local commercial fields. Remote spore catches from the local air spora are helpful as a "check sample".

2.1.2 Sample Size

The sample size required will mainly depend on the purpose of the study. If rare (new) virulence factors or small frequency changes are to be documented, large samples will be needed.

Since virulence is treated as a binary trait, its distribution in the population should be binomial. Therefore $\binom{n}{r} p^r q^{n-r}$ gives the probability of r successes in a sample of size n

and can be applied to sampling for a virulence with a given frequency p with q being $1-p$. More informative is the zero term of the binomial distribution which indicates the chance to miss a certain virulence in a sample:

$$P(0) = q^n. \tag{11.1}$$

The complement $1-P(0)$ indicates the chance to find a certain virulence at least once in a sample. For infrequent virulence genes in large samples ($p<0.1$ and $np<5$) the zero term of the Poisson formula would be more appropriate (Sokal and Rohlf 1981):

$$P(0) = e^{-np}. \tag{11.2}$$

The probability, for example, to miss a virulence of $p=10^{-4}$ in a sample of $n=500$ isolates would be 0.951. Another general problem is to determine the sample size that is required to test the significance of the difference between two virulence or race frequencies. Sokal and Rohlf (1981, p. 766) give an equation to determine n. An example presented in their book underlines the need for large sample sizes. If one wishes to be $(1-\beta)=0.80$ certain of detecting a true difference between two samples $p_1=0.65$ and $p_2=0.55$ at the $\alpha=0.05$ level of significance, $n=396$ observations would be needed in each sample. For further information, see Snedecor and Cochran (1967) and Sokal and Rohlf (1981).

2.1.3 Frequency and Timing of Sampling

In epidemiology it is essential to monitor temporal and spatial changes in virulence patterns *alongside* the disease progress. Hence, a number of dates for sampling and disease readings during the growing season are required to quantify selective changes (see Chap. 3). They should, wherever possible, relate to defined growth stages and start with at least one date during the exodemic (Robinson 1976). As the selection process may be more pronounced during the esodemic, i.e. when the canopy is closed, at least three to four dates during this period have to be considered for sampling. The disease readings should also include an estimate of the number of colonies per unit area of field so that absolute frequencies for certain virulences or races can be derived. In case of remote sampling of the "overall population" with volumetric spore traps, like the jet trap, absolute numbers of infectious virulent spores per volume unit of air can be calculated and used as an important parameter in forecasting.

2.2 Testing the Sample

Although direct testing with mobile nurseries or jet spore traps may be sufficient in some cases, there is more flexibility in indirect virulence tests. In most studies single spore or single colony isolates were employed for the virulence test. Alternatively, it has been suggested to inoculate bulk isolates and to count the number of colonies on differential and control cultivars. Both methods shall be discussed in some detail.

2.2.1 Design of Virulence Tests

Storage of Isolates

Cereal rust spores can easily be stored in liquid nitrogen (−45° to −50 °C) for long periods. Fresh uredospores are produced on susceptible hosts treated with maleic hydrazide to enhance spore production (see Bushnell and Roelfs 1984). Powdery mildews have to be kept on living plants. Inoculated seedlings that are grown in large test tubes in vermiculite (Bronson 1981) or 0.5% water agar amended with a low concentration of liquid fertilizer can be stored at temperatures just above 0 °C for more than two months.

Plant Material

Resistant reactions are expressed on leaf segments or whole plants. However, general experience has shown that the distinction of phenotypes is clearer with whole plants. On the other hand, detached leaf cultures save space and are easily protected from contamination during incubation. The plants must be grown under fixed environmental conditions and to a defined age to reduce variability. It is recommended to test for the purity of seed batches with reference isolates. A good reference isolate would be one that is virulent only for the resistance of the differential line in question.

When whole plants are used, three seeds of each differential can be sown as a circle with a susceptible control in the center (Bronson 1981). The pots are covered with a chimney. Detached leaves are cut into 2–3-cm-long segments and placed on 0.5% water agar with 15 ppm benzimidazole (Limpert and Schwarzbach 1981). Two segments per tester are arranged in polystyrene petri dishes or rectangular boxes.

Inoculation

When single colony isolates of rust are being employed, the spores of single uredia are collected with small cyclone spore collectors. The spores are suspended in light-weight mineral oil and sprayed onto the plants. Mildew spores are usually inoculated dry through settling towers (200–300 conidia cm^{-2}). Single colony isolates can be established with small, rubber-tipped pencils. However, to save this step of colony transfer, the conidia of a discrete colony can be sucked into a Pasteur pipette and blown directly into a small settling tower (Limpert and Schwarzbach 1981).

Incubation

The incubation period depends on temperature. It usually ranges between 7 and 14 days. Since temperature may have a strong influence on the infection type produced, constant environmental conditions are extremely important for reproducibility of results. Only "a definite parasite genotype interacting with a definite host genotype in a definite environment results in a definite phenotype" (Browder 1985).

Differential Host Varieties

In order to study virulence in relation to specific-host genes, a set of differentials in which each member possesses a single, unique resistance gene is most efficient and informative (Roelfs 1984). This rationale is applied to the selection of host lines used for virulence surveys of *P. graminis* f. sp *tritici* (Luig 1983). Isolines with a uniform genetic background can be particularly useful for genetic, physiological or epidemiological studies of virulence (Moseman 1972; Kølster et al. 1986).

Other workers have emphasized the use of cultivars that are of some significance in practical plant production (Scott et al. 1980). They argued that more information relevant to commercial crops would be gained from such differentials. However, these workers were not only interested in the presence or absence of certain virulence genes but also in the quantitative degree of adaptation to those cultivars.

Table 11.1 Infection types (IT) of *Erysiphe graminis* (After Torp et al. 1978)

IT[a]	Mycelium growth	Sporulation	Development of chlorosis/necrosis
0	None	None	–
1	Weak	None	+
2	Moderate	Weak	+
3	Strong	Moderate	+
4	Strong	Strong	–

[a] In connection with IT 1, 2, 3 the following letters denote:
n = small necrotic spots; N = large necrotic spots;
c = small chlorotic spots; C = large chlorotic spots.
Intermediate IT classifications are possible (0–1, 1–2, etc.)

2.2.2 Reading Infection Types or Counting Colonies

Infection-Type Readings

Gene-for-gene interactions are generally characterized by the infection type that is produced by an individual isolate together with a host plant. The typical infection type conferred by a certain avirulence gene is generally listed in descriptions of sources of resistance genes or is "stored" in type-culture collections (Roelfs 1984).

In powdery mildew five basic infection types are distinguished (Table 11.1). If infection types are to be specified very precisely, e.g. in descriptions of host resistance genes, extra letters can be added to the infection type number (e.g. 2 Cn). Infection type 0/4 was not considered in Table 11.1. It corresponds to a more or less "immune" reaction with a few normally developing colonies on barley hosts with the *ml-o* gene. Usually infection types 3 and 4 are read as a susceptible reaction (Wolfe 1972).

As for wheat stem rust there is an extended range of classifications. An infection-type description given by Luig (1983) regarding necrotic, green island, chlorotic or heterogeneous expressions of infection types 0 through 4 lists as many as 41 different classifications.

Colony Counts

A somewhat different approach was suggested by Wolfe and Schwarzbach (1975) for cereal powdery mildews. These authors argued that more accurate estimates of virulence frequencies would be obtained from analyses of bulk isolates. Their method implies the counting of "successful" colonies on the differential cultivars and for comparison on susceptible controls. The relative number of colonies produced per leaf area on a cultivar with resistance gene A is taken as a direct estimate of the frequency of virulence against A (Wolfe and Schwarzbach 1975). This estimate can be biased: any quantitative effect of the resistance gene itself or of the cultivar's background which affects the number of colonies produced will contribute to the error of the estimate. For example, unrecognized resistance genes in the control cultivars may lead to virulence frequencies that exceed 100% (see Schwarzbach 1978). While usual infection-type readings regard virulence only, the method of colony counts implies two components of fitness: virulence (in the narrow, "major" gene sense) and disease efficiency, which is an attribute of aggressiveness. Therefore, colony counts do not measure virulence alone, but rather the combined effects of virulence and aggressiveness. To eliminate the non-systematic influence of different host background genes, the use of a test set composed of near-isogenic lines is recommended (Munk, in preparation). Most accurate estimates of virulence frequencies can only be obtained with single colony isolates.

Race frequencies cannot be determined with bulk isolates. Sequential tests of bulk isolates on several test varieties (Wolfe and Schwarzbach 1975) may provide an estimate of certain, simple combinations of virulence genes, but for complex combinations the testing of individual isolates is indicated (Scott et al. 1980). Of course, it is also not possible to detect variation within a bulk sample. On the other hand, indirect tests of bulk isolates or direct analyses with mobile nurseries have proved very useful. They provide fast information on the virulence spectrum of a population which may be sufficient for many purposes. Direct or indirect analyses of bulk isolates seem advantageous in estimating frequencies of very rare virulence genes, simply because a larger sample of the population will be tested, thus outweighing the errors or disadvantages mentioned above.

Limpert and Schwarzbach (1981) tested single colonies but only considered interactions of what they termed "high virulence". These are by definition interactions conferring infection type 4 and at least half as many colonies as on the susceptible check. It may be argued that high virulence is a parameter confounding virulence and aggressiveness as well. However, it is a useful approach to exclude the quantitative effect of some resistance genes with few but necrosis-free colonies which were otherwise treated as matching interactions. On the other hand, a new mutant virulence having low parasitic fitness might go unmentioned in such an analysis.

All the various methods to estimate virulence frequencies still contain some pragmatic elements. An improved input of epidemiology into host-parasite genetics will be discussed in Section 3.

2.2.3 Measuring Aggressiveness

Aggressiveness is an important component of fitness. However, aggressiveness itself is made up of several components. The appropriate methods for analyses of aggressiveness may be more analytic or more synthetic (see Hau, Chap. 19). The latter could be simply restricted to the measurement of a single defined parameter like the "cumulative spore production per inoculated spore 10 days after inoculation". The former would require separate measurements of each component. Techniques for studying aggressiveness in the pathogen are also used to study quantitative resistance in the host.

The *latent period* (time from inoculation to begin of sporulation) should not be measured with a single observation. It can be expressed by the time (in hours) between inoculation and that time when 50% of the sporulating colonies have appeared on a marked leaf area. The *disease efficiency* (number of colonies per inoculated spore per unit leaf area) is easily determined in combination with the latent period: trap the inoculated spores on one or more marked slides and count the number of spores on 1 cm^2 under a dissecting microscope. The *colony growth* (colony size) can be measured (in mm^2) with a vernier caliper, e.g. 10 days after inoculation. The *sporulation* is expressed as a *rate* (spores per day) or as a *capacity* (cumulative number of spores per infectious period). Values may be given per colony or per unit area of a colony. Techniques were described by Rotem (Chap. 2). The *infectious period* (period of sporulation) can also be determined in combination with the latent period: calculate the time from the end of latency until that time when 50% of the colonies have ceased to form new spores.

Several factors will influence the aggressiveness of an isolate: host cultivar, leaf insertion and leaf age of the host plant, or abiotic factors (temperature, light). For some components a pathogen population may show more variation than for others. In powdery mildew of barley, latent period and disease efficiency vary little among compatible races, whereas spore production and colony growth appear to be good discriminatory variables (Stähle 1986a). In leaf rust of barley, however, the latent period may well be race-specific (Parlevliet 1976).

Of crucial importance is to have standardized conditions throughout a study of aggressiveness. The inoculum density must be uniform and defined. This can be attained by spraying spore suspensions with fluorochemical carriers (Bushnell and Rowell 1967) using a quantitative inoculator (Knudsen 1984).

2.3 Analyzing the Data

2.3.1 Virulence Frequencies and Associations

The knowledge of individual virulence gene frequencies is regarded as the most important information yielded by virulence or race surveys (Van der Plank 1975; Wolfe and Schwarzbach 1975). As explained above, these frequencies can be calculated from colony counts or more precisely from infection-type readings. For many purposes a simple descriptive presentation of the results will suffice. However, more detailed analyses of frequency changes require some statistical tests.

When comparisons of different samples from different locations or sampling dates are made, it is a common problem to test the significance of deviating frequencies. The G-test of independence is recommended as the appropriate method to solve this problem, since it is easy to compute and/or reduces the observed type I error close to the intended one (Sokal and Rohlf 1981, p. 765). In Section 2.1.2 it was emphasized that large sample sizes may be needed to detect small differences.

A special topic is the analysis of virulence associations (Van der Plank 1975; Wolfe and Schwarzbach 1975; Browder and Eversmeyer 1977; Alexander et al. 1984). That is "whether individuals in the population are more likely to have particular combinations of virulence or avirulence to a given pair of cultivars than expected by chance" (Alexander et al. 1984). Virulence associations which depart from their expected value are most interesting to plant breeders and pathologists. They provide clues for the choice of resistance genes which could be combined in cultivars or cultivar mixtures to minimize pathogen reproduction. However, it must be outruled that the non-random association of virulence genes was not simply due to sampling errors, to a lack of recombination in the pathogen or to selection of certain associations (Wolfe and Knott 1982). When the association of two virulence factors is to be tested at a time, two-way contingency tables are to be calculated. Of course, one can also use multiway tables for analyzing reactions on three or more test varieties simultaneously. This, however, requires more computational work and time.

With a differential set of n lines, an analysis of an association regarding k differentials at a time,

$$\frac{n!}{k!\,(n-k)!}$$

comparisons can be done. The value of this term is maximal for $k = \frac{n}{2}$.

The appropriate statistical test for comparison of observed and expected frequencies (again) is the G-test of independence (Alexander et al. 1984). It is considered to be more reliable than the traditional chi-square test used by Wolfe und Schwarzbach (1975) and Royer et al. (1984).

Browder and Eversmeyer (1977) have proposed a somewhat different approach to study "associations of pathogenicity", a "pathogenicity association coefficient" (PAC) and a "virulence association coefficient" (VAC). However, the names of the two coefficients do not fit their definition very luckily. While a high VAC means a high proportion of isolates being virulent on both resistance genes, a high PAC means a high proportion of the sum of isolates being either virulent or avirulent on both resistance genes. Thus, VAC and PAC have quite a different meaning, while virulence and pathogenicity can be regarded as rather similar attributes of a pathogen.

2.3.2 Race Designation

Although races represent the natural entities (spores with one or more virulences) which interact with a host plant, the concept of races and race designations has been disapproved by a number of authors (e.g. Browder et al. 1980; Vanderplank 1983). This objection is partly based on the large number of races to be expected (2^n) when the number n of identified virulences increases. Also, the abundance of designation systems (see Roelfs 1984) and the introduction of extra differentials to previous "standard differential sets" has led to this criticism. On the other hand, the term race has proved very useful in pathosystems with stable populations of virulence phenotypes, e.g. the North American wheat stem rust population (Jeger and Groth 1985).

One of the disadvantages of most earlier coding systems has been that one cannot infer an isolate's virulence combination from its "race name".

More recently, a mathematical approach to the coding of races has led to short, unequivocal race names (i.e. numbers). These can be automatically computed from dual numbers which represent virulent and avirulent reactions of an isolate on a closed set of differential varieties. Two coding systems of this kind been proposed. One designates decimal (Habgood 1970), the other, octal code numbers (Gilmour 1973) to specific virulence combinations. The former system has been applied to *P. striiformis* (Johnson et al. 1972) and *E. graminis* f. sp. *hordei* (Welz and Kranz 1987), the latter to a number of diseases (Engel 1981; Jones and Clifford 1984).

The statistical analysis of race frequencies is very similar to that of virulence frequencies.

2.3.3 Measuring Virulence and Race Dynamics

The change of frequencies over time is a typical problem for regression analysis. Based on the assumption of logistic increase or decrease of virulence frequencies, Leonard (1969) proposed a discrete selection model that was designed to calculate the relative fitness of virulent vs avirulent subpopulations of oat stem rust. The same model has also been used in powdery mildew investigations (Meah et al. 1982; Welz, unpublished). Bronson (1981) extended the model to an indefinite number of subpopulations. The models assume that generations are discrete and that selection coefficients are constant.

If q is the frequency of a less fit strain/virulence and p is the frequency of the reference strain/virulence with a fitness of $W_p = 1.0$, then after n generations:

$$\frac{p_n}{q_n} = W_q^n \frac{p_0}{q_0}. \quad (11.3)$$

Given a population of k strains or virulence factors, $k-1$ selection coefficients can be estimated. If the data are plotted as $\ln(q/p)$ vs generation, a straight line is yielded with a slope of $\ln W$. A one-tailed t-test can be used to test whether this slope is significantly smaller than zero (Leonard 1969). Genetic change in pathogen populations can also be expressed by continuous models. Groth and Barrett (1980) showed that when q_t is the frequency of a less fit phenotype after t days, the relative fitness of that phenotype per day can be calculated from the equation:

$$\frac{p_t}{q_t} = W_q^t \frac{p_0}{q_0}. \quad (11.4)$$

Since generation times are difficult to determine in field experiments, continuous selection models may become more important in epidemiological modelling. Comparing the two classes of models it is evident that relative fitness, be it expressed per generation or per day, can be solely calculated from the change of frequencies over time. Kranz (1987) therefore suggested to quote additionally to fitness the size $y_t f_{it}$ of a subpopulation having the relative frequency p_i at time t when disease severity is y_t.

The linear regression analysis that is performed with the log-transformed frequency data will usually employ the method of least squares. For some data sets, however, it may be preferable to base the estimation of the regression on the maximum likelihood method (Mood et al. 1974, p. 276). It will be especially more accurate when sample sizes vary strongly and when the marginal values 0.0 and 1.0 are to be considered among the frequencies.

A problem that should also be addressed in further studies of virulence shifts is the choice of the regression model (MacKenzie, cited in Royer et al. 1984). Is it justified to assume constant selection coefficients and hence apply a linear regression model? At least there is some evidence for density-dependent selection in powdery mildews (Rouse et al. 1984; Stähle 1986b) and for temperature dependence in cereal rusts (Browder 1985) and mildews (Eckhardt 1987). Thus, if selection in pathogen populations is "soft" and not "hard" (Leonard and Czochor 1980), a non-linear regression model might be more suitable (see Madden and Ellis, Chap. 13).

2.3.4 Measuring Pathogen Diversity

Investigations of stabilizing selection have often focussed their interest on the number of virulence genes per isolate (Luig and Watson 1970; MacKey 1981). Welz and Kranz (1987) called this parameter the virulence complexity of an isolate. Under the assumption of independence of virulence loci it should be binomially distributed among the population. However, this expectation is barely met with pathogens that reproduce primarily asexually (Roelfs and Groth 1980; Welz und Kranz 1987). The departure of observed from expected frequencies can easily be tested by the G-test or the chi-square method. It must be kept in mind that both directional selection and the degree of sexuality within a population will have a great impact on the distribution of complexity. Either of the two forces can only be discerned when the other one is kept equal among the samples.

When genetic variation is to be compared among samples from differently selecting host populations, the frequency distribution of the number of gene differences will be a better measure (Roelfs and Groth 1980). It can be expressed by "the number of gene differences in each of all possible paired combinations of genotypes, weighted according to the product of the frequencies of the two genotypes being compared. The expected distribution based on random gene association is a Poisson distribution" (Roelfs and Groth 1980).

A simplified approach to analyzing virulence difference was suggested by Welz and Kranz (1987). They compared the virulence factors of each isolate with those of the most frequent race in the sample. The differences from the most frequent race are summed over the number of possible gene differences and plotted against their frequency. The expected values of this distribution should as well follow a Poisson distribution if gene association is random.

Indices of diversity are a ready measure to describe genetic or other variations of a population. They are frequently used in ecology but equally applicable to plant pathology. Although there are quite a number of different indices (Lebeda 1982), it is sufficient to consider only a few, since many of them are intercorrelated. A very simple and yet informative measure is the *dominance index* of a population (May 1976) which is the proportion of the most frequent phenotype.

The *percentage of distinct phenotypes* (Lewontin 1974) is another simple parameter. If, for instance, 30 different races were detected among 200 isolates, the "variation" was 15%. *Simpson's measure of diversity* (Pielou 1969) expresses both the number of phenotypes and the evenness of the distribution. It is defined as:

$$D = 1 - \sum_i \frac{n_i (n_i - 1)}{n (n - 1)}, \tag{11.5}$$

in which n_i = the number collected of the i-th phenotype and
n = the sample size.

The *Index of Shannon* (May 1976) is frequently used because it includes richness as well as evenness and is considered to be less sensitive to changes in sample size (Groth and Roelfs 1982). This index is calculated as:

$$H' = -\sum_i p_i \ln p_i, \tag{11.6}$$

in which p_i is the frequency of the i-th phenotype.

Groth and Roelfs (1982) pointed out that intra- or interspecific comparisons should not be attempted between pathogen populations unless similar differential host sets are used. To aid comparisons in cases where the host sets are rather dissimilar, a uniform number of differentials should be selected for which virulence occurs at intermediate frequencies (Groth and Roelfs 1982).

Each type of virulence analysis contains some pragmatic elements which limit the usefulness of results. In the conventional approach virulence is treated as a binary character although there is increasing evidence that virulence and avirulence are rather the extreme points of a continuum in the pathogen related to a continuum between immunity and susceptibility in the host (Wolfe 1972; Browning 1980). Clearly, this simplification arose from a genetic point of view defining the aegricorpus (Loegering 1966) as an interaction of single "major" genes. It ignores the quantitative level of adaptation which is most interesting to epidemiologists. Sometimes "old" races aquiring higher fitness can be more dangerous than "new" races of a pathogen (Nelson

1979). With increased exploitation of partial resistance more information on fitness components will also be welcomed by plant breeders. Therefore, some measurement of aggressiveness should be added to virulence analyses. Further studies on the variation of fitness components at the population level are needed. There is also a lack of understanding in the relation of aggressiveness and fitness. High aggressiveness is often taken as equivalent to high fitness. This may be true for a single epidemic or a controlled selection experiment, but at least in natural host-pathogen systems maximum aggressiveness must not always be correlated with best survival (Browning 1980).

There are claims that the concept of races was largely unnecessary in modern plant pathology so that interest should be centered on virulence gene frequencies (Wolfe and Schwarzbach 1975; Browder and Eversmeyer 1977; Vanderplank 1983). Referring to this point of view, it must not be forgotten that the theory of gene frequencies was largely based on the biology of species with sexual reproduction. In pathogen populations which reproduce predominately, asexually the genotype (e.g. the individual conidium, spore, bacterial cell) rather than the gene is the unit of selection.

9 References

Alexander HM, Roelfs AP, Groth JV (1984) Pathogenicity associations in *Puccinia graminis* f. sp. *tritici* in the United States. Phytopathology 74: 1161–1166

Bronson CR (1981) The influence of unnecessary virulence genes on the reproductive fitness of *Erysiphe graminis* f. sp. *tritici*. Thesis, Michigan State University

Browder LE (1985) Parasite: host: environment specificity in the cereal rusts. Annu Rev Phytopathol 23: 201–222

Browder LE, Eversmeyer MG (1977) Pathogenicity associations in *Puccinia recondita tritici*. Phytopathology 67: 766–771

Browder LE, Lyon FL, Eversmeyer MG (1980) Races, pathogenicity phenotypes, and type cultures of plant pathogens. Phytopathology 70: 581–583

Browning JA (1980) Genetic protective mechanisms of plant-pathogen populations: their coevolution and use in breeding for resistance. Tex Agric Exp St Misc Publ 1451: 52–75

Bushnell WR, Roelfs AP (eds) (1984) The cereal rusts, vol 1. Academic Press, London

Bushnell WR, Rowell JB (1967) Fluorochemical liquid as a carrier for spores of *Erysiphe graminis* and *Puccinia graminis*. Plant Dis Rep 51: 447–448

Eckhardt H (1987) Pathotypendynamik von *Erysiphe graminis* DC. f. sp. *hordei* Marchal in Abhängigkeit von biotischen und abiotischen Faktoren. Z Pflanzenkr Pflanzenschutz 94: 169–177

Engel KH (1981) Pathotypen-Hierarchien und ihre Anwendung. Arch Phytopathol Pflanzenschutz 17: 139–145

Eyal Z, Yurman R, Moseman JG, Wahl I (1973) Use of mobile nurseries in pathogenicity studies of *Erysiphe graminis* f. sp. *hordei* on *Hordeum spontaneum*. Phytopathology 63: 1330–1334

Gilmour J (1973) Octal notation for designating physiologic races of plant pathogens. Nature 242: 620

Groth JV, Barrett JA (1980) Estimating parasitic fitness: a reply. Phytopathology 70: 840–842

Groth JV, Roelfs AP (1982) The effect of sexual and asexual reproduction on race abundance in cereal rust fungus populations. Phytopathology 72: 1503–1507

Habgood RM (1970) Designation of physiological races of plant pathogens. Nature 227: 1268–1269

Jeger MJ, Groth JV (1985) Resistance and pathogenicity: epidemiological and ecological mechanisms. In: Fraser RSS (ed) Mechanisms of resistance to plant disease. Martinus Nijhoff, Dordrecht

Johnson R, Stubbs RW, Fuchs E, Chamberlain NH (1972) Nomenclature for physiologic races of *Puccinia striiformis* infecting wheat. Trans Br Mycol Soc 58: 475–480
Jones ERL, Clifford BC (1984) Rhynchosporium of barley. UK Cereal Pathogen Virulence Survey Ann Rep 1983, Trumpington, Cambridge, pp 60–63
Kiyosawa S (1980) On the virulence analysis of pathogen race frequencies. Ann Phytopath Soc Jpn 46: 582–593
Knudsen JCN (1984) Selection for partial resistance to powdery mildew in barley. Vortr Pflanzenzüchtg 6: 32–43
Kølster P, Munk L, Stølen O, Lohde J (1986) Near isogenic barley lines with gene for resistance to *Erysiphe graminis* f. sp. *hordei*. Crop Sci 26: 903–907
Kranz J (1987) Raten für die Dynamik von Subpopulationen bei Krankheitserregern. Z Pflanzenkr Pflanzenschutz 94: 225–229
Lebeda A (1982) Measurement of genetic diversity of virulence in populations of phytopathogenic fungi. Z Pflanzenkr Pflanzenschutz 89: 88–95
Leonard KJ (1969) Selection in heterogeneous populations of *Puccinia graminis* f. sp. *avenae*. Phytopathology 59: 1851–1857
Leonard KJ, Czochor RJ (1980) Theory of genetic interactions among populations of plants and their pathogens. Annu Rev Phytopathol 18: 237–258
Lewontin RC (1974) The genetic basis of evolutionary change. Columbia University Press, New York
Limpert E, Schwarzbach E (1981) Virulence analysis of powdery mildew of barley in different European regions in 1979 and 1980. Proc 4th Int Barley Genetics Symposium, Edinburgh, pp 458–471
Limpert E, Schwarzbach E, Fischbeck G (1984) Influence of weather and climate on epidemics of barley mildew, *Erysiphe graminis* f. sp. *hordei*. Prog Biometeorol 3: 146–157
Loegering WQ (1966) The relationship between host and pathogen in stem rust of wheat. Hereditas 2: 167–177
Luig NH (1983) A survey of virulence genes in wheat stem rust, *Puccinia graminis* f. sp. *tritici*. Fortschritte in der Pflanzenzüchtung, vol 11. Parey, Berlin
Luig NH, Watson IA (1970) The effect of complex genetic resistance in wheat on the variability of *Puccinia graminis* f. sp. *tritici*. Proc Linn Soc NSW 95: 22–45
MacKey J (1981) Alternative strategies in fungal race-specific parasitism. Theor Appl Genet 59: 381–390
May RM (ed) (1976) Theoretical ecology. Blackwell, Oxford
Meah MD, Hermansen JE, Jørgensen JH (1982) Interactions between powdery mildew isolates and compatible barley cultivars. Phytopathol Z 105: 45–50
Mood AM, Graybill FA, Boes DC (1974) Introduction to the theory of statistics, 3rd edn. McGraw-Hill, Tokyo
Moseman JG (1972) Isogenic lines for reaction to *Erysiphe graminis* f. sp. *hordei*. Crop Sci 12: 681–682
Munk L (in prep) A survey of barley powdery mildew in Denmark 1978–1983
Nelson RR (1979) The evolution of parasitic fitness. In: Horsfall JG, Cowling EB (eds) Plant disease, vol 4. Academic Press, London, pp 23–46
Parlevliet JE (1976) Evaluation of the concept of horizontal resistance in the barley / *Puccinia hordei* host-pathogen relationship. Phytopathology 66: 494–497
Pielou EC (1969) An introduction to mathematical ecology. Wiley Interscience, New York
Robinson RA (1976) Plant pathosystems. Springer, Berlin Heidelberg New York
Roelfs AP (1984) Race specificity and methods of study. In: Bushnell WR, Roelfs AP (eds) The cereal rusts, vol 1, Academic Press, London
Roelfs AP, Groth JV (1980) A comparison of virulence phenotypes in wheat stem rust populations reproducing sexually and asexually. Phytopathology 70: 855–862
Roelfs AP, Casper DH, Long DL (1984) Races of *Puccinia graminis* in the United States and Mexico during 1983. Plant Dis 68: 902–905
Rouse DI, MacKenzie DR, Nelson RR (1984) Density dependent sporulation od *Erysiphe graminis* f. sp. *tritici*. Phytopathology 74: 1176–1180
Royer MH, Nelson RR, MacKenzie DR (1984) An evaluation of the independence of certain virulence genes of *Erysiphe graminis* f. sp. *tritici*. Phytopathology 74: 1007–1010

Schwarzbach E (1978) Monitoring airborne populations of cereal mildew. In: Scott PR, Bainbridge A (eds) Plant disease epidemiology. Blackwell, Oxford
Schwarzbach E (1979) A high throughput jet trap for collecting mildew spores on living leaves. Phytopathol Z 94: 165–171
Scott PR, Johnson, R, Wolfe MS, Lowe HJB, Bennett FGA (1980) Host-specificity in cereal parasites in relation to their control. Appl Biol 5: 349–393
Snedecor GW, Cochran WQ (1967) Statistical methods, 6th edn. Iowa State University Press, Ames, Iowa
Sokal RR, Rohlf FJ (1981) Biometry, 2nd edn. Freeman, San Francisco
Stähle U (1986a) Untersuchungen zum Einfluß der Virulenzstruktur von Rassen des Gerstenmehltaus (*Erysiphe graminis* DC. f. sp. *hordei* Marchal) auf ihre Fitness. Z Pflanzenkr Pflanzenschutz 93: 177–182
Stähle U (1986b) Untersuchungen zum Einfluß der Inokulumdichte auf die Fitness von Erregerrassen des Gerstenmehltaus (*Erysiphe graminis* DC. f. sp. *hordei* Marchal). Z Pflanzenkr Pflanzenschutz 93: 172–176
Torp J, Jensen HP, Jørgensen JH (1978) Powdery mildew resistance genes in 106 northwest European spring barley varieties. Royal Vet Agric Univ Copenhagen Yearbook 1978, pp 75–102
Vanderplank JE (1975) Principles of plant infection. Academic Press, London
Vanderplank JE (1983) Durable resistance in crops: should the concept of physiological races die? In: Lamberti F, Waller JM, Van der Graaf NA (eds) Durable resistance in crops. Plenum, New York, pp 41–44
Welz G, Kranz J (1987) Effects of recombination on races of a barley powdery mildew population. Plant Pathol (Lond) 36: 107–113
Wolfe MS (1972) The genetics of barley mildew. Rev Plant Pathol 51: 507–522
Wolfe MS (1984) Trying to understand and control powdery mildew. Plant Pathol (Lond) 33: 451–466
Wolfe MS, Knott DR (1982) Populations of plant pathogens: some constraints on analysis of variation in pathogenicity. Plant Pathol (Lond) 31: 79–90
Wolfe MS, Minchin PN (1976) Quantitative assessment of variation in field populations of *Erysiphe graminis* f. sp. *hordei* using mobile nurseries. Trans Br Mycol Soc 66: 332–334
Wolfe MS, Schwarzbach E (1975) The use of virulence analysis in cereal mildew. Phytopathol Z 82: 297–307
Wolfe MS, Schwarzbach E (1978) The recent evolution of barley powdery mildew in Europe. In: Spencer DM (ed) The powdery mildews. Academic Press, London, pp 129–157
Wolfe MS, Slater SE, Minchin PN (1981) Mildew of barley. UK Cereal Pathogen Virulence Survey Ann Rep 1980, Trumpington, Cambridge, pp 42–56
Wolfe MS, Slater SE, Minchin PN (1983) Mildew of barley. UK Cereal Pathogen Virulence Survey Ann Rep 1982, Trumpington, Cambridge, pp 45–59
Wolfe MS, Slater SE, Minchin PN (1984) Mildew of barley. UK Cereal Pathogen Virulence Survey Ann Rep 1983, Trumpington, Cambridge, pp 42–49

Chapter 12

Analysis of Epidemiological Components in Yield Loss Assessment

P. S. Teng and K. B. Johnson[1]

1 Introduction

Plant disease epidemics are multicomponent systems resulting from the dynamic interaction between pathogen populations, host populations, and the physical environment, often influenced by man's activities (Kranz 1974). Epidemics, like most ecosystem phenomena, possess a hierarchical property which reflects the many levels of complexity in biological organization. Plant disease epidemics have commonly been analyzed from a "top-down" approach using the disease progress curve as a starting level, or a "bottom-up" approach using monocyclic processes as the starting level (Teng 1985a). Regardless of the approach used, it is generally accepted that the dynamics of epidemics can only be understood by the use of quantitative techniques, foremost of which is modeling. Many models have been published which describe the disease progress curve or alternatively, simulate the effect of environment on pathogen-host interactions. The top-down or bottom-up approaches for studying epidemics are also reflected in studies of disease effects on host plants, in particular, the disease effects which result in yield loss. Assessment of yield/crop losses would be impossible without a means to realistically quantify the dynamics of epidemics.

Disease results from pathogen activities on a host, but commonly measurements are made on the host. With the diseased population, all or part of host plants have been used to denote the intensity of infection by a pathogen. The term "disease incidence" is used to describe the proportion of infected individuals within a host population, while "disease severity" is used to describe the proportion of host tissue showing symptoms (James and Teng 1979). Disease intensity is used to mean either incidence or severity or an index of both (see Kranz, Chap. 3). Pathogen populations have been measured using variables such as urediospore numbers (Burleigh et al. 1972) or fungal mass (Harrower 1978) and the relationship between pathogen variables and disease variables has been shown to be significant and positive. Whether the pathogen or disease is measured will influence how loss assessments are made. For example, if disease severity is determined, then the data may be input to empirical, regression disease-loss models and to coupled disease-crop yield models, for estimating loss. If only the pathogen or pathogen processes are measured, then the literature shows that the data may be used only in conjunction with a coupled pathogen-crop yield model for loss assessment (see Sect. 3.2).

Using the convention of MacKenzie (1983), a conceptual distinction is made between yield loss and crop loss. Yield loss is used to describe losses caused by single pathogens

[1] Department of Plant Pathology, University of Minnesota, St. Paul, Minnesota 55108, USA

or single sets of disease symptoms, and is measured using a relatively small experimental unit, such as a sample of plants. It assumes that other variables do not have a significant effect on the particular loss equation. Crop loss is used to describe losses caused by multiple pests and is estimated using relatively large experimental fields which reflect the effect of several biotic and abiotic variables. This distinction facilitates the discussion on which epidemiological components need to be considered when moving from small-sized, single pathogen/disease units (yield loss assessment) to farm-sized, multiple pest units (crop loss assessment).

Robinson (1976) suggested the term "pathosystem" to denote biological systems characterized by a parasite-host interaction, as in plant diseases. To accurately assess yield or crop losses in pathosystems therefore requires measurement of the pathogen (or disease) and the host. Progression dynamics of the disease epidemics need to be characterized in order that the rate, intensity, and duration of interaction are accounted for in loss assessment. The environment becomes a consideration because of genotype-environment interactions which greatly affect yield and the disease polycycle or monocycle. The environment is also a consideration when loss estimates are aggregated over a region. In this chapter, we will discuss the above issues and make suggestions for dealing with them.

2 Components Analysis in Yield Loss Assessment

Regardless of whichever model type is used for yield loss assessment, a pathogen or disease intensity value is commonly the input variable. This input variable needs to realistically reflect not only the effect of the disease at the specific time that the disease is assessed, but also the cumulative effect of the disease as the crop develops and grows. How the dynamics of the epidemic, as it affects crop yield, is represented in a model also influences the subsequent use of the model as well as the type of experiments that must be conducted to provide data for modeling.

2.1 Analysis of the Disease Progress Curve (DPC)

ADPC is plotted using sequential disease observations over time. It is a cumulative curve, commonly *S*-shaped when untransformed disease values are used (Fig. 12.1) (see also Chap. 3). Disease values on the vertical axis reflect the particular assessment method for a disease, and may be a proportion, an index, or a rating (James and Teng 1979), although a percentage scale is the most common. The horizontal axis is time in physical units, crop development stage in numerics, or a cumulative unit such as degree day.

Van der Plank (1963) provided a means to linearize the DPC (Fig. 12.1) using the logit transformation, $\ln y/(1-y)$. The untransformed *S*-shaped DPC is describable through the logistic and other equations, although it is difficult to use the parameters of these equations for comparison of epidemics, and the subsequent relationship to yield. Linear regression of logit y on time allows the removal of environmental and experimental variation from the data, and the determination of r, the apparent infection rate (slope of line L, Fig. 12.1). Linearization of the DPC further allows an evaluation of the data through the F, t, s, and r^2 statistics for each epidemic. At the same time, the initial disease intensity (c) and terminal disease intensity (T) are determined.

Another parameter that has been used for describing epidemics in relation to loss assessment is the area under the disease progress curve (AUDPC), in Fig. 12.1, the area under U. A similar area under the disease progress line, L, can be calculated. The AUDPC is a useful concept as it integrates the amount of disease that a crop is exposed to over a given time period. See also Berger, Chapter 9, in this Volume.

For the purposes of disease-loss empirical modeling, only a portion of the DPC is sometimes addressed. It is known that the same disease intensity at different crop growth stages will result in different amounts of yield loss, especially in indirect-loss pathosystems (Zadoks and Schein 1979). These are pathosystems in which the plant part affected by disease is different from the plant part considered the economic yield, such as in cereal foliar diseases (Teng and Gaunt 1981). Also, with crops which are exposed to long-duration epidemics and have a relatively long period for yield accumulation, such as in potato, it may be desirable to partition disease effects into distinctive intervals (James et al. 1972). The DPC has been broken up into actual disease assessments for a specific time or growth stage, $y(t_1)$, into rates for specified time intervals, e.g., $r(t_4-t_2)$, and into AUDPC for known intervals, A and B in Fig. 12.1. With a DPC, the AUDPC is commonly calculated by linear interpolation of daily increments of disease and then cumulating the daily values over the desired interval. For example, James et al. (1972) calculated r for 7-day intervals, while Teng and Bissonnette (1985) calculated AUDPC for 10-day intervals. The area under the epidemic line, L is easier to calculate than the DPC, by multiplying one-half the horizontal axis by vertical axis values. Robinson (1976) distinguished between exodemics and endodemics in pathosystems; these concepts in practice present problems for loss assessment because they suggest that on individual plants, a separate epidemic occurs on each leaf in the case of foliar pathogens and that it may not be valid to represent disease occurring on a plant by a single DPC.

2.2 Relating DPC Components to Empirical Loss Models

Empirical models for estimating yield loss caused by a single disease were categorized into the following types: single point (SP), multiple point (MP), integral, or AUDPC and response surface (Teng 1985b).

Single point models utilize one independent variable to estimate loss; this variable has to reflect the entire epidemic. The term "independent variable" is used here to mean the input variable for a disease-loss model and is also equivalent to a "mathematical predictor" (Madden 1983). Early in crop-loss assessment literature, it was thought that the effect of a disease epidemic on yield could be quantified through indentifying a "critical point" of the crop's growth (James 1974). Many models for cereal diseases reflected this idea, for example, the estimation of loss due to wheat stem rust by

$$\% \text{ Loss} = -25.53 + 27.17 \ln(X) \tag{12.1}$$

where X is percent stem rust severity at three-fourths berry stage (Romig and Calpouzos 1970). The model for estimating yield loss due to neck blast on rice was another critical point model:

$$\% \text{ Loss} = 0.57\,X, \tag{12.2}$$

where X is percent blasted neck nodes 30 days after heading (Katsube and Koshimizu 1970). Both these models were derived using linear regression and explained a high proportion of the variance in loss due to disease, suggesting that there was validity in representing the entire epidemic by means of a single disease assessment. Work on other diseases has not led to the identification of a critical stage and James and Teng (1979) suggested that often, the limited data base used for modeling disease loss may lead to a single crop growth stage showing good statistical correlation for loss.

Other critical point models utilize the concept of a disease-free period or a critical disease threshold (Teng 1985b). The potato late blight model of Olofsson (1968) estimated loss as a function of blight-free days, in which the crop was considered no longer blight-free if disease could be detected. Thus, the longer the blight-free period, the lower the loss in tuber yield. Large (1966) used a 75% blight severity as the cutoff point for causing losses; if it was exceeded in mid-September then yield loss equaled 4%, if it was exceeded in late July then yield loss equaled 50%, with intermediate losses in between. The model types of Olofsson (1968) and Large (1966) have parameters which are very locality-specific, making a strong assumption that epidemics develop with relatively constant rates and onset times between seasons. The data used for developing these models would lend itself to predicting loss from blight severity at a given time in the growing season. With potato early blight, Teng and Bissonnette (1985) could not derive any SP models capable of explaining more than 26% of the variation in loss caused by disease. These authors felt that in a crop such as potato, with a relatively long period of tuber bulking, no single time could be selected to represent an entire epidemic. Even with cereals which have distinctive yield components developing over different growth stages, Teng et al. (1979) found that the regression parameters of the single point models for estimating barley leaf rust losses were very unstable between locations and seasons. It would appear from the literature that SP models of disease loss make gross assumptions on epidemiology (Zadoks and Schein 1979).

SP models in general have been to estimate loss caused by short-duration epidemics with relatively stable infection rates, and in which most of the yield loss is due to one or two yield components (Teng 1985b). The inherent weakness in using a single point model is that it assumes a constant infection rate between epidemics encountered in experiments and in subsequent application.

Multiple point models estimate yield loss from several disease assessments made during the growing season. These assessments have been represented as actual disease intensities at specific times/growth stages (Burleigh et al. 1972; Teng et al. 1979), or as disease increments over known intervals (James et al. 1972; Teng and Bissonnette 1985).

With barley leaf rust, Teng et al. (1979) developed the following model:

$$\begin{aligned}\%\ \text{Yield Loss} = {} & -0.07 - 4.35\,X3 + 3.26\,X9 + 0.72\,X4 - 1.82\,X10 \\ & + 0.62\,X5 + 0.42\,X11 - 0.03\,X6 + 0.22\,X12,\end{aligned}$$

in which the X variables were, respectively, disease severities for leaf position 1 and 2 for the growth stages of 58/59, 64/65, 73/74, and 83/83 on a decimal code (Zadoks et al. 1974). Like the wheat leaf rust model of Burleigh et al. (1972), the barley leaf rust model explained over 79% of the loss variation and both groups of workers felt that the models could account for variable infection rates of the disease during a growing season. With cereals, a sampling problem arises because each plant commonly has several tillers, each of which has been exposed to disease for a different duration. The distinction of exodemics and endodemics (sensu Robinson 1976) is particularly apt here. In a population of cereals, any field epidemic comprises subepidemics, each subepidemic is characterized by the host subpopulation of equivalent age. When measuring mean disease in this population for loss estimation, the sampling should be done on all main tillers or tillers of approximately the same age. Many SP models use disease intensity assessed on one plant part as the input for loss estimation in cereals, while the MP models commonly use two or more parts of a plant. With cereal MP models, using actual disease severity at selected growth stages as input variables (Burleigh et al. 1972; Teng et al. 1979), partial regression coefficients often have negative values, which biologically can be interpreted as a disease-decreasing loss. While Gaunt (1980) has notes that this max be possible, others have considered them to be a statistical artifact with no biological meaning (James and Teng 1979).

In another type of MP model, using disease increment per week as the independent variable (in Fig. 12.1, the difference between y on week 3 minus y on week 2, if time is in weeks, and so on), James et al. (1972) divided the potato-growing season into nine periods. The partial regression coefficient for each period indicated the potential loss for that period, such that total % yield loss = sum of weekly % yield loss. The MP model for potato late blight explained 98% of the loss variation and had a standard error of loss estimate of 3%. James et al. (1972) used disease increments per week because they considered late blight epidemics in Canada to have highly variable infection rates each season and that there was no single period which could dominate in estimating loss. When Teng and Bissonnette (1985) divided the growing season in Minnesota into five 10-day periods and tested both actual disease severity and disease increment per 10-day period as the independent variables for regression, they found that disease increment MP models were not significantly better than actual disease severity MP models for estimating loss due to potato early blight. The use of a disease increment per period is conceptually appealing as it tries to account for wide fluctuations in disease progress observed under some natural and experimental conditions. Teng and Bissonnette (1985) also tested both actual disease severity and disease increments in the same equation for loss estimation, but could not find an improvement over the MP models which used either variable. As with the James et al. (1972) work, Teng and Bissonnette (1985) concluded that separate MP models were required for "early" and "late" season pathosystems. Generally, the addition of sequential disease assessments as input variable to a model increases the predictive ability of the model (Teng and Shane 1984). However, the statistical technique commonly used for developing SP and MP models, least squares regression, is subject to abuse especially if the number of observations in a data set does not justify the inclusion of an inordinately large number of independent variables (Teng 1981). For example, the r^2 and F-statistics to evaluate regression models are influenced by n, the number of observations.

Integral models predict loss from input variables that represent disease for a defined epidemic duration, such as the AUDPC (see also Chap. 9). Yield loss is estimated from AUDPC, in arbitrary units, for a fixed epidemic duration. Schneider et al. (1976) developed a model for Cercospora leaf spot on cowpea, using the AUDPC from flowering to terminal severity. Teng and Bissonnette (1985) could only explain 19%

and 55%, respectively, of the loss variation in short and long season potato cultivars, due to early blight with AUDPC models.

It has been noted that the ratio between different AUDPCs is independent of infection rate (Van der Plank 1963). Therefore, it would not be possible to define the relative susceptibilities of different crop growth stages to disease effects using the AUDPC. Two DPCs with the same AUDPC could have different onset times, different onset severities, different terminal severities, and different infection rates. Thus, AUDPC models appear applicable only to short-duration epidemics with stable infection rates which occur relatively late in the season (James 1974; Teng 1985b).

The AUDPC model type was modified to account for the growth stage effect in loss assessment by Hills et al. (1980), who weighted disease intensity by week number. This enabled them to distinguish between short- and long-duration powdery mildew effects on sugar beet yield and to derive models explaining 84% of the loss variation. Meronuck and Teng (1984) were unsuccessful in adding the infection rate and epidemic onset growth stage as regressors in AUDPC models for estimating bean rust loss.

The relationship between disease and loss is inherently a nonlinear one even though the majority of empirical disease-loss models have been developed using linear regression. A nonlinear model for relating loss to either an SP or AUDPC, based on the Weibull distribution, was proposed by Madden et al. (1981). Epidemiologically the model is very appealing as it allows the definition of a minimal threshold of disease, below which no loss is incurred. The problems of experimentally proving this minimal threshold are real, as in the early phase of an epidemic not all plants are infected and sampling a diseased population with low mean disease generally shows high variability in the data (Teng 1985b).

Response surface models estimate yield loss from two different types of input variables derived from the epidemic or host. Teng and Gaunt (1981) presented a conceptual model for predicting yield loss (Y) from disease severity (X) and crop growth stage (T), which pictorially may be represented as a three-dimensional response surface with Y as the vertical axis and X, T as the two horizontal axes. These workers

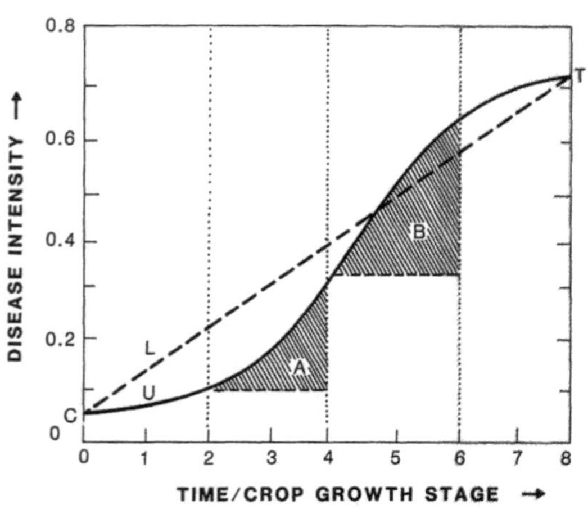

Fig. 12.1 Diagram of a disease progress curve showing curve elements important for yield loss assessment

felt that a response surface best represented the physiological basis for yield loss, in which the same loss could be caused by different disease severity–growth stage combinations. Further, the surface could account for epidemics of various characteristics. Calpouzos et al. (1976) predicted wheat stem rust losses from the slope of an epidemic line (infection rate of L, Fig. 12.1) and the crop growth stage of epidemic onset. In this model, DPCs are implicitly transformed through the use of equal divisions to plot a nonlinear growth stage scale on the horizontal axes. As with other model types, the data base for developing response surface models has to contain the range of epidemics commonly encountered under natural conditions, otherwise spurious predictions of loss will result. Teng and Gaunt (1981), using data from Teng et al. (1979), showed that, with barley leaf rust, early season CP models with good statistical acceptance were biologically unsound as part of a response surface model because the epidemics generated during the early part of the season had very low severities.

Synoptic models are multivariate models estimating yield loss which incorporate all the independent variables in a single equation (Teng 1985b). Data for model development are collected by sampling natural field plots and the repeatability or general application of any synoptic model depends largely on there being a wide range of disease intensities in the data for each disease. If the plots are all under natural conditions, this may not be the case, with the result that insufficient "levels" of each yield-reducing factor being present in the data to allow development of generalized models (Teng 1985b).

Any analysis of epidemiological data for modeling yield loss should contain a determination of the magnitude of representational errors, especially of interplot interference (James 1978). Van der Plank (1963) had warned of the "cryptic error" in field experiments with differential epidemics, in which epidemics in experiments do not have the same characteristics as those developing in farmers' fields. In yield loss experiments, which commonly contain treatments with a heathy, a natural epidemic, and several intermediate levels of disease, plots with higher disease can increase the amount of disease and infection rate in adjacent plots with lower disease, a phenomenon called positive interference (James et al. 1973). Similarly, negative interference can occur (Bowen et al. 1984). The amount of interference, and its related representational error, increases as experimental units decrease in size. The dynamics of an epidemic occurring on a single plant cannot therefore be equated to those of a field epidemic (Teng and Shane 1984). This point is important as single plants have been used to provide the data for developing empirical disease-loss models, notable examples are those for cereal diseases (Richardson et al. 1975; King 1976). Apart from the epidemiological implications of using single plants to represent fields, James and Teng (1979) have noted that there are statistical problems in the models derived from single plants in that these commonly account for a very proportion of the loss variation and have high standard errors and F-statistics.

3 Components Analysis in Crop Loss Assessment

In this section we will discuss the estimation of losses caused by multiple diseases, and over large areas.

3.1 The Problem of Scale and Aggregation

Yield loss models have commonly been derived from data collected in small plots, from single tillers or plants, and from parts of farmer's fields. It is common knowledge in statistics that experimental plots need to be of a predetermined size to allow valid comparisons of treatment effects on crop yield, since the need is for plot results to be comparable to those of actual farming conditions. With regional loss assessment, in which many production systems, climatic zones, soil types, etc. all contribute to a diversity in epidemiological and crop yield potential, the problem becomes one of how to derive a representative sample of fields in a given area. Most plant disease surveys (King 1976) attempt to make assessments at one or more crop growth stages common to all fields, with the assumption that there is comparable progression dynamics in the fields. Without this assumption, one or two disease variable models would not be applicable. Furthermore, the ecological boundaries for epidemics with certain characteristics may not always correspond to the criteria for stratifying a region in loss assessment. The problem of aggregating yield loss estimates from many fields to represent the mean loss for a region, has no simple solution and is an area requiring much research (Teng and Oshima 1983).

3.2 Crop-Pest Loss Models

Recently, use of integrative crop growth simulation models as tools for studying pest-induced crop losses has received increased attention (Boote et al. 1983; Gutierrez et al. 1983; Johnson et al. 1986a; Loomis and Adams 1983; Rouse 1983; and Seem, Sect. 4.3.2). This interest in mechanistic modeling of crop growth comes about with the realization that empirical (statistical) approaches to modeling crop loss are limited by being insensitive to environment. Also, empirical models are not well suited for biological explanation or for studying pest control optimization (Rouse 1983). In contrast, mechanistic models, which integrate over several levels of biological organization, are well suited for explanation of biological phenomena (Loomis and Adams 1983). Furthermore, crop growth models are driven by environmental variables, which makes them responsive to changing environmental conditions. Crop growth simulation models should ideally be able to accurately predict potential yield in the absence of pests.

Most crop-pest simulation models developed for studying yield loss address the tissue and whole-plant levels. The output from these models is usually expressed as an average plant or as unit area of crop. Nonetheless, these models also include both lower and higher levels of biology, and include phenomena at the process level (e.g. photosynthesis, respiration, water utilization) and inter- or intraplant competition for solar radiation, nutrients, and water. Within a crop model, dry matter is dynamically partitioned to the various plant tissues based on the relative ability of each plant part to compete for photosynthetic assimilates. Thus, as part of the partitioning process, it is possible to include compensatory mechanisms that may result in response to pest-induced stress.

The effects of a disease may be coupled to a crop growth model in three ways: (1) levels of disease severity may be directly input from a file that contains daily or hourly environmental information, (2) disease severity may be determined from a mathematical function such as a logistic, and (3) a disease simulation model which is itself responsive to environment, may be used to produce severity levels linked to the crop model. The first two approaches are simpler than the third and are useful when yield or other disease effects on crop growth are of prime interest. The third approach is useful when prediction of disease is also of interest; this offers the most potential for understanding pathogen-host interrelationships. Coupled models utilizing all three approaches are useful for examining the dynamics of loss due to changing disease severities during the season.

The desired level of detail and the nature of the disease influence which of the above three approaches is used. Boote et al. (1983), who were interested in detailed interactions, identified seven mechanisms by which disease influences crop growth: stand reduction, reduction in photosynthetic rate, acceleration of leaf senescence, direct utilization of plant assimilates, tissue consumption, and turgor reduction. For simpler models, which may not model all plant processes at high levels of detail, Johnson et al. (1986a) suggested that pest effects be viewed in terms of Monteith's (1977) hypothesis. This states that crop growth is the product of two independent factors, the proportion of solar radiation intercepted (dependent on green leaf area), and the efficiency of its use. Functional equations can then be developed for how a disease affects these two components of growth.

Development of coupled crop-pest models is still in its infancy. Gutierrez et al. (1983) coupled the effects of verticillium wilt to a cotton model, from which yields were estimated as influenced by inoculum density, pathogen virulence, and plant spacing. Adams and Loomis (1983) used a powdery mildew-sugar beet growth model to show that loss was caused mainly through reduction in photosynthesis of the canopy and that mildew infection rate was affected by plant spacing. Johnson et al. (1986a) developed a simple potato growth model to couple the effects of early blight, verticillium wilt, and potato leafhopper. Early blight effects were modeled by reducing green leaf area through lesion expansion and defoliation. Verticillium wilt was modeled by accelerating the aging rate of leaf tissue. Potato leafhoppers reduced both green leaf area through hopperburn and the photosynthetic efficiency of remaining green tissue. Simulation modeling and field experiments showed that the combined effects of these three pests were less than the additive effects of solitary pest populations (Johnson et al. 1986b).

4 Concluding Remarks

The science of crop loss assessment, inter alia, attempts to provide some understanding of the effect of disease epidemics on crop yield, from which improved disease management may result. Although much of the work on modeling disease losses has been influenced by Van der Plankian logic, serious limitations have become evident and have led workers to examine models with more explanatory ability than empirical models.

The future use of crop-pest models is very bright given their responsiveness to changing environment and the ability to translate varying pest intensities at different times of the season into crop loss. An important feature of crop-pest models over statistical models is that model output (e.g., tuber mass, leaf mass) can be directly compared with field situations. Crop-pest models may also be the only realistic way to understand or predict the effects of multiple pests on yield. In the future, crop-pest models may even be used to estimate regional losses from defined distributions of epidemiological parameters and environmental conditions. Concurrently, these models may be used to determine the best method for developing simpler empirical yield loss models.

5 References

Adams SS, Loomis RS (1983) Explanatory simulation of sugar beet yield losses from powdery mildew using coupled models of host and pathogen. Phytopathology 73: 827 (Abstr.)

Boote KS, Jones JW, Mishoe JW, Berger RD (1983) Coupling pests to crop growth simulators to predict yield reductions. Phytopathology 73: 1581–1587

Bowen KL, Teng PS, Roelfs AP (1984) Negative interplot interference in field experiments with leaf rust of wheat. Phytopathology 74: 1157–1161

Burleigh JR, Roelfs AP, Eversmeyer MG (1972) Estimating damage to wheat caused by *Puccinia recondita tritici*. Phytopathology 62: 944–946

Calpouzos L, Roelfs AP, Madson ME, Martin FB, Welsh JR, Wilcoxson RD (1976) A new model to measure yield losses caused by stem rust in spring wheat. Minn Agric Exp Stn Tech Bull 307: 1–23

Gaunt RE (1980) Physiological basis of yield loss. In: Teng PS, Krupa SV (eds), Crop loss assessment. Minn Agric Exp Stn 7: 98–111

Gutierrez AP, De Vay JE, Pullman GS, Frieberthauser GE (1983) A model of verticillium wilt in relation to cotton growth and development. Phytopathology 73: 89–95

Harrower KM (1978) A novel technique for assessing the reaction of wheat cultivars to *Septoria tritici*. In: Close RC et al. (ed) Epidemiology and crop loss assessment. Lincoln College Press, pp 3-1—3-6

Hills FJ, Chiarappa L, Geng S (1980) Powdery mildew of sugar beet: disease and crop loss assessment. Phytopathology 70: 680–682

James WC (1974) Assessment of plant diseases and losses. Annu Rev Phytopathol 12: 27–48

James WC (1978) Importance of interplot interference in field experiments involving plant diseases. In: Close RC et al. (ed) Epidemiology and crop loss assessment. Lincoln College Press, pp 29-1—29-15

James WC, Teng PS (1979) The quantification of production constraints associated with plant diseases. Appl. Biol 4: 201–267

James WC, Shih CS, Hodgson WA, Callbeck LC (1972) The quantitative relationship between late blight of potato and loss in tuber yield. Phytopathology 62: 92–96

James WC, Shih CS, Callbeck LC, Hodgson WA (1973) Interplot interference in field experiments with late blight of potato. Phytopathology 63: 1269–1275

Johnson KB, Johnson SB, Teng PS (1986a) A simple potato growth model for crop-pest management. Agric Syst 16: 189–209

Johnson KB, Teng PS, Radcliffe EB (1986b) The effect of interacting populations of early blight, verticillium wilt and the potato leafhopper on potato yield. Phytopathology 76: 1046–1052

Katsube T, Koshimizu Y (1970) Influence of blast disease on harvests in rice plant. I. Effect of panicle infection on yield components and quality. Bull Tohoku Natl Agric Exp Stn (Morioka) 39: 55–96

King JE (1976) Relationship between yield loss and severity of yellow rust recorded on a large number of single stems of winter wheat. Plant Path 25: 172–177

Kranz J (1974) Comparison of epidemics. Annu Rev Phytopathol 12: 355–374

Large EC (1966) Measuring plant disease. Annu Rev Phytopathol 4: 9–28

Loomis RS, Adams SS (1983) Integrative analyses of host-pathogen relations. Annu Rev Phytopathol 21: 341–362

MacKenzie DR (1983) Toward the management of crop losses. In: Kommedahl T, Williams PH (ed) Challenging problems in plant health. American Phytopathological Society, St. Paul, USA, pp 82–92

Madden LV (1983) Measuring and modeling crop losses at the field level. Phytopathology 73: 1591–1596
Madden, LV, Pennypacker SP, Antle CE, Kingsolver CH (1981) A loss model for crops. Phytopathology 71: 685–689
Meronuck RD, Teng PS (1984) Economic thresholds for chemical control of bean rust. Phytopathology 74: 796 (abstr.)
Monteith JL (1977) Climate and the efficiency of crop production in Britain. Philos Trans R Soc Lond B Biol Sci 2 (1): 277 294
Olofsson B (1968) Determination of the critical injury threshold for potato blight. Medd Waxtskyddsanst Stockholm 14: 81–93
Richardson MJ, Jacks M, Smith S (1975) Assessment of loss caused by barley mildew using single tillers. Plant Path 24: 21–26
Robinson RA (1976) Plant pathosystems. Springer, Berlin Heidelberg New York
Romig RW, Calpouzos L (1970) The relationship between stem rust and loss in yield of spring wheat. Phytopathology 60: 1801–1805
Rouse DI (1983) Plant growth models and plant disease epidemics. In: Kommedahl T, Williams PH (eds) Challenging problems in plant health. American Phytopathological Society, St. Paul, USA, pp 387–398
Schneider RW, Williams RJ, Sinclair JB (1976) Cercospora leaf spot of cowpea: models for estimating yield loss. Phytopathology 66: 384–388
Teng PS (1981) Use of regression analysis for developing crop loss models. In: Chiarappa L (ed) Crop loss assessment methods. FAO/CAB 3: 51–55
Teng PS (1985a) A comparison of simulation approaches to epidemic modeling. Annu Rev Phytopathol 23: 351–379
Teng PS (1985b) Construction of predictive models. II. Forecasting crop losses. Adv Plant Pathol 3: 179–206
Teng PS, Bissonnette HL (1985) Developing equations to estimate potato yield loss caused by early blight in Minnesota. Am Potato J 62: 607–618
Teng PS, Gaunt RE (1981) Modeling systems of disease and yield loss in cereals. Agric Syst 6: 131–154
Teng PS, Oshima RJ (1983) Identification and assessment of losses. In: Kommedahl T, Williams PH (eds) Challenging problems in plant health. American Phytopathological Society, St. Paul, USA
Teng PS, Shane WW (1984) Crop losses caused by plant pathogens. CRC Crit Rev Plant Sci 2: 21–47
Teng PS, Close RC, Blackie MJ (1979) Comparison of models for estimating yield loss caused by leaf rust on Zephyr barley in New Zealand. Phytopathology 69: 1239–1244
Van der Plank JE (1963) Plant diseases: epidemics and control. Academic Press, London
Zadoks JC, Schein RD (1979) Epidemiology and plant disease management. Oxford Univ Press, London, New York
Zadoks JC, Chang TT, Konzak CF (1974) A decimal code for the growth stages of cereals. Weed Res 14: 415–421

Chapter 13

How to Develop Plant Disease Forecasters

L. V. Madden and M. A. Ellis[1]

1 Introduction

As stated by Fry (1982) in his recent text, use of disease management techniques "when they are not needed is inefficient at best because their use results in unneeded cost to growers, consumers, and perhaps to the environment." One approach for determining when or if to apply disease control techniques is the use for forecasting systems. The intensive development and use of plant disease forecasters is a relatively new and exciting application of epidemiology.

Disease forecasters operate by predicting an outbreak or increase in disease intensity based on weather, host, or pathogen conditions. Many popular and functional forecasters are based on identifying past infections; few predict future infections. As such, it has been argued that the term forecaster is a misnomer because decisions are usually based on past, or at best current, conditions. Zadoks (1984) preferred to use "disease warning" instead of disease forecasting system or predictor. We prefer to use the terms interchangeably with the knowledge that all of the systems are "predicting" general disease increase based on completed components of a disease cycle.

Disease forecasting systems have been reviewed and discussed several times over the last 2 to 3 decades (e.g., Miller 1959; Waggoner 1960; Bourke 1970; Krause and Massie 1975; Shrum 1978; Pennypacker and Stevenson 1982; Fry 1983, Chap. 6; Zadoks 1984; Thresh 1986; Fry and Fohner 1985). Forecasters have been classified by these authors in various ways, including whether: (1) crop, disease, pathogen, environment, or a combination of these were used for making forecasts; (2) forecasts were made pre- or postplanting; (3) forecasts were based on initial inoculum, secondary inoculum and rate of disease increase, or combinations of both; and (4) forecasts were based on empirical or fundamental information. All of these classifications may be useful depending on the circumstances. The emphasis in this chapter is on point (4).

No attempt is made in this chapter to review all currently and previously used forecasting systems. Extensive reviews can be found in the articles cited above. Instead, we present the common experimental approaches for developing disease forecasters and validating them in commercial fields. We draw heavily on the systems we have or are currently working with. The approaches described are not all-inclusive but represent techniques we find most useful. Because a reliable forecasting system will not be used without grower acceptance, we also discuss different ways of using forecasters in commercial fields.

[1] Department of Plant Pathology, The Ohio State University, Ohio Agricultural Research and Development Center, Wooster, OH 44691, USA

2 Fundamental Forecasting Systems

We consider *fundamental disease forecasting systems* as those developed from experiments in the laboratory, controlled-environment chamber, greenhouse, or field and describe one or more aspects of the host-parasite relation as influenced by environment. Implicit in this definition is that the experiments permit "cause and effect" to be determined. Mechanistic and "derived" are two other terms for this approach. Our definition agrees closely to that presented by Krause and Massie (1975), although we would not call some of their examples fundamental.

Fundamental forecasters in use today are basically simple and generally based on one component of the disease cycle. In fact, we believe that to develop an accurate and usable forecaster, one must keep it as simple as possible. In this chapter we emphasize the infection disease cycle component with the understanding that other components can also be very important.

2.1 Infection

The most successfully used disease cycle component in forecasting is infection. Obviously, without infection disease outbreaks will not occur. Many environmental factors can influence infection; with plant pathogenic fungi, temperature (T), and wetness duration (W) are important variables that control infection. Some forecasting systems, such as BLITECAST for late blight of potatoes, use periods of high relative humidity (RH), e.g., hours of RH > 90%, instead of W. When ambient RH is greater than 90 or 95%, there is often free moisture on at least some plant parts. In the past it was much easier to measure and record RH than it was W. Today, it is much easier to detect and record free moisture on either plant parts or sensors (Sutton et al. 1984 and Chap. 7).

The time required by a spore to proceed from the ungerminated state to the state of an established infection we consider to be the *infection period* (IP). IP is usually a function of at least T and probably other factors as well. Plotting IP vs T usually produces a *U*-shaped curve (Fig. 13.1 A). For instance, it takes a minimum of 12 h of leaf wetness at 13 °C for infection of grape leaves by *Guignardia bidwellii* conidia, causal agent of black rot (Spotts 1977). A simple forecasting system can be developed that utilizes measurements of T and W to determine if an IP has occurred.

Experience has shown that the IP for many fungi can be very short. For instance, between 6 and 30 °C, at least some sporangia of *Phytophthora cactorum,* causal agent of strawberry leather rot, can germinate and cause strawberry fruit infection at W < 5 h (Grove et al. 1985). In these situations, one must describe the relative level (i.e., proportion) of infection at various combinations of T and W. These proportions are estimates of the probability of infection. Then one can produce curves depicting hours of W vs T for specified predicted levels of infection (Fig. 13.1B). Scheduling fungicide applications based on predicted infection would be practical if environmental conditions were *not* favorable for high infection daily. A forecasting system based on no or low infection (e.g., 10%) would be completely impractical when conditions were favorable *daily* for ~100% infection. If disease epidemics are periodic or discontinuous in the latter case, then low levels of infection probably are not the controlling factor in disease development. One might then test the possibility of using high predicted infection probabilities (e.g., >75%) together with other disease cycle components to forecast disease development.

2.1.1 Experimental Procedures

Controlled environment chambers are extremely valuable for relating infection level to T and W (see also Aust and Kranz, Chap. 1, and Rotem, Chap. 2). Despite the controversy concerning chambers, we, as well as others, have been very successful in relating controlled environment studies to field situations. First, the relation between number of infections and inoculum level should be established. A level of inoculum high enough to produce countable infections (e.g., lesions), but low enough so that competition among spores for infection sites is not great, should be used. Ideally, one should use an inoculum level from the (nearly) linear portion of the disease inoculum.

A wide range of T's, including the maximum and minimum T at which infection can occur, and wetness durations should be tested. Investigators may find that a minimum and maximum T will be difficult to determine if they use long enough W. In these cases, a wide range of T's that demonstrate the functional relationship between infection level and T is sufficient. At least five or six T's need to be tested, preferably several more. At least five or six W's also have to be tested at each of the T's. The range of W's should be chosen to give a potential range of infection from 0 to 100%. Preliminary experiments

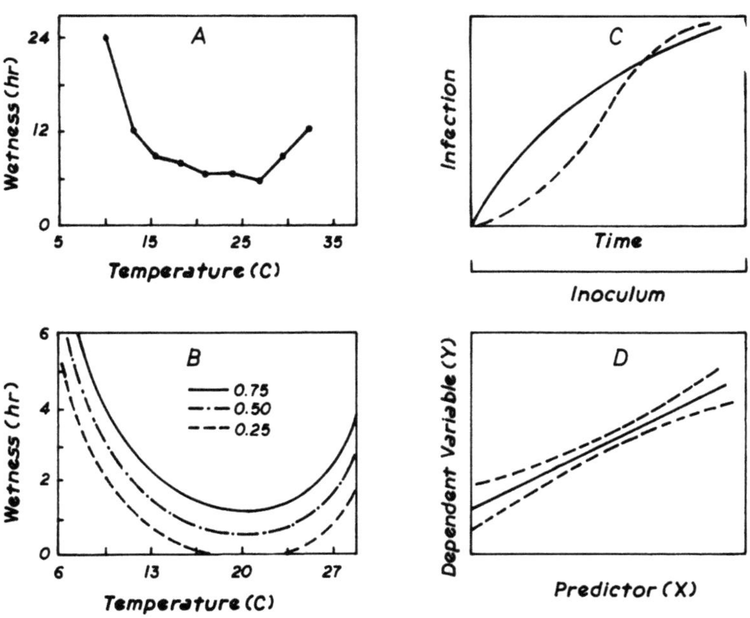

Fig. 13.1 a–d. Observed and theoretical disease responses. **A** Minimum hours of wetness at selected temperatures required for infection of grape leaves by *Guignardia bidwellii* conidia (Spotts 1977). **B** Calculated hours of wetness at a range of temperatures required for 0.25, 0.5, and 0.75 of strawberry fruit to be infected by *Phytophthora cactorum* sporangia (Grove et al. 1985). **C** Theoretical number of infections (or probability of infection) in relation to (1) time that the host is exposed to inoculum (or inoculative vectors), or (2) to the level of inoculum. **D** Simple linear relationship between a dependent variable (*Y*) and an independent (predictor) variable (*X*), together with an 95% confidence band around the predicted values (*dashed line*) (Neter et al. 1983)

should enable the researcher to choose the minimum and maximum W. The extremes in infection may not be obtainable but one should attempt to determine them if possible. If there are time or space limitations, one might be able to eliminate the shortest W's at the T's farthest from the optimum, if no infection is expected. For example, a requirement of at least 8 h of wetness at T = 28 °C would suggest that W's less than 8 need not be tested at 30 °C.

Few investigators have the controlled-environment facilities to test all temperatures simultaneously. We generally are limited to two chambers. Temperatures to be tested are chosen in a random order. Both chambers are set at the same desired T. One chamber is a modified dew chamber to ensure continuous free moisture. Uniform plants or plant parts (e.g., fruit) are placed in the dew chamber and incubated. A specified number of plants or plant parts are then removed at the preassigned W values, e.g., 1, 2, 4, and 8 h, and placed in the second (drying) chamber. In the second chamber, a wetness sensor can be gently misted to simulate plant wetness and monitored to determine drying time. If a very short W is desired, inoculating in the drying chamber can be done. If sensor drying does not simulate plant drying, then electrical sensors attached directly to the plant parts to measure plant wetness can be used. Visual assessment of drying is also possible but can be subjective and is also inconvenient with many experiments. We have found that varying the surface coating of electrical wetness sensors and using different cut-off voltages can acceptably simulate plant wetness and subsequent drying. After drying, plant material can be moved to another chamber or elsewhere (e.g., greenhouse). After the appropriate incubation period, the level of infection can be determined.

The above experiment can be expanded to include different cultivars that potentially differ in susceptibility to the pathogen. Host plant age (maturity) and predisposition, as well as position of the inoculated plant part (low vs high fruit) can be studied. If the chambers are large enough, one temperature run could include more than one cultivar or plant condition. Usually, several runs are necessary.

2.1.2 Data Analysis

In some situations, an investigator will discover exact combinations of T and W in which infection will not occur. For instance, at T = 29 °C, infection of grapes by *Guignardia bidwellii* will not occur at W < 9 h. In these situations, a simple IP curve can be generated and used (Fig. 13.1), no complicated statistical analysis would be required. Unfortunately, results are usually not this precise and invariable for most diseases.

Regression and analysis of variance (ANOVA) are the techniques of choice for analyzing data obtained from the above described experiment(s). With ANOVA, one must be careful to use the appropriate linear model and error terms. As typically done, there is a restriction on randomization, i.e., all wetness durations are tested with the chamber set at a single temperature. This restriction results, basically, in a "repeated measures" experimental design (e.g., Gill 1978).

A more productive approach to ANOVA in many cases is to use regression analysis to determine the functional relationship between Y and W and T (i.e., Y = f(T, W)). In general, a "bell-shaped" curve will be needed to describe Y as a function of T and a monotonically increasing curve to describe Y as a function of W. All models need to be evaluated using standard statistical techniques (e.g., Neter et al. 1983).

2.1.3 *Other Physical Factors and Nonconstancy*

There is a wide range of environmental components other than T and W that affect infection. With fungi, atmospheric moisture and radiation can be very important. Moisture is typically represented as relative humidity (RH, %) although vapor pressure deficit (VPD, bars) or absolute humidity (AH, gm^{-3}) may be more appropriate. Radiation, due primarly to sunlight, also can affect the infection process. In many climates, however, there is not sufficient variability in radiation from day-to-day to be useful in disease forecasting. In climates with highly variable cloud cover, radiation could be an important predictor of disease outbreaks. Radiation, or more correctly, radiant flux incident on a surface (irradiance) should be measured in watts per square meter (or equivalent units), or if photosynthetically active radiation is desired, micro-Einsteins per square meter per second. Hours of irradiance greater than zero (or hours of light) could be used as a predictor variable as well.

If one can accurately predict infection based on variables such as T and W, then we see little reason to try to relate radiation to infection. If infection variability is high, and preliminary experiments indicate that radiation (light) affects infection, then certainly this factor should be considered in developing a forecasting system.

Environmental conditions are seldom constant in nature, yet most of us conduct experiments with constant T and RH, as well as continuous W. The above described experiments can, however, be expanded to include variability. In modeling, one can divide a fluctuating environmental factor (e.g., T) into a number of smaller units such as 1-h or 30-min intervals, and assume that conditions are approximately constant during these intervals. The rate of progress toward infection can then be estimated, and the actual predicted infection is determined by summing the predictions from each short-time interval. For more discussion on this topic and other approaches, one should read Hau et al. (1985).

The experiment for ascertaining the effect of T and W on Y can be expanded to include the additional physical components of the environment. With controlled environment chambers, it will generally be easiest to set T and one or more of the additional factors for one run of the experiment; within the run, various durations of wetness, or perhaps RH greater than a certain value, can be investigated. To avoid confounding of T and the other fixed factor(s), one must run combinations of T with the other factor(s).

2.1.4 *Vectors*

Viruses, spiroplasmas, and mycoplasma-like organisms (MLO's), generally require insect vectors for plant infection (see Raccah and Irwin, Chap. 14). Vector numbers can be used to predict disease levels (Thresh 1986). With persistently transmitted pathogens, exposure time to the plant is analogous to W. The functional form of Y vs vector exposure time, at a single T, will be a concave or sigmoid curve as in Fig. 13.1 C. This curve corresponds to either a Poisson or binomial probability distribution of infections.

One can conduct vector/infection experiments by varying either the duration of plant exposure or the number of vectors per plant (or plant part). If T was also of concern, chambers could be set at single T's in a random order, and the complete set of the vector factor could be run at each T. The experimental design would then be analogous to the T/W experiment. Probability models (e.g., Poisson distribution) then should be used to describe and predict Y.

2.1.5 *Inoculum Level*

Van der Plank (1975) discussed at great length infection as a function of inoculum level. He considered the probability of infection to be described by a Poisson distribution. Rouse et al. (1985), working with epiphytic populations of bacteria, considered infection to be best described by a log-normal distribution. If one could readily measure inoculum levels in the field, e.g., *Pseudomonas syringae* pv. *syringae* numbers on bean leaves, then a statistical model could be used to predict infections. Those interested in this area should read Rouse et al. (1985) for more details.

2.2 Other Disease Cycle Components

All components of a disease cycle could be used in developing a forecasting system. Relying on infection, although common, may not be sufficient for accurate predictions. For instance, environmental conditions might be favorable for 100% fruit infection, but if there is no inoculum on the fruit, obviously no infection will occur. Survival of inoculum, sporulation and dispersal are all necessary for viable inoculum to reach a previously disease-free fruit and infection to occur (see Rotem, Chap. 5). Relying on infection models alone when inoculum is absent or the host is not susceptible would result in excessive control applications.

Methods described above for infection can generally be used to determine the relationship between weather variables and other factors such as sporulation. Linking all aspects of a disease cycle is complex and almost always requires simulation. Most disease forecasters do not utilize all components of the disease cycle. Nevertheless, consideration of the entire disease cycle, linked by computer simulation can be productive. Fry et al. (1983), for instance, have successfully used a complicated simulator incorporating both disease, environment, and fungicide effects, to improve on BLITECAST forecasts. For aspects of the host relevant to the development of forecasts, see Seem (Chap. 4, Sects. 4.2.1, 4.2.2, 5.1, and 5.2).

3 Empirical Forecasting Systems

Empirical forecasting systems are developed from the observation and analysis of current and historical data on disease levels and other biotic and abiotic factors. Experimental results are correlative in nature, although sometimes cause and effect can be inferred from the results. Although many would feel that fundamental forecasting systems are preferable to empirical ones because of their explanatory capability, many useful and functional forecasting systems are empirical.

Empirical forecasters can be classified as: (1) dealing with a single prediction per year (season); or (2) involving multiple predictions throughout a growing season. Forecasters of the first type are most useful when it is important to predict the initial inoculum or early level of disease, or when only a limited use of a control technique (e.g., fungicide application) is practical. Forecasting the seasonal maturation of *Venturia inaequalis* ascospores (MacHardy and Gadoury 1985) and stripe rust of winter wheat, caused by *Puccinia striiformis* (Coakley et al. 1982), exemplify the former and latter reasons for this type of prediction. Development of these systems usually requires observations from several years and/or several locations.

Forecasting systems that involve multiple predictions (e.g., weekly) within a season are most useful when a disease can increase rapidly over much of the season, or when the economic value of the crop is high enough to justify multiple applications of a control technique. Examples of this type of forecast system include those used for potato late blight (caused by *Phytophthora infestans*) and leaf spot of peanuts (caused by *Cercospora* spp.) (Krause and Massie 1975; Parvin et al. 1974). Development of these systems requires observations made throughout the growing season on disease, weather, and/or other factors. In general, more than 2 years or locations need to be studied to develop such a forecaster, and more years for validation.

Empirical forecasters can be developed in one of two ways. The first approach, which we call qualitative, involves the development of prediction criteria ("rules") without any formal statistical analysis. The second, quantitative approach is based on statistical analysis and modeling of the observed data. Despite the inclination of many plant pathologists to reject the qualitative approach, both approaches can be successfully used.

3.1 Single Season Predictions

Predicting that a disease outbreak is likely or that a control technique, usually a pesticide application, is necessary to avoid disease intensity buildup and resulting yield loss, is highly beneficial, especially with crops that do not routinely receive many, or even any, pesticides. Forecasters can be developed using qualitative and quantitative methods.

3.1.1 Qualitative Development of Forecasters Based on Weather

As Stevens (1935) showed more than 50 years ago, one can subjectively relate disease intensity, e.g., Stewart's bacterial wilt, to the previous winter weather conditions and develop a useful forecasting scheme. Predictions are based on a simple sum of the average ambient air temperatures in December, January, and February.

It may be possible to develop similar systems for other important pathogens that survive in or near the fields where the crop is planted. Survival can either be in alternate hosts, vectors, plant debris, or soil. An investigator needs data from several years to obtain a wide range of disease/weather values. An exact number cannot be given because this depends on the variability from year-to-year. At first, forecasts can be based on a very susceptible cultivar; refinements could then include less susceptible cultivars. Patience and persistence of the investigator will hopefully lead to some simple relations that could accurately predict disease or the need to apply a control.

3.1.2 Qualitative Development of Forecasters Based on Disease Intensity

Predicting future disease intensity, likely yield losses, or the need to use a control, based on current disease level is highly desirable. In fact, some very successful forecasters of this type have been developed. For instance, the first fungicide application to control *Botrytis* leaf blight of onions can be based on a so-called critical disease level (CDL) of one *B. squamosa* lesion per ten leaves (Shoemaker and Lorbeer 1977). The CDL can be determined with a series of plots in which sprays are initiated corresponding to several different disease intensities. Repeating this study in time or at different locations could result in finding a CDL in which initiating a fungicide schedule at the CDL results in final disease and/or yield that are little different from a standard full application schedule.

This approach also can be used with crops that are sprayed only once or, at most, a few times. At Ohio State University, P. E. Lipps has developed an effective system for recommending the application of a sterol-inhibiting fungicide to control powdery mildew of wheat based on a CDL (*unpublished*). A spray is only recommended if 1% disease severity is reached on the flag or second leaf by the boot stage or 1% of the flag leaf at flowering. The CDL was obtained by experimenting with a range of disease severities at different growth stages and then observing final yield.

Using a CDL in forecasting is becoming increasingly more popular. Disease levels are now routinely used in some European countries to make pesticide recommendations; indeed, EPIPRE is based on this approach (Zadoks 1984). There are, however, two inherent problems in using CDL's. Firstly, many diseases are difficult to assess, even for plant pathologists. Large errors in assessment could be made (see Kranz, Chap. 3). The second problem is less easily solved. Because of the incubation period, visible disease is only a fraction of total disease. One percent visible severity on a leaf may actually represent a much higher total severity. If visible disease always represented a constant percentage of total disease, then no problem would exist. With diseases that progress at a variable rate dependent on the environment, the proportion of "invisible" disease will vary. Therefore, spraying when 1% visible disease is reached may be too late in some years. Despite this cautionary note, using a CDL has been successful and will continue to be so. Investigators need to realize that for some diseases such a system may not work. We also refer to Sect. 4.3.3 of Chap. 9 (R. D. Berger).

3.1.3 Quantitative Development of Forecasters

Formal statistical techniques are valuable in forecaster development. Predicting stripe rust and *Septoria* blotch of wheat as a function of winter/spring weather variables are two good examples of this type of forecaster (Coakely et al. 1982, 1985). So too, is the prediction of ascospore maturatin of *V. inaequalis* (MacHardy and Gadoury 1985). Forecasts are made using linear or nonlinear models.

Regression analysis is the most popular technique for model building. Many articles are available for plant pathologists to learn the basics of regression (Butt and Royle 1974; Madden 1983; Neter et al. 1983); only relevant points can be discussed here.

An example of a linear regression model used for disease forecasting is the following equation of Coakley et al. (1985):

$$Y = 147.5 - 3.0\,X_1 - 2.1\,X_2, \qquad (13.1)$$

in which Y represents disease severity of *Septoria* blotch of wheat on adjusted Julian day 170 (assuming that heading occurred at day 144 of the year), X_1 is the total consecutive days without precipitation between 26 March and 4 May, and X_2 is the total consecutive days between 4 April and 3 May that minimum temperature was $\leqq 7\,^\circ\mathrm{C}$.

Developing a regression model as above is not a trivial matter. Every potential model must be evaluated using several statistics as well as with graphic techniques. The coefficient of determination (R^2) must be acceptable using a subjective criteria, e.g., >0.50 or >0.90. The mean square error (MSE), or error variance, must be reasonably low. There must be an overall significant relation between Y and the X's as determined by an F-test ($P<0.05$ or 0.01). Usually, overall significance at $P<0.05$ is easy to obtain; utility of the model is not guaranteed simply because the calculated F was significant. All estimated parameters, except B_0 (Y-intercept) should be significant as determined by t-tests. The X's should not be highly correlated. For instance, the number of days with an average T below 7 °C will be highly correlated with the number of days with an average T below 6 °C; one would not want to use both terms in the model. Correlations among the X's can be easily determined prior to the regression analysis. With each regression model, the degree of correlation among the X's can be assessed with the variance inflation factor (VIF) calculated for each X. Many of the statistical assumptions that must be satisfied to properly use a regression-derived model can be evaluated graphically. In particular, the residuals (differences between observed and predicted Y's, which are estimates of the error terms [e's]) can be plotted and examined in various ways to properly evaluate a model (Neter et al. 1983).

Choosing the X's is subjective and open to controversy. With a situation like *Septoria* glume blotch of wheat, there are virtually an infinite number of weather variables that could be calculated. The starting (or ending) time (e.g., day 140) for calculating the weather variables as well as the time duration for the variables must be decided upon. Coakley et al. (1985) presented a formal and automated procedure for making these decisions. Arbitrary starting and ending periods, e.g., 1 week or 1 month, may be sufficient during early model building.

A mechanism for using only a small subset of the possible X's must be utilized. Randomly selecting X's to insert in a model will not be productive. There are many *stepwise* procedures for variable selection; many of them can be performed automatically by statistical computer packages. With correlated X's, there is not a unique result from stepwise regression. In fact, one should not accept the "final" model from a stepwise procedure as the end product of the analysis. Alternative models must be considered. With the help of a statistician, plant pathologists should attempt to use more than one stepwise regression procedure, possibly even the recently popular interactive variable selection method. Each model should be evaluated based on R^2, R_a^2 (R^2 adjusted for degrees of freedom), MSE, significant estimated parameters, correlation among the X's, acceptable residuals, presence of outliers or other data points with a large influence on the results (high "leverage" points), and ability to accurately predict Y's used and *not* used in the model.

A model can be *verified* by its ability to predict the Y's used in developing the model. For instance, consider a simple linear relation between Y and a single X. The predicted Y's, based on regression analysis, can be determined together with a confidence band (e.g., 95%) around the predicted line (Fig. 13.1 D). Overall agreement between observed and predicted disease values can be ascertained with R^2 (or MSE); individual agreements can be determined as the difference between each observed and predicted value. Equivalently, the observed Y's should be close to the predicted line and within a narrow confidence band.

A model can be *validated* by its ability to predict Y's not used in developing the model. This data can come from another year or location, or be deliberately "held back" in the initial model development and parameter estimation. There is an alternative to not using a proportion of the data which has been called the "leave one out", PRESS (Predicted Error Sum of Squares), or the "jacknife" procedure. The PRESS procedure can best be visualized as an iterative method. Initially, the first observation or value in the data set (both disease and X's) is omitted from the data and the regression parameters are determined. Then the omitted Y is predicted based on the omitted X's and parameters calculated *without* using the observation. The omitted observation is returned to the data set and the second observation is omitted. The regression analysis is performed and the whole procedure then continues. At the end, there is a set of predicted Y's that were determined without using the observed Y and X's in the parameter estimation. The various regression statistics can then be calculated such as the R^2 for the PRESS situation (R_p^2).

Investigators may wish to use nonlinear models for predicting disease based on environmental or other variables. For empirical model building as discussed here, we do not recommend the use of nonlinear models. Estimation procedures are exceedingly more difficult than in the linear case. Selection of the predictors does not easily lend itself to nonlinear models. Model evaluation is less clear-cut and most calculated statistics are only approximate. Only after extensive linear model building is complete should investigators consider nonlinear models for this type of data.

Sometimes, a long-term data base may exist in which disease is only represented as "mild" or "severe", or continuous disease intensity data are available but the investigators cannot develop a precise regression model because of high variability. This latter situation is not unexpected if one is trying to predict disease in mid- to late-season based on weather conditions the previous winter. Fortunately, it is possible to predict relative disease classification (e.g., mild or severe) based on weather or other variables by means of discriminant analysis. The objective of a discriminant analysis is to assign an observation to one of two or more groups with a low error rate, on the basis of a set of (measured) discriminating variables (Lachenbruch 1975; or other multivariate text). The discriminating variables are directly analogous to the predictors (X's) of regression. A linear model is produced in which a new variable, the discriminant score (D) replaces Y of regression.

Discriminant models can and should be evaluated just like a regression model. Overall significance can be measured with a statistic called Wilks' lambda which can be approximated with an F-distribution. Significance of the individual X's can be determined in a similar fashion. Stepwise procedures can be employed for variable selection. An R^2, as such, is not calculated, but an error rate (0–1) for misclassifications is.

Madden et al. (1983) successfully used discriminant analysis to predict "mild" and "severe" levels of maize dwarf mosaic in early season maize trap plants in southern Ohio, USA, based on winter and spring weather variables. Data from 13 years were analyzed and a stepwise procedure resulted in the following model:

$$D = 0.23\,X_1 + 0.97\,X_2 + 0.43\,X_3 - 14.69, \tag{13.2}$$

in which X_1 is the number of days in January, February, and March with an average air temperature $<0\,°C$; X_2 is the average daily minimum air temperature in December, January, and February; and X_3 is the number of days with rain in March and April. The X's for each year were used to calculate D, and all years were then correctly classified into the appropriate group.

Although the data used for 13 years are extensive, we do not recommend discriminant analysis for much smaller data sets, especially if more than two groups are used. When three or more groups are to be classified, the explanation for discriminant analysis is based on matrix algebra and becomes much more difficult, and alternate statistical methodology is possible.

3.2 Multiple Season Predictions

With crops that are heavily managed throughout the growing season, it is often beneficial to predict at many times disease increase or the need for pesticide applications. A forecaster of this type may reduce the number of pesticide applications and should result in better timing of applications.

3.2.1 Qualitative Development

The many systems for forecasting potato late blight, most based on cumulative weather conditions, attest to the reliability of certain qualitatively derived forecasters (Zadoks 1984; Krause and Massie 1975). As with the single season forecasters, observations on disease, weather, and possibly other factors are needed from several seasons and possibly more than one cultivar and/or location. A patient and persistent investigator needs to compile the data in such a form that results are obvious and uncomplicated. Hyre's (1959) diagram for developing a forecaster of lima bean downy mildew (caused by *Phytophthora phaseoli*) serves as a good example (Fig. 13.2.).

Favorable temperatures and rainfalls, empirically derived, are used to schedule fungicides. Weather data are summarized in the form of arbitrary, ordinal values (e.g., severity values of BLITECAST) that represent the overall favorableness of environmental conditions (Krause and Massie 1975). Control applications usually are determined when the accumulation of the so-called severity values reaches some critical level. The critical level can be determined from observations based on (many) previous seasons or on field experiments in which treatments correspond to spraying at different critical values (Madden et al. 1978).

Fungicide sprays can be recommended on the basis of a biological factor. For instance, one could use number of spores trapped as a predictor of disease increase (Berger 1969; Harrison et al. 1965).

3.2.2 Quantitative Development

Understanding the relationship between disease at a given time and a series of possible predictor variables can be dramatically improved by using formal statistical procedures. The classic studies of Burleigh et al. (1972), Royle (1973), and Schrödter and Ullrich (1965) on predicting disease intensities with regression equations demonstrate the useful quantitative development of within-season forecasting systems. Following their methods will likely result in forecasters for other diseases and crops.

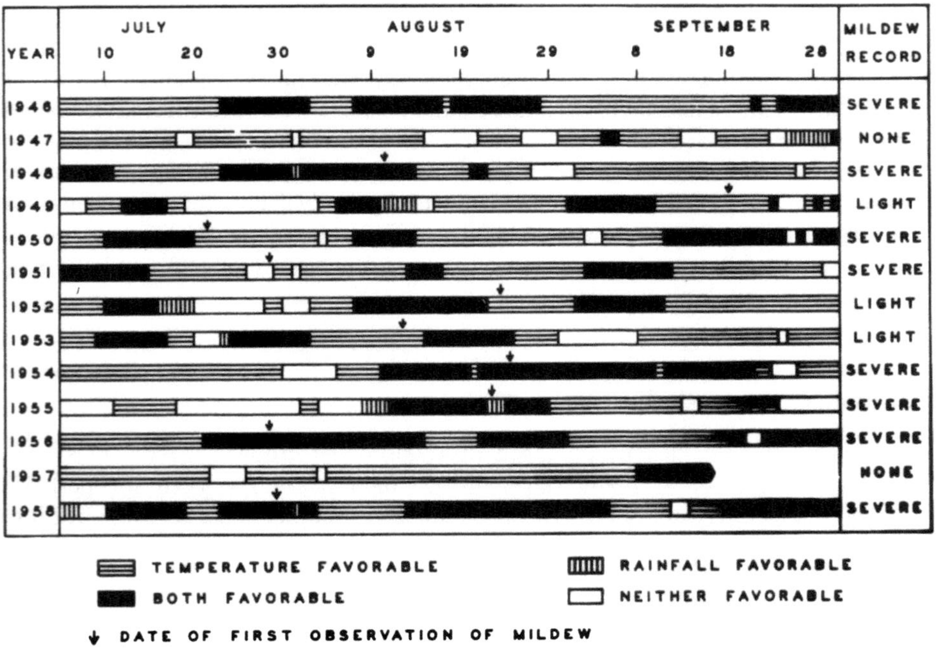

Fig. 13.2 Summary of temperature and rainfall data in relation to the importance of downy mildew of lima bean from 1946–1958 in Bridgeton, NJ, USA (Hyre 1959)

The previous discussion on using regression and discriminant analysis holds in the present situation and will not be repeated here. The set of observations, however, may come from *t* weeks during a single season. A separate analysis could be performed on each year and the results compared. We can almost guarantee that there will be differences among years or locations or cultivars, both in the significant predictor variables and in the estimated parameters. There are two overriding reasons for the differences. Firstly, investigators will seldom consider all of the factors that influence disease. Indeed, disease development is related to environmental factors in much more complex ways than can be depicted by empirical models. Sometimes simple equations will be able to predict disease, but sometimes they will not. Secondly, the high degree of correlation among many weather variables makes it difficult to determine the best relationship between disease and a set of predictors. This is a statistical problem, not a biological one. As discussed previously, in-depth variable selection and model development, with the aid of a statistician, can lead to useful empirical models.

One special situation needs to be addressed regarding predicting the increase in disease intensity throughout a season. Even with invariable environmental conditions, the change in disease will not be constant. Therefore, one should not attempt to relate the absolute change in disease intensity with a set of variables. One needs to use a transformation of disease that provides an additive scale for relating disease to the predictors. If one had information on the mode of disease increase (e.g., logistic or Gompertz), then the proper linearizing transformation could be used (Madden 1980). Even if the mode of increase is not known, the logistic transformation still eliminates certain statistical problems.

3.3 Combining Fundamental and Empirical Forecasting Systems

The classification of forecasters used in this chapter can break down in application. Many empirical forecasters have a fundamental basis. Likewise, fundamental forecasters usually have many empirical elements. Fundamental knowledge and models of a disease can, sometimes, be incorporated into a previously developed empirical forecaster. This was done with potato late blight in which the empirical BLITECAST was modified by incorporating computer simulation results, specifically host resistance and fungicide effects (Fry et al. 1983). The classification is not meant to be absolute but rather to serve as a practical guide to view and develop such systems. Investigators will need to balance many factors in developing disease forecasters. Knowledge of the pathosystem, historical data, facilities for controlled-environment research, computers, and modeling expertise will all influence the investigation.

4 Field Testing

No matter how a forecaster is developed, it must be functional and reliable in the field under practical crop management practices. Disease management strategies based on a forecaster should not be recommended without proper validation.

If a forecaster is used to predict disease intensity at a particular time (or crop growth stage), then one needs to obtain data on disease and the predictors (e.g., T) that were *not* used in model development. These data may or may not be obtained from field studies in which the crop and/or environment is manipulated to potentially achieve varying levels of disease. For instance, one could alter irrigation to change early season moisture in some treatments to see how well final disease can be predicted. Often it will not be possible to effectively vary conditions in a small plot area to obtain a range of disease intensities. One then needs to use data from other locations or years to properly validate the model(s).

Validation of a model cannot be defined in absolutes. Although predicting 0% disease intensity when there actually was 90% would be considered unacceptable. predicting 80% would probably be considered an acceptable prediction. (Discussion of these predictions does not pertain to acceptable disease control.) Deciding on the acceptable error level is subjective, although statistics can help. Ideally, the actual disease value should lie within a *prediction* interval (e.g., $P = 0.05$) determined by the regression model. Or with a discriminate model, the posterior probability of classification into the correct group should be greater than 0.50. With enough new observations, all predictions will not meet the above constraints. Then, a judgment must be made as to the usability of the forecaster. If imprecise predictions make a difference in disease management, model modifications are necessary. If, however, relatively large errors still result in the same management (i.e., spray or no spray), then the forecaster is operating acceptably.

Many forecasters are used specifically for scheduling fungicides. Proper validation then consists of a comparison of the forecaster-generated spray schedule with other treatments. These other treatments should include the commonly used commercial or standard schedule(s) and possibly a no-fungicide control. There is danger in using a treatment with a high level of disease (e.g., control) due to interplot interference.

Sprayed plots (plants) near the controls will have a higher level of disease than if they were near low-disease plots (plants). Providing a large separation between plots or having barrier plants (crops) can be beneficial. Our treatments or experimental units are generally in close proximity. This results in a conservative test of a forecaster. If the system is successful under these severe conditions, we feel that it will work under other circumstances. Investigators should be aware that the failure of a forecaster in an experimental plot may be due to unrealistic conditions.

In a field trial for validating a forecaster, one needs to compare among the treatments: the amount, type, and frequency of fungicide applications; disease intensity at one or more times; possibly the rate of disease increase; and yield. Fohner et al. (1984) believe that a forecaster should suppress disease more effectively than the commercial or standard schedule. We believe that a forecaster should be used to achieve optimal control, obtaining a given level of control most *efficiently*. In our opinion, a forecaster that results in a somewhat higher level of disease and a slightly lower yield, but schedules fewer sprays and results in greater profit to a grower than a commercial control schedule, would be very successful. A forecaster that schedules more control applications but results in high yields would also be successful. Determining optimal management practices concerns many factors beyond the scope of disease forecasting. For the narrower topic discussed herein, we feel that an acceptable and practical forecaster results in disease levels statistically equal to or less than the commercial schedule. Information on the economics of reducing (increasing) the number of spray applications in relation to yield is lacking and studies need to be conducted.

5 Implementation–Delivery and Operation

Having fundamental or empirical predictive models will not result in better disease management if they are not used by growers. Unfortunately, there is not always an automatic transfer of information and technology from the researcher to the extension agent to the growers. In many ways, work on forecaster implementation is as important as work on model development.

Butt and Jeger (1985) list four reasons why growers do not use predictive models. These are: "*research*, in that the models (forecasters) are not available; *synthesis*, in that the models have not been simplified and 'packaged'; *dissemination*, in that there is no mechanism for communicating to the grower; and *reception*, in that the model is neither clear nor usable by the growers." The previous sections of this chapter have discussed the "research" of model development and testing. The last three areas involve a large commitment in extension-related activity, both by university personnel and the private sector.

As discussed by others (e.g., Butt and Jeger 1985; MacKenzie 1981), there is a natural range of grower acceptance and ability (inclination) to use forecasting systems. Many growers are risk adverse, meaning that they will use excess pesticide (rates or frequency) to provide *insurance* against disease epidemics. Other growers simply prefer to follow a fixed chemical schedule due to the common practice of contracting a company to spray their crop at preassigned times. Nevertheless, there are still a substantial number of growers who are estremely eager to use new and innovative approaches.

The level at which forecasting systems are used varies greatly. Although many of the popular systems such as BLITECAST and the apple scab forecaster use data collected within the grower's field, others such as the stripe rust of winter wheat forecaster in the northwestern USA use data collected at weather stations and make forecasts for whole regions. No overall principle exists for the proper level of operation. One normally would expect that individual field-based systems are more accurate than a regional system. This may be true, but if a regional system works, then obviously it is accurate enough. Generally, if a forecaster was developed with data (disease, weather, etc.) from individual fields, then the system will work best at this level, and vice versa.

There are many ways of collecting and manipulating weather data collected from within fields. Traditionally, nonelectronic sensors were used, such as hygrothermographs placed in weather shelters. Data then had to be transcribed by hand and processed in some fashion. BLITECAST gained great popularity when growers could call in their weather data by phone and obtain their recommended spray schedule, without charge, from an operator who ran a mainframe computer program. When a charge was made for the service, interest declined. Efforts have been made by various workers to eliminate the need for a mainframe computer. Forecasting programs have been incorporated into microcomputers and even calculators. Simplified charts for making calculations have also been used. All of these methods, however, depend on a considerable investment of growers' time in transcribing weather sensor data and inputing the data into a computer/calculator/chart scale.

Advances in microprocessor technology have resulted in an alternative approach to data monitoring and disease forecasting. We are using a microprocessor-based electronic device to monitor and store weather data, run disease prediction programs, and display suggested control measures (Ellis et al. 1984, 1986; Jones et al. 1984). Models for additional diseases and other pests are easily incorporated using a plug-in module.The work required by a grower is thus substantially reduced compared to the hygrothermograph approach.

There are still some problems with this method. Cost is considered excessive by many growers considering that complete personal computers can be purchased for less. Quality control and postsale service of damaged or defective units can also be a severe problem. A final problem concerns grower attitudes. If the electronic unit is not used properly or fails, due to a lightning strike or other natural causes, the grower may decide to revert back to a commercial schedule without even asking for help. Nevertheless, microprocessors and electronic sensors will continue to be used to monitor the physical environment (Chap. 7). Instead of processing data in the field, a more logical approach is to transmit the collected data via phone line or telemetry to a receiving microcomputer in an office or home. There the computer can use the transmitted data to determine disease forecasts. New programs are easily incorporated into the receiving computer, a machine that can also be used to receive regional forecasts. We are proceeding in the development of several of these components.

Reception of this technology by growers will be dependent on the number and risk probability (see Weltzien, Chap. 14) of important pests that are predicted. Many growers are reluctant to purchase this equipment to predict only one disease. Incorporating models to predict two or more diseases drastically increases grower acceptance. Ultimately, prediction models for arthropods and weeds need to be incorporated into electronic units to better integrate pest management strategies.

6 Summary and Conclusions

Plant pathologists have developed many useful empirical and fundamental forecasting systems. Some are used to predict disease outbreaks and resulting losses once per season, while others are used throughout a growing season to schedule fungicide applications. Predictors are usually weather variables, but disease and/or inoculum levels are also utilized. Traditionally, disease forecasters were used to reduce the number of chemical sprays. More recently, forecasters have been found useful for recommending highly selective sprays in crops that are not routinely protected by fungicides.

Disease forecasters can be developed quantitatively and qualitatively. The latter approach is older, yet has resulted in some of the most successful prediction systems. We find it likely that additional forecasters will be qualitatively derived. However, we believe that most new forecasters will be quantitatively derived using statistical techniques such as regression and discriminant analysis. The analysis will be based on empirical field studies and also fundamental controlled-environment experiments.

Not all growers will use forecasting systems. Those who are risk-adverse will not wish to use a system that reduces the number of spray applications. These growers may, however, use a system that recommends new, additional chemical applications. Growers who cannot tolerate *any* disease, such as seed producers (MacKenzie 1981), should probably not use forecasters that reduce chemical treatments. Finally, growers who do not wish to closely monitor their crops will not find forecasting to be of benefit. Nevertheless, there is a large grower clientele that will eagerly use disease forecasting to more efficiently manage their crops. This latter grower will find forecasting most useful when: the disease is important and variable in level; effective and legal controls are available; and appropriate techniques are available to record the predictor variables (Bourke 1970). Forecasters will continue to be developed for important plant diseases and the number of growers using forecasters will increase, but only through the continuing basic and applied research by epidemiologists and extension specialists.

7 References

Berger RD (1969) A celery early blight spray program based on disease forecasting. Proc Fla State Hortic Soc 82: 107–111

Bourke PMA (1970) Use of weather information in the prediction of plant disease epiphytotics. Annu Rev Phytopathol 8: 345–370

Burleigh JR, Eversmeyer MG, Roelfhs AP (1972) Development of linear equations for predicting wheat leaf rust. Phytopathology 62: 947–953

Butt DJ, Jeger MJ (1985) The practical implementation of models in crop disease management. In: Gilligan CA (ed) Advances in plant pathology, vol 3. Mathematical modelling of crop disease. Academic Press, London, pp 207–230

Butt DJ, Royle DJ (1974) Multiple regression analysis in the epidemiology of plant diseases. In: Kranz J (ed) Epidemics of plant diseases. Mathematical analysis and modeling. Springer, Berlin Heidelberg New York, pp 78–114

Coakley SM, Boyd WS, Line RF (1982) Statistical models for predicting stripe rust on winter wheat in the Pacific Northwest. Phytopathology 72: 1539–1542

Coakley SM, McDaniel LR, Shaner G (1985) Model for predicting severity of *Septoria tritici* blotch on winter wheat. Phytopathology 75: 1254–1251

Ellis MA, Madden LV, Wilson LL (1984) Evaluation of an electronic apple scab predictor for scheduling fungicides with curative activity. Plant Dis 68: 1055–1057
Ellis MA, Madden LV, Wilson LL (1986) An electronic grape black rot predictor for scheduling fungicides with curative activity. Plant Disease 70: 938–940
Fohner GR, Fry WE, White GB (1984) Computer simulation raises question about timing protectant fungicide application frequence according to a potato late blight forecast. Phytopathology 74: 1145–1147
Fry WE (1982) Principles of plant disease management. Academic Press, London
Fry WE, Fohner GR (1985) Construction of predictive models. I. Forecasting disease development. In: Gilligan CA (ed) Advances in plant pathology, vol 3. Mathematical modelling of crop disease. Academic Press, London, pp 161–178
Fry WE, Apple AE, Bruhn JA (1983) Evaluation of potato late blight forecasts modified to incorporate host resistance and fungicide weathering. Phytopathology 73: 1054–1059
Gill JL (1978) Design and analysis of experiments in the animal and medical sciencies, vol 2. The Iowa State Univ Press, Ames Grove GG, Madden LV, Ellis MA, Schmitthenner AF (1985) Influence of temperature and wetness duration on infection of immature strawberry fruit by *Phytophthora cactorum*. Phytopathology 75: 165–169
Harrison MD, Livingston CH, Oshima N (1965) Control of potato early blight in Colorado; II. Spore traps as a guide for initiating applications of fungicides. Am Potato J 42: 333–340
Hau B, Eisensmith SP, Kranz J (1985) Construction of temporal models: II. Simulation of arial epidemics. In: CA Gilligan (ed) Mathematical Modelling of crop disease. Adv plant pathology vol 3, pp 31–65
Hyre RA (1959) The development of a method for forecasting downy mildew of lima bean. Plant Dis Rep 257: 179–180
Jones AL, Fisher PD, Seem RC, Kroon JC, Van de Motter PJ (1984) Development and commercialization of an in-field microcomputer delivery system for weather-driven predictive models. Plant Dis 68: 458–463
Krâuse RA, Massie LB (1975) Predictive systems: modern approaches to disease control. Annu Rev Phytopathol 13: 31–47
Lachenbruch PA (1975) Discriminant analysis. Hafner, New York
MacHardy WE, Gadoury DM (1985) Forecasting the seasonal maturation of ascospores of *Venturia inaequalis*. Phytopathology 75: 381–385
MacKenzie DR (1981) Scheduling fungicide applications for potato late blight with BLITECAST. Plant Dis 65: 394–399
Madden LV (1980) Quantification of disease progress. Prof. Ecology 2: 159–176
Madden LV (1983) Measuring and modeling crop losses at the field level. Phytopathology 73: 1591–1596
Madden L, Pennypacker SP, MacNab AA (1978) FAST, a forecast system for *Alternaria solani* on tomato. Phytopathology 68: 1354–1358
Madden LV, Knoke JK, Louie R (1983) Classification and prediction of maize dwarf mosaic intensity. In: Gordon DT, Knoke JK, Nault LR, Ritter RM (eds) Proc Int Maize Virus Disease Colloq and Workshop, 2–6 August 1982. The Ohio State Univ, Ohio Agric Res and Devel Center, Wooster, pp 238–242
Miller PR (1959) Plant disease forecasting. In: Holtan CS (ed) Plant pathology problems and progress. Univ of Wisconsin Press, Madison, pp 557–565
Neter J, Wasserman W, Kutner MH (1983) Applied linear regression models. Irwin, Homewood, IL
Parvin DW Jr, Smith DH, Crosby FL (1974) Development and evoluation of a computerized forecasting method for Cercospora leaf spot of peanuts. Phytopathology 64: 385–388
Pennypacker SP, Stevenson RE (1982) The management of plant pathogens. In: Hatfield JL, Thomason IJ (eds) Biometerology in integrated pest management. Academic Press, London, pp 243–265
Rouse DI, Nordheim EV, Hirano SL, Upper CD (1985) A model relating the probability of foliar disease incidence to the population frequencies of bacterial plant pathogens. Phytopathology 75: 505–509
Royle DJ (1973) Quantitative relationships between infection by the hop downy mildew pathogen, *Pseudoperonospora humuli*, and weather and inoculum factors. Ann Appl Biol 73: 19–30

Schrödter H, Ullrich J (1965) Untersuchungen zur Biometeorologie und Epidemiologie von *Phytophthora infestans* (Mont.) de By auf mathematisch-statistischer Grundlage. Phytopathol Z 54: 87–103
Shoemaker PB, Lorbeer JW (1977) Timing initial fungicide applications to control *Botrytis* leaf blight epidermics on onions. Phytopathology 67: 409–414
Shrum RD (1978) Forecasting of epidemics. In: Horsfall JG, Cowling EB (eds) Disease, an advanced treatise, vol. 2. How disease develops in populations. Academic Press, London, pp 223–238
Spotts RA (1977) Effect of leaf wetness duration and temperature on the infectivity of *Guignardia bidwellii* on grape leaves. Phytopathology 67: 1378–1381
Stevens NE (1935) Experimental forecast of the incidence of bacterial wilt of corn in 1935. Plant Dis Rep 19: 69–70
Sutton JC, Gillespie TJ, Hildebrand PD (1984) Monitoring weather factors in relation to plant disease. Plant Dis 68: 78–84
Thresh JM (1986) Plant virus disease forecasting. In: McLean GD, Garrett RG, Ruesink NG (eds) Plant virus epidemiology: monitoring modelling, and predicting outbreaks. Academic Press, London, pp 359–386
Van der Plank JE (1975) Principles of plant infection. Academic Press, London
Waggoner PE (1960) Forecasting epidemics. In: Horsfall JG, Dimond AE (eds) Plant pathology, an advanced treatise, vol 3. Academic Press, London, pp 291–312
Zadoks JC (1984) A quarter century of disease warning, 1958–1983. Plant Dis 68: 352–355

Chapter 14

Techniques for Studying Aphid-Borne Virus Epidemiology

B. Raccah[1] and M. E. Irwin[2]

We dedicate this chapter to the late Dr. Warren C. Adlerz, who devoted his life to the epidemiology of aphid-borne viruses.

1 Introduction

Epidemics caused by fungi pivot on complex interactions among the host plant, its pathogen, and the environment. Epidemics of plant-pathogenic viruses include a fourth element, the vector, which is usually an insect. Although being a specific case amongst dispersal mechanisms (see Rotem, Chap. 5), only the vector element compounds the complexity of the relationship by necessitating an understanding of the interactions between the pathogen and its vectors and between the vectors and the host plant of the virus. By adding the vector element to the system, studying plant-pathogenic virus epidemiology becomes more tightly interwoven. Even when plant-pathogenic fungi are transmitted by insects, the association between pathogen and vector is relatively loose, whereas the association between virus and vector is almost always intricate and complex.

We restrict this chapter to virus groups that are associated with aphids as vectors. Epidemics of these viruses are notorious and well documented (McLean et al., 1986; Plumb and Thresh 1983). Other important virus diseases are transmitted by several additional groups of arthropods (leafhoppers, planthoppers, whiteflies, mealybugs, thrips, beetles, bees, eriophyid mites, and others), nematodes (Barker and Noe, Chap. 15), and fungi. Certain procedures described herein relating disease incidence to vector presence are or might be applicable to these additional vector groups. However, differences in their life cycles, flight, and feeding behavior mandate procedures and techniques that are group-specific. These will not be discussed in this chapter; the reader is referred to McLean et al. (1986) and Plumb and Thresh (1983).

The five parameters that must be ascertained to understand the spread of aphid-borne viruses are: identity of the virus that causes the infection; source plants from whicht the virus is acquired; role of aphids in the transmission process; movement of infective aphids; role of aphids; and environmental conditions impinging on the above parameters.

Each of these parameters should be determined or estimated before a valid assessment of its probable contributions to spread can be made. Previous studies and reviews provide considerable information on available techniques (McLean et al. 1986; Plumb and Thresh 1983). Here we describe the methods currently in use to collect the above data and add recently developed techniques and procedures.

---

[1] Department of Virology, ARO, Volcani Center, Bet Dagan, Israel
[2] University of Illinois and Illinois Natural History Survey, Urbana, Illinois, USA

2 Detecting Plant-Pathogenic Viruses in Plants and Vectors

Various methods have been devised to detect and quantify viruses in plants and vectors. The methods for a given situation should be chosen according to research objectives, the degree of accuracy required, and the practicality of the methods in epidemiological studies. In many instances, disease symptoms are sufficiently obvious to detect the presence of a virus in a host plant. However, where the disease is symptomless or where more than one virus is likely to be present, visual inspection can result in erroneous information. Measures are often needed that are more sensitive or more specific than visual detection. We refer the reader to comprehensive reviews on the subject by Clark and Bar-Joseph (1984), Dodds et al., (1984), and Milne and Luisoni (1977). In this chapter we emphasize procedures that are commonly in use in plant virus epidemiology, with particular focus on those that are applicable to the detection of viruses in aphids.

2.1 Visual Inspection of Infected Plants

Visual inspection continues to be the most common and simplest way to monitor virus infection in plants in the field (see Kranz, Chap. 3). A skilled inspector can rapidly field-screen hundreds of plants by walking along the plant rows under study. While rapid and simple, this procedure fails to detect all of the infected plants in a plot for several reasons: (1) Certain viruses do not cause symptoms in their host plants in the field; (2) some symptoms are expressed only during the latter stages of virus infection, long after the host plant serves as a source for virus acquisition by aphid vectors; (3) some infected plants do not produce clear symptoms when the temperature is above optimum, and (4) symptoms only appear on flowers, fruits, or subterranean parts of the plant.

Before modern detection techniques were available, potentially diseased plants where tested by mechanical inoculation of a series of test plants that would produce visible symptoms. Grafting is done for some viruses that are not transmitted easily by mechanical inoculation. Although simple, these procedures are laborious and require an abundance of space to obtain the desired results.

2.2 Laboratory Methods for Detecting Viruses in Plants and Vectors

In addition to indexing for disease progress in the field, laboratory procedures are most common for detecting plant pathogenic viruses in plants and vectors.

Electron Microscope Procedures (EM):

The procedure is too laborious to be used for the mass indexing needed in virus epidemiological studies involving plants or vectors. However, the leaf-dip method has been used successfully to detect citrus tristeza virus (CTV) in citrus trees (Bar-Joseph et al. 1974). Milne (1984) provides a detailed description of electron microscope methods.

Immunoelectron microscopy (IEM) can combine the advantages of EM and serology. The technique has special merit when two distinct disease-causing agents are present in the same host tissue. Milne and Luisoni (1977) detail the procedures in IEM.

Two IEM techniques are useful for separating viruses of more than one type from the same host tissue: trapping and decoration. Trapping increases the probability of finding virus particles in EM even when they are in low concentration in host tissues.

The procedure is based on trapping virus particles on grids that were previously coated with specific antibodies. The decoration procedure complements trapping. It is based on the creation of an aura of antibody molecules which surround each virus particle trapped on the grid. This makes each particle highly visible and more easily identifiable.

Serological Methods

Serological methods are relatively inexpensive and highly specific. They are based on the affinity that exists between the virus and an appropriate antibody. Many techniques have been developed to take advantage of this reaction, including: immunodiffusion reactions in gels (Ouchterlony 1964) and enzyme-linked immunosorbent assays (ELISA) (Clark and Bar-Joseph 1984).

ELISA is probable the most widely used procedure for detecting and identifying viruses for epidemiological studies, second only to visual inspection. ELISA is simple, inexpensive, and adaptable to large-scale monitoring. It allows quantitative and qualitative determinations of a virus. The technique is based on adsorption of the antibody to a solid support (in most cases a microtiter polystyrene microplate) and visualization of the virus presence by a label which is conjugated to the antibody. For principles and descriptions of various ELISA techniques and their uses for detecting viruses, see Clark and Bar-Joseph (1984).

ELISA has been used also to detect viruses in aphids. The small size of most vectors of plant viruses and the low concentration of the viruses in the vectors demand a very sensitive procedure to detect the presence of the virus in a given vector. Indeed, the high sensitivity of ELISA has made it possible to detect nonpersistent (Gera et al. 1978) and persistent (Tamada and Harrison 1981) viruses in their aphid vectors. It was, however, determined that the detection of noncirculative viruses is erratic and inconsistent with the transmission capability of the vectors (Carlebach et al. 1982).

The sensitivity of ELISA for detecting viruses in aphids is influenced by several factors. The following are recommendations for maximizing the sensitivity of ELISA through careful preparation of material: Aphids should be extracted in a small volume of buffer (0.05–0.1 ml) using an all-glass small homogenizer. Tamada and Harrison (1981) used a micromortar and 600-mesh carborundum to grind the aphids for testing with ELISA. One of the major problems for detecting viruses in vectors by ELISA is the undesirable, highly nonspecific reaction found in virus-free aphids. One approach that has been adopted to increase the specific reaction was to reduce the pH and increase the molarity of the conventional ELISA extraction buffer (Carlebach et al. 1982) Nonspecific reaction was also reduced by omitting polyvinyl pyrrolidone (PVP) (Carlebach et al. 1982; Tamada and Harrison 1981), and by adding a reducing agent, diethyl dithiocarbamic acid (DIECA), to the extraction buffer (Carlebach et al. 1982). Tamada and Harrison (1982) were able to detect potato leafroll virus (PLRV) by extending the duration of incubation with the substrate. Berger et al. (1985) recently used the dot blot immunosorbent assay (DBIA) in an attempt to detect tobacco etch virus in aphids. This procedure allowed them to detect ca. 0.5 pg virus. However, they were unable to detect the virus in viruliferous aphids either by DBIA or by ELISA.

Use of Viral Genome for Detecting Viruses

These procedures are based on the extraction of replicative RNA (double-stranded RNA) (Dodds et al. 1984) or on virus detection using complementary DNA (cDNA)

(Gonda and Symons 1978). In many epidemiological studies, serological tests have insufficient specificity to detect and distinguish among the components in a complex of related viruses (e.g., PLRV, beet western yellow virus, and barley dwarf virus). Detection differentiation methods based on the virus genome could provide the needed sensitivity. If these procedures could be simplified and altered to be used with nonradioactive lables, they would be most useful in future epidemiological studies.

Jayasena et al. (1984) used cDNA to detect the presence of subterranean clover red leaf virus in single aphid vectors. Boulton et al. (1984), using a slightly different technique, detected maize streak virus in single leafhopper vectors. They squashed viruliferous insects in the presence of alkali on nitrocellulose paper. The detection limit was 2.5 pg DNA or 12.5 pg virus per insect. Using the same system, attempts were unsuccessful to detect cauliflower mosaic virus (CaMV) in *Brevicoryne brassicae, Myzus persicae,* and *Acyrthosiphum pisum* (B. Raccah, R. Hull and P.G. Markham, unpublished results) or potato virus Y (PVY) in *Myzus persicae* (B. Raccah and A. Rosner, unpublished results).

2.3 Indexing for Disease Progress in the Field

Indexing plants in the field at specific incidence intervals provides a cumulative record of the number of plants infected over a period of time. This record is most accurate and valuable during the earlier stages of spread when multiple infections account for only a small fraction of the total infections in a field. Indexing is easily accomplished by labeling the plants that are and are not infected in the field or by marking those plants on maps of the field at specific intervals during the epidemic. Another, less laborious, approach is to test the percentage of infected plants (i.e., disease incidence) in samples that are periodically and randomly taken from the field. Serological methods, as described earlier in this chapter, allow fast and reliabel indexing of plant material for the presence of specific viruses (Bar-Joseph et al. 1979).

Disease progress, as a function of time or space, has been presented in numerous studies (e.g., see reviews cited in McLean et al. 1986). Virus spread can be expressed (1) as the distribution of the infected plants on maps of fields, (2) as progress curves that described earlier in this chapter, allow fast and reliable indexing of plant material for mathematical expressions (Raccah 1986).

Indexing for disease progress is a way of testing hypotheses and predictive models regarding the relative importance of different aphids as vector species. Relating virus spread to vector presence and activity is not always easy to accomplish by indexing for disease progress. When the incidence of disease is high and the rate of spread rapid, there is a considerable amount of multiple infections, i.e., inoculations of plants that are already infected. This has a dampening effect on the number of new infections and accounts for the levelling off of disease progress curves as the incidence approaches 100%.

Indexing for disease progress is sufficient to track disease incidence through time or space. However, when the objective of the study is to dissect the disease progress into its component parts (e.g., to differentiate among the components that produce the disease progress), it is important to relate the actual numbers of new potential infections over time to vector activity. Three ways have frequently been employed to assess this information: (1) use of bait trap plants, (2) repetitive plantings of field plots; and (3) infectivity studies with field-captured aphids as potential vectors (as described later in this chapter).

2.4 Indexing Incoming infection in the Fields

Trap Plants

Trap plants are virus-free potted plants that are placed in field plots for predetermined lengths of time (see also Aust and Kranz, Chap. 1, and Rotem, Chap. 5). Exposure of trap plants permits one to substantiate the infection pressure in the field during the period of exposure. Use of trap plants was suggested by Broadbent et al. (1950), who exposed potted, glasshouse-grown potato plants to natural aphid infection, in a potato plot, for a specific time period, and then indexed for infection to potato viruses. The percentage of trap plants infected represents the infection pressure during the time of exposure. Schwartz (1965) exposed test plants to record the spread of tristeza virus in South Africa. The relationship between the timing of virus spread in the field and the vectors responsible for the spread was ascertained. When properly correlated, trap plants were able to relate the infection pressure of the two viruses to the incidence of aphids in the field. The peaks of CMV infection in trap plants have correlated well with vectors species that were active during trap plant exposure (Raccah 1983; Raccah et al. 1985).

Trap plants are usually exposed in precise arrangements of specific age categories that favor optimal infection under field conditions. Thus, the infection pressure indicated by trap plants is generally higher than that actually present in the plot of field. Plant density and spatial patterning are also known to influence the landing rate of aphid vectors (M. E. Irwin, unpublished data) and must be considered when selecting where and how to place trap plants in the field. To minimize the discrepancy between percentage infection of trap plants and actual field infection pressure, we suggest that anyone using trap plants to determine infection pressure should mimic field plant type, age, size, and spacing as closely as possible. When carefully thought out and executed, experiments using trap plants can complement virus epidemiological information obtained with other techniques.

Repetitive Plantings of Field Plots

Repetitive plantings in isolated plots can also be used to correlate infection pressure with vector activity. This allows normal field growth of plants and insures that similarly aged plants are exposed to potential vectors throughout the season. Schultz et al. (1985) used this technique to help determine which aphid species were important vectors at different times during the growing season. The technique works well with nonpersistently transmitted viruses where the primary inoculum is from seedlings that contained virus as seeds of infected parent plants. Infector plants can be inoculated at a given date with the knowledge that they will become sources for virus acquisition about 1 week after inoculation. Schultz et al. (1985) exposed field-sown test plants for 2 weeks, after which all infector plants were removed and adjacent rows read a week later. This insured minimal secondary spread except from infector plants. All four repetitive plantings during the season were treated in the same manner. Two variables that were not controlled were vector activity and weather conditions (different parts of the season). Timing of sowing, inoculating, and symptom indexing are critical for the success of this technique. The results can be misleading if any virus is moving into the plots from outside. However, it allows one to conduct the experiment almost entirely in the field and to reduce reliance on glasshouse tests. It also eliminates the potential differences between infector plants and test plants in the field, because both are part of the same plot.

3 Sources of Plant-Pathogenic Viruses

The intensity and distribution of virus infection in the field and the timing of infection are correlated with the number of virus sources and their distance from cultivated crop. When the virus is seedborne or vegetatively propagated, the source plants are often the same variety and make up a portion of the crop. Weeds that can serve as virus sources and that are present in the field behave similary to the source plants found in the crop. However, many viruses are acquired by aphids from source plants located outside of the field. These source plants can be wild or cultivated. The contribution of wild plants to virus epidemiology was reviewed by Bos (1981). Once the virus source is known, it is possible to calculate the contribution of the source to the rate of infection in the field (Ruesink and Irwin 1986). To determine the contribution of a known or unknown source of the virus, it ist necessary to identify the virus in the various source plants and to prove that each can serve to inoculate the crop plants. It then becomes necessary to identify which aphids are present and, of those, which can act as vectors of the virus. Their capabilities to transmit from the sources (wild, weed, or crop) to the crop under study must be determined. This process, recently described by Raccah and Gal-On (1984), was used to locate source that contributed to the epidemiology of PVY in a pepper field. It was found that *Hyoscyamus desertorum* was an excellent source in the laboratory. Later, it was also proved to be the principal source of PVY in the pepper-growing region of Arava in southern Israel (S. Cohen, O. Okko, and B. Raccah, unpublished data).

4 Assaying for Vector Potential

Determining vector inocultativity is important in quantifying vector potential (Irwin and Ruesink 1986; Marcus and Raccah 1986; Ruesink and Irwin 1986) because different vectors species have different inherent capabilities to transmit. Irwin and Ruesink (1986) devised the term "vector propensity" to describe the actual probability that a given vector specimen will transmit a virus under field conditions. They define vector propensity as the probability that a single vector specimen, having had the opportunity to acquire a virus in nature by landing on an infected plant, will subsequently transmit the virus, provided it lands on a noninfected host plant of the virus in nature. Vector propensity deserves special attention because the term encompasses the probabilities that culminate in the transmission and infection process in an agroecosystem. Assaying potential vectors to determine if they are inoculative is an important part of quantifying vector propensity.

Infectivity tests are the most reliable methods by which the presence of virus can be ascertained in aphids. Several approaches have been developed for assaying vector inoculativity and for detecting the presence of virus in aphids. Many of these approaches are common for detection of viruses in plants. The ones that are specific or advantageous for detecting virus in aphids can be grouped into laboratory assays and field live assays.

4.1 Laboratory Assays

A common approach to the study of the role of aphid vectors in plant virus epidemiology is to determine which aphid species are present on the vegetation or in flight during the epidemic. These species are then raised in the laboratory and tested for their capacity to transmit the virus that caused the epidemic. The most common method is through timed probe assay (Sylvester 1954). This procedure measures the innate ability of the aphid species, or biotype, to transmit the virus isolate to a given test plant. The method is very laborious and requires considerable patience on the part of the investigator.

Acquisition access tests require the manual transfer of aphids from the source plant to the target hosts. In many cases two chambers are required, one for acquisition-access feeding and another for inoculation-access feeding. Aphid feeding time is not monitored in either the acquisition-assess or inoculation-access feeding. These procedures measure the capability and efficiency of the aphid species, and in particular the laboratory aphid biotype, to transmit the virus. If tests the transmissibility of the virus isolate that is present in the laboratory and does not necessarily reflect the transmissibility of other isolates of virus which are, often concurrently, present in a field situation. However, this simple procedure permits one to test the transmissibility of various isolates that can be field collected. This test lacks accuracy but greatly increases the ability to process a large number of specimens in a reasonably short time period, something that is esssential for epidemiological studies.

These two transmission tests do not actually measure the probability of a given vector species to acquire and inoculate virus under field conditions. Therefore, we use a method termed the "arena test" that attempts to take into consideration important vector behavioral components (Irwin and Ruesink 1986; Schultz et al. 1985). In this procedure, aphids are allowed to move freely among infected and noninfected hosts. Laboratory clones of aphids are often used in preference to field-collected or "wild" aphid specimens. This tends to lower the variance of the results of the test somewhat from the situation in the field. In the arena test, a specific isolate of the virus is standardized by growing the isolate-containing source plants under specific conditions in growth chambers.

The arena test allows a fast and close-to-reality determination of vector propensity with a specified virus isolate. The system allows free movement between source and test plants, thus permitting multiple acquisitions and multiple inoculations by each specimen. Furthermore, aphids are given the opportunity to probe frequently and for as long as they choose. Movement from plant to plant and the potential for serial transmissions are accounted for and only slightly hindered by the limited number of test seedlings. However, the following disadvantages should be noted: (1) movement of the aphids is quite restricted, especially the aspect of flight; (2) a laboratory clone of the aphid and a cultivated isolate of the virus are used instead of wild specimens and isolates, although this could be corrected simply by using field-collected specimens; and

(3) the test is spatially and temporally confining, a fact that probably alters the behavioral responses of the vectors to their environments. Nevertheless, it gives more accurate values of vector propensity than those recorded through conventional transmission studies.

4.2 Field Live Assays

Field-oriented assays of aphids as vectors of a given plant virus are based on capturing wild specimens and placing them individually on test seedlings for the required period of time (inoculation-access feeding). If any of these aphids are inoculative, they will have the potential to transmit the virus, providing a measure of propensity. This technique has worked particularly well with persistently transmitted plant viruses where the vector retains infectivity for a long period of time (e.g., A'Brook and Dewar 1980; Plumb 1983). The technique also works well with nonpersistently transmitted viruses; however, successes with this system are dependent on rapid placement of captured specimens on test plants (Halbert et al. 1981; Labonne et al. 1982; Raccah et al. 1985). Evaluation of vector importance can be done by monitoring the rate of aphids alighting on crops (a measure of vector activity) and determining the proportion of the population that actually transmits the virus. Information derived from such assays is used directly, like the index (*sensu* Plumb 1983), as a measure of transmission in the field at that particular time.

Adlerz (personal communication) has found that aphids which were trapped alive in yellow water traps, retain their infectivity. This procedure is now being used by him in Florida in order to compare the transmission of cucurbits viruses by aphids trapped alive in yellow water traps versus those trapped by suction. However, the possible loss of virus by aphids which are maintained in water should be ruled out experimentally.

The obtained value depends on the amount of inoculum present in the field and on the vector propensity. It varies not only with species composition of vectors but also with the change in amount of inoculum present in the field through time. Such a value has great potential in describing disease progress, particularly on a simulative basis, but the work and expertise involved to provide these values on a daily or even weekly basis is staggering. Indeed, most plant virus epidemiological programs have insufficient resources to pursue this course of action. Furthermore, the described test cannot lead to sound methods for predicting epidemics because the value derived, while probably more accurate than the one currently used for vector propensity, is dependent upon the amount of inoculum present in the field and not a near constant that can be fit into a mathematical model.

An alternative approach is based on determining vector propensity to obtain a value that is independent of the amount of inoculum present in a field. This means that once measured in a given locality, each aphid species will have a constant vector propensity value. This parameter can be multiplied by the probability that a given specimen of a given species will alight on an infected plant (a measure of the amount of inoculum present) and, when summed over all specimens of all species, will provide an estimate of infectivity (*sensu* Plumb 1983). Irwin and Ruesink (1986) infected all plants in a soybean plot with soybean mosaic virus (SMV), so that the potential of a vector landing on an infected plant was very high. Once captured on a vertical net set downwind in the field, each aphid specimen could be placed on a test plant and tested for its ability to inoculate. In this way, it was possible to obtain a close approximation of the field potential of landing aphids to transmit the virus. This method assumes natural acquisition-access feeding and a reasonable amount of natural movement, leaving only the vector-inoculation phase somewhat confined and restricted. With SMV, this assay has provided the needed data to calculate the total vector contribution to the infection process (Halbert et al. 1981; Ruesink and Irwin 1986). Schultz et al. (1983) showed that neither plant age at the time of inoculation nor length of time a plant had been infected has a significant effect on an aphid's ability to acquire SMV from a soybean plant. These authors concluded that capturing aphids at different times during the season should not affect the relationship between soybean plants and SMV through time. These conclusions have yet to be tested in other virus-aphid systems.

The described technique has its drawbacks: at least one species of aphid, *Aphis citricola,* was found to fly against the wind when wind velocities were low; and although aphid movement activity (measured by aphid catch per net per hour) was high on several successive days, vector propensity varied greatly among days of much activity. These variations in data most likely resulted from not taking into account certain key behavioral responses of the aphids, among which probing, or the lack of probing, seems most important.

5 Vector Activity

5.1 Vector Abundance and Activity

When monitoring aphids as vectors of plant viruses, researchers have generally tended to quantify various aspects of vector abundance. Abundance denotes numbers per unit of area or, in the case of aerial abundance, numbers per volumetric unit. In either context, abundance can be synonymized with density. Population densities vary with time, producing a change in the density and morphs of any given vector species. The question that must be asked, however, is how important is it to measure the abundance of a vector species? If the objective is to understand, and perhaps model, the epidemiology of a given virus, then the vector density may indirectly affect virus spread by influencing the amount of vector movement and associated activity. Vector movement is a very important element in virus spread. Thus, monitoring population density cannot help resolve the amount and timing of virus spread unless the relationship between vector abundance and vector movement are at least partially understood.

Increases in vector abundance within a crop can precipitate a change in vector activity. With aphids, the development of winged morphs and the process of emigration usually result from stress factors imposed on the host plant, often due to overcrowding of aphids. Thus, it is important to monitor charges in abundance of a vector species on its host plant when such data lead to predicting the timing and quantity of movement. To correlate the phenomenon of movement with that of abundance implies more knowledge than we presently possess about how the two phenomena are linked. Irwin (1980) described several methods for assessing vector abundance. It is also important to monitor environmental parameters (e.g., plant-water stress) because environmental stimuli may be of key importance in the timing of aphid movement.

Many types of activity are related to the behavior of specific vector species. Among the various activities , feeding, dispersal, and host-seeking (including the landing on nonhost species) play roles in virus spread. Without feeding activity (and here we include probing), the virus cannot be acquired from an infected plant, nor can an infective vector inoculate a noninfected plant. Without dispersal activity, the virus cannot move from an infected host plant, and without host-seeking activity by the vector, the virus cannot be carried to potential noninfected host plants of the virus. Host-seeking activity involves finding a suitable host plant for the vector, landing, probing, and often feeding, even on vector-noncompatible plants. Thus, vector activity is the key to virus spread, and each of the components of activity mentioned above should be considered when assessing an epidemiological process, or predictors for forecasting (Madden and Ellis, Chap. 13).

5.2 Monitoring Movement and Host-Seeking Activity

Aphids can move great distances in the air (Taylor 1986); they also move shorter distances between and within fields. Taylor (1986) suggests a continuum between the two. An aphid can get from one plant to another within a field by flying (i. e. aerially) or by walking. Alate (winged) aphids can easily walk among plants when the canopy closes. Irwin (unpublished data) has observed considerable early season spread of SMV across as well as along rows, indicating that a great deal of the spread, especially early in the season, can be attributed to movement by flight. Spread in other aphid and crop systems could be influenced to a much greater degree by walking.

A simple experiment described by Irwin and Ruesink (1986) can provide valuable information about the relative importance of walking activity on virus spread. It requires two treatments in both of which the individual plants are isolated from one another so that the aphids cannot pass from one plant to another by walking. In one treatment the stems of all plants should be ringed with a sticky substance lined to form a sticky barrier (Broadbent and Tinsley 1951), or surrounded with a liquid-filled moat. The other treatment is identical, with the exception that it lacks the barrier to walking among plants. A large percentage (e. g., 50 %) of the plants should be inoculated. For those systems that have colonizing vectors, vectors should be placed on the infected plants to begin colonies at the start of the experiment. If the resulting difference in incidence of virus infection is significant between the two treatments, then walking vectors had an effect on spread. Consequently, when such an experiment reveals that walking adds significantly to the incidence of disease, walking activity should be monitored. We know of no way, other than direct observation, of monitoring walking in a field in which the canopy has closed and the vectors can walk from plant to plant without having to cross the soil. Early in the season, when plants are spatially separated for a long period of time, walking can be assessed by placing pitfall traps into the soil among plants. This has recently been done to assess the importance of apterae in the spread of barley yellow dwarf in Washington State (Pike and Halbert 1983). Depending on the situation and vectors involved, some care is needed to assure that the species in question cannot detect and avoid the trap.

Much attention has focused on the aspect of monitoring aphid landings (Irwin and Ruesink 1986; Schultz et al. 1985). The goal was to monitor near absolute numbers of aphids alighting in soybean fields in central Illinois. Coupled with vector propensity data, this information could provide a good estimate of vector intensity. The conventional traps for aphids may provide erroneous data because of nonproportional catches. The Moericke yellow pan trap is widely used, but various species of aphids are differentially attracted to its color. Therefore, actual and relative alighting activity cannot be derived from these trap catches. Another common aphid trap is a clear, sticky, vertically-oriented cylinder placed 1–1.5 m above the ground. Such traps measure relative abundance of all aphid species at a given point in time. This objection to their use in measuring landing rates is that they are in fact measuring aerial density, and that the catch is dependent on wind speed: at any given density of aerial aphids, the faster the wind, the more that are likely to be caught. Suction traps of the type described by Taylor (1986) are very good for determining absolute aerial densities, and may provide data concerning migrations over medium and long distances. However, it is difficult to predict landing rates from data of aerial densities. The complexities of cues eliciting alighting behaviour for aphids are great and varied. Thus, these and other conventional aphid traps were found to be inadequate for measuring landing rates of aphids.

Irwin (1980) surmised that a trap was needed that would behave exactly like the crop canopy on which the aphids were landing naturally. Such a trap would need the following properties: (1) horizontal attitude, (2) position over and relatively close to the top of the canopy, and (3) the same spectral reflectivity and characteristics as the canopy (Irwin 1980). If the trap looked just like the canopy, its size and shape should make little difference to the catch, especially after canopy closure.

The trap described below appears to provide near-absolute landing rates of all aphid species in a field; it has consistently provided numbers that match the progress of viral epidemics (Schultz et al. 1985). The horizontal mosaic green pan trap consists of a square (12 × 12 cm) ceramic tile (Cambridge 815 from the Cambridge Tile Co., PO Box 15071, Cincinnati, OH 46215, USA) of rugose texture and a color that closely resembles that of mature soybean leaves, based on data from reflectance spectrophotometry. The trap is mounted on a vertical metal rod driven into the ground, to which is attached a two-ended chemistry clamp. The distal arm of the clamp is attached to a plastic sandwich box (Sterilite Corporation, Townsend, MA 01469, USA), within which the tile is placed. The sandwich box is maintained in a horizontal position at a height even with the top of the soybean canopy. The box is filled with a 50% solution of ethylene glycol in water. The use of liquid-filled traps reduces labor costs and preserves the aphids in an easily identifiable condition. Ethylene glycol also minimizes evaporation.

The efficiency of the trap has yet to be proven, it appears to be catching near-absolute landings of aphids. However, during rainfall, most of the aphids are lost because the rain drops propel the aphids out of the trap. Placing vertical baffles in the sandwich box did not reduce this problem (M. E. Irwin, S. E. Halbert and G. E. Kampmeier, unpublished data). Therefore, all of the trapped aphids must be collected daily before the onset of a rain storm.

While the horizontal mosaic green pan trap was designed specifically to collect alate aphids in soybean canopies, the differences in spectral characteristics among chlorophyll-producing plants are minimal, making the trap generally adaptable to most row-crop and low-growing agroecosystems.

6 Epilogue

The parameters regulating plant-pathogenic virus spread by arthropod vectors are more complex and interrelated than those regulation other plant-pathogenic disease spread. Furthermore, many of these parameters are difficult to measure and almost impossible to monitor on a regular basis. We have cataloged the techniques and procedures currently available and, in our opinion, most critical to the study of plant virus epidemiology.

Because of the interwoven complexities of the parameters governing plant-virus epidemics, the logical way to approach an understanding of them is to develop simulative computer models. While this has been reasonably well achieved for other plant disease systems, it is in its infancy with regard to plant-virus epidemics. The spread of aphid-borne viruses has only recently received attention by simulation modelers. Madden et al. (1983) developed a model that describes the correlation between virus incidence and aphid species prevalent at the time. Models that take into account the distribution of infection (clumped or random) in relation to aphid species and aphid activity have recently been described (Marcus and Raccah 1986; Ruesink and Irwin 1986).

The models developed to date have been useful in helping plant virus epidemiologists to understand the important factors that regulate virus spread. They have also helped the search for parameters that insure good simulation and validation, and eventually prediction, of actual spread. The near-term goal of modeling epidemics of plant viruses, in our opinion, is to determine the parameters that will predict the onset of epidemics and assess the utility of various control measures.

7 References

A'Brook J, Dewar AM (1980) Barley yellow dwarf virus infectivity of alate aphid vectors. Ann Appl Biol 96: 51–58

Bar-Joseph M, Loebenstein G, Oren Y (1974) Use of electron microscopy in eradication of tristeza sources recently found in Israel. In: Weathers LG, Cohen M (eds) Proc 6th Conf Int Org Citrus Virologists. Univ of Califormia Press, Riverside, CA

Bar-Joseph M, Garnsey SM, Gonsalves D, Moscovitz, Purcifull DE, Clark ME, Loebenstein G (1979) The use of enzyme-linked immunosorbent assay for the detection of citrus tristeza virus. Phytopathology 69: 190–194

Berger PH, Thornbury DW, Pirone TP (1985) Detection of picogram quantities of potyviruses using a dot blot immunobinding assay. J Virol Methods 12: 31–39

Bos L (1981) Wild plants in the ecology of virus diseases. In: Maramorosch K, Harris KF (Eds) Plant diseases and vectors: ecology and epidemiology. Academic Press, London, pp 1–34

Boulton MI, Markham PG, Davies JW (1984) Nucleic acid hybridization techniques for the detection of plant pathogens in insect vectors. Proc Crop Prot Conf 3: 181–186

Broadbent L, Chaudhurk RP, Kapica L (1950) The spread of virus diseases to single potato plants by winged aphids. Ann Appl Biol 37: 355–372

Broadbent L, Tinsley TW (1951) Experiments on the colonization of potato plants by apterous and by alate aphids in relation to spread of virus diseases. Ann Appl Biol 38: 411–424

Carlebach R, Raccah B, Loebenstein G (1982) Detection of potato virus Y in the aphid *Myzus persicae* by enzyme-linked immunosorbent assay (ELISA). Ann Appl Biol 101: 511–516

Clark MF, Bar-Joseph M (1984) Enzyme immunoassay in plant virology. Methods 7: 51–87

Dodds JA, Morris TJ, Jordan R (1984) Plant viral dsRNA. Annu Rev Phytopathol 22: 151 – 168

Gera A, Loebenstein G, Raccah B (1978) Detection of cucumber mosaic virus in viruliferous aphids by enzyme-linked immunosorbent assay. Virology 86: 542–545
Gonda TJ, Symons RH (1978) The use of hybridization analysis with complementary DNA to determine the RNA sequence homology between strains of plant viruses: its application to several strains of cucumoviruses. Virology 88: 361–370
Halbert SE, Irwin ME, Goodman RM (1981) Alate aphid (Homoptera: Aphididae) species and their relative importance as field vectors of soybean mosaic virus. Ann Appl Biol 97: 1–9
Irwin ME (1980) Sampling aphids in soybean fields. In: Kogan M, Herzog DC (Eds) Sampling methods in soybean entomology. Springer, Berlin Heidelberg New York, pp 239–259
Irwin ME, Ruesink WG (1986) Vector intensity: a product of propensity and activity. In: McLean GD, Garrett RG, Ruesink WG (eds) Plant virus epidemics: monitoring, modelling and predicting outbreaks. Academic Press, London, pp. 13–33
Jayasena KW, Randles RJ, Barnett OW (1984) Synthesis of a complementary DNA probe specific for detecting subterranean clover red leaf virus in plants and aphids. J Gen Virol 65: 109–117
Labonne G, Fauvel C, Leclant F, Quiot JB (1982) Description d'un piege a suction: son emploi dans la recherche des aphides vecteurs de virus transmis sur le mode nonpersistent. Agronomie (Paris) 2: 773–776
Madden LV, Knoke JK, Louie R (1983) The statistical relationship between aphid trap catches and maize dwarf mosaic virus inoculation pressure. In: Plumb RT, Thresh JM (eds) Plant virus epidemiology the spread and control of insect-borne viruses. Blackwell, Oxford, pp 159–168
Marcus R, Raccah B (1986) A model for spread of non-persistent virus disease. J Appl Stat 13: 167–175
McLean GD, Garrett RG, Ruesink WG (1986) Plant virus epidemics: monitoring, modelling and predicting outbreaks. Academic Press, London pp 550
Milne RG (1984) Electron microscopy for identification of plant viruses in in vitro preparations. Methods Virol 7: 88–121
Milne RG, Luisoni E (1977) Rapid-immune electron microscopy of virus preparations. Methods Virol 6: 265–281
Ouchterlony O (1964) Gel-diffusion techniques. In: Ackroyd JF (ed) Immunological methods. Blackwell, Oxford, pp 55–78
Pike KS, Halbert SE (1983) Spread of barley yellow dwarf virus in the winter. Abstr. 4th Int Congr of Plant Pathology. (Melbourne, Australia), no 874
Plumb RT (1983) Barley yellow dwarf virus — a global problem. In:Plumb RT, Thresh JM (eds) Plant virus epidemiology: the spread and control of insect-borne viruses. Blackwell, Oxford, pp 185–198
Plumb RT, Thresh JM (eds) (1983) Plant virus epidemiology: the spread and control of insect-borne viruses. Blackwell, Oxford, pp 377
Raccah B (1983) Monitoring insect vector populations and the detection of viruses in vectors. In: Plumb RT, Thresh JM (eds) Plant viruses epidemiology: the spread and control of insect-borne viruses. Blackwell, Oxford, pp 51–60
Raccah B (1986) Non-persistent viruses: epidemiology and control. Adv Viruses Res 31: 387–429
Raccah B, Gal-On A (1984) The role of flying aphid vectors in the transmission of cucumber mosaic virus and potato virus Y to peppers in Israel. Phytoparasitica 12: 209 (abstr.)
Raccah B, Gal-On A, Eastop VF (1985) The role of flying aphid species in the transmission of cucumber mosaic virus and potato virus Y to peppers. Ann Appl Biol 106: 451–460
Ruesink WG, Irwin ME (1986) Soybean mosaik virus: a model and some implications. In: McLean GD, Garrett RG, Ruesink WG (eds) Plant virus epidemics: monitoring, modelling and predicting outbreaks. Academic Press, London, pp 295–313
Schultz GA, Irwin ME, Goodman RM (1983) Factors affecting acquisition and transmission of soybean mosaik virus. Ann Appl Biol 103: 87–96
Schultz GA, Irwin ME, Goodman RM (1985) Relationship of aphid landing rates to the field spread of soybean mosaic virus. J Econ Ent omol 78: 143–147
Schwartz RE (1965) Aphid-borne virus diseases of citrus and their vectors in South Africa. A. Investigations into the epidemiology of aphid-transmissible virus diseases by means of trap plants. S Afr J Agric Sci 8: 839–852
Sylvester ES (1954) Aphid transmission of nonpersistent plant viruses with special reference to the *Brassica nigra* virus. Hilgardia 23: 53–98

Tamada T, Harrison BD (1981) Quantitative studies on the uptake and retention of potato leafroll virus by aphids in laboratory and field condition. Ann Appl Biol 98: 261–276
Taylor LR (1986) The distribution of virus diseases and the migrant vector Aphid. In: McLean GD, Garrett RG, Ruesink WG (eds) Plant virus epidemics: monitoring, modelling and predicting outbreaks. Academic Press, London, pp 35–57

Chapter 15

Techniques in Quantitative Nematology

K. R. Barker and J. P. Noe[1]

1 Introduction

Wide ranges in the kinds and numbers of plant-parasitic nematodes occur in all agroecosystems, including those with annual and perennial crops, ornamentals, forests, and even greenhouse crops (Freckman and Caswell 1985; Norton 1978). Although parasites may be present in low numbers to billions per hectare, only certain species cause major epidemics and crop losses.

Specific symptoms associated with nematode-induced epidemics are limited to a few diseases. Therefore, the methods of investigation focus on sampling and characterizing the dispersal and population dynamics of nematodes themselves rather than disease symptoms. Other problems in methodology arise from the typically polyspecific communities of nematodes and their ability to predispose plants to attack by other pathogens.

2 Nematode-Community Characteristics

Interactions between plant-parasitic nematodes and their hosts range from neutral ecological associations to various levels of parasitism and pathogenesis. In the nematode-host relationships, nematodes act as aggravators, vectors, incitants, or pathogens. Thus, the microniche and feeding habits of the predominant taxa must be considered in investigating nematode-induced epidemics.

The majority of plant-parasitic nematodes occur in the soil and feed on the roots of plants. Knowledge of the basic nematode feeding habits is a prerequisite to an investigation of their role in plant disease. Assessment of associated disease should include quantification of soil and plant components of the nematode populations as well as measurement of host responses. The capacity of nematodes to move on their own is limited to a few centimeters per day, but this may be influenced by soil texture and other factors (Prot and Netscher 1979). Although long-distance dispersal of nematodes occurs frequently via plant parts and infested soil, it is passive and by chance (Wallace 1973).

Generally, the preplant numbers of a pathogenic nematode can be correlated negatively to the yield of an annual crop (Seinhorst 1965; Oostenbrink 1966; Barker et al. 1985). Lower than expected yields occurring erratically in sections of a field are often

[1] Departments of Plant Pathology, North Carolina State University, Raleigh, North Carolina, and University of Georgia, Athens, Georgia, USA, respectively.

a recognizable indication of nematode damage. Nematode-infested fields typically show a patchy or undulating pattern of plant growth, the result of the contagious spatial patterns of these pests.

2.1 Spatial and Temporal Population Patterns

Because of the limited host-specific symptoms and signs associated with most plant-parasitic nematode species, the spatial and temporal population dynamics of these pests have become a focal point for investigating the epidemiology of related diseases. Other methods utilized experimentally for assessing nematode epidemics include aerial infrared photography, bioassays, and relative adenosine triphosphate (ATP) content in cysts of *Globodera* spp. (Barker 1985a).

The vertical and horizontal dispersal patterns of species in a given community vary greatly, and numerous techniques have been utilized for characterizing spatial patterns of plant-parasitic nematodes. Methods of spatial pattern analyses based on quadrat-count data have been divided into four approaches (Campbell and Noe 1985). These include the mapping of population densities in two or three dimensions for visual assessment and interpretations; the fitting of certain probability distributions to indicate possible nonrandomness in the spatial pattern of a population and to infer aspects of the biological basis for observed patterns; the computation of various indices of aggregation that measure the degree of nonrandomness and spatial patterns; and methods based on spatial autocorrelation, including quadrat-variance techniques, spectral analyses, and lag correlation.

The aggregated or clustered spatial patterns of various nematodes (e.g., Fig. 15.1) have been depicted in different types of maps drawn manually or computer-assisted (Barker and Campbell 1981; Campbell and Noe 1985). The close relationship of certain nematode levels, including those of *Helicotylenchus digonicus* and *Merlinius brevidens*

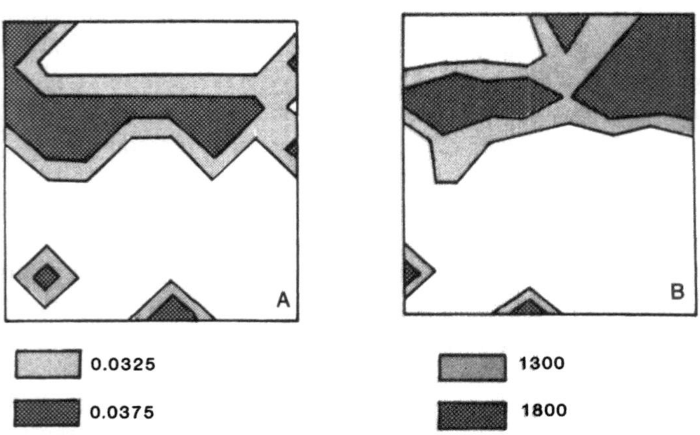

Fig. 15.1 a and b. Contour maps showing relative concentrations of **A**, sodium (MEQ/100 cm^3 soil); and **B**, *Meloidogyne incognita* population density (second-stage juveniles per 500 cm^3 soil) in a field previously planted with tobacco. All three clusters of higher densities of *Meloidogyne incognita* coincide with areas of the field which have higher levels of sodium. (After Noe and Barker 1985)

with soil texture (Goodell and Ferris 1980), and *Meloidogyne incognita* with specific edaphic factors (Noe and Barker 1985) can be characterized through mapping. All mapping methods require that data be collected in a systematic-grid pattern.

Nematode-count frequency data obtained from grids sampled to construct maps also may be fitted to discrete probability distributions which then may be utilized to estimate the degree of clustering or randomness of the nematode-dispersal patterns. The negative binomial probability distribution, described by the mean and a dispersion parameter k, is suitable for such investigations. Population data that can be fitted to the negative binomial would be characterized as nonrandom or clustered. In contrast, data fitting a Poisson distribution would indicate a random-spatial pattern. Additional discrete distributions are available (Campbell and Noe 1985), but have been utilized less frequently for nematode investigations. For information on related subjects, see Gilligan (Chap. 6).

Although the use of indices of dispersion and autocorrelation methods has received little attention in nematology, these techniques can be useful in the description and comparison of nematode-dispersal characteristics. Indices of dispersion can be used to compare the impact of various management practices on dispersal, or to describe changes in dispersal over time. Spatial autocorrelation provides an even more detailed description of spatial patterns, allowing analysis of changes in the size and orientation of nematode clusters over time (Noe and Campbell 1985).

The vertical-spatial dispersal of nematode species also varies greatly. Most plant-parasitic nematodes are found in the upper 30 cm of soil on typical annual crops. Generally these pests are closely associated with a vertical distribution of host roots (Wallace 1973). Although the vertical-spatial distribution of nematodes has been characterized less well than that of the horizontal dispersal, one must be cognizant of how these organisms are distributed, both vertically and horizontally, when relating them to the development of plant disease.

2.2 Sampling

The timing and methods of sampling are dependent on the objectives of the assay and the population dynamics of the species involved (Barker 1985a). Sampling with primary objectives of simple detection (surveys) is best performed when maximum population numbers occur. Nematicide efficacy and resistant-cultivar assessments should be based on samples collected when the nematode numbers on the untreated susceptible plants are near maximum, but before the effects of the control treatments diminish (near mid-season for annual crops). Although the ideal time for sampling to relate nematode numbers to crop performance is at planting, assays for advisory purposes often are conducted in the fall to minimize detection failures. Extrapolations from near-maximum fall population levels to the subsequent spring may be based on various types of models. Research should include samples collected at planting (pi), one or more during mid-season (pm), and one at harvest (pf) (Barker 1985a). Ideally, the rate of nematode development and decline should relate to degree days and the influence of various environmental factors (Ferris 1985).

The spatial patterns of nematodes are useful in developing more reliable sampling procedures (Barker and Campbell 1981). For example, the very conservative model described by Karandinos (1976) facilitates the calculation of the theoretical number of samples needed to estimate mean density with a given degree of precision where the population mean and dispersion parameter (k) are available. The components of this model are as follows:

$$n=(Z\alpha/2)^2\left(\frac{1/\bar{x}+1/k}{E^2}\right) \tag{15.1}$$

with n = number; Z = upper $\alpha/2$ point of the standard distribution; $\bar{x}$ = population mean; k = empirically determined dispersion parameter of the negative binomial; and E = level of precision indicated as a predetermined standard error expressed as a decimal of the mean. Other models such as Taylor's power law and Morisita's index of dispersion also have potential for estimating required samples sizes in contagious populations (Barker and Campbell 1981; Taylor 1971). For other aspects of sampling, see Kranz (Chap. 3) and Gilligan (Chap. 6).

3 Role of Laboratory and Greenhouse Research

Greenhouse and laboratory studies are useful in elucidating the relative host suitability for a given nematode species, including the resistance, tolerance, or susceptibility of new cultivars. They are essential for developing information on how environmental parameters such as temperature and/or moisture and various edaphic factors impact nematode development, reproduction, and survival. Such information is essential for simulating plant-nematode interactions (Ferris 1976) as will be discussed later. There is still a need for fundamental information on nematode biology, including life tables for these purposes.

4 Use of Artificial Infestations in Fields

Because of the high variability of nematode dispersal in natural field infestations, field microplots and larger sections of fields may be artificially infested to characterize various aspects of nematode-related epidemics. This approach typically includes the use of a general biocide to largely eliminate indigenous nematodes and other plant pests. Selected plots of appropriate size are then infested with known numbers and kinds of nematodes as well as beneficial organisms. Nitrogen-fixing bacteria should be introduced for legumes and mycorrhizal fungi where they have an important function on the target crop. Temporary plant-growth problems still may be encountered with biocides; these problems can only be overcome by treating plots several months prior to infestation.

Microplots made of various materials, including fiberglass, and tile pipes of different sizes, are useful for determining the damage and potential loss effected by nematodes on numerous crops (Barker 1985b). Nematode inoculum for these purposes may be increased in monoxenic culture (Martin et al. 1983) or on greenhouse plants. This approach has considerable potential for determining plant responses to single and multiple nematode species and their role in disease complexes (Martin et al. 1982). Because plant responses to nematodes in microplots may not agree with those in naturally infested fields, results from such tests should be evaluated in parallel, especially as related to damage potential (Barker et al. 1981).

Small field plots also may be infested with nematodes with procedures similar to those used for microplots. For example, a dispenser that delivers standard inoculum of *Globodera rostochiensis* has been developed (Marks and McKenna 1981). An agar solution of 0.125% facilitates a more uniform introduction of *Meloidogyne* eggs into field plots (Ball and Ferris 1982). These types of artificially infested field plots offer the advantage of a more uniform nematode distribution. Nevertheless, problems of invasion of pathogenic fungi as well as nontarget nematodes typically develop after one to two growing seasons.

5 Utilizing Natural Field Infestations

Most epidemiological elements of nematode host-parasite relationships are density-dependent, so it is necessary to establish a wide range of densities in field plots before studies are initiated. Crop-loss functions and nematode reproductive curves then can be derived over an appropriate range, which should include tolerance levels, economic thresholds, and equilibrium densities. It is usually not feasible to add nematode inoculum to large field plots at the required rates, as may be done with foliar pathogens, because of the highly buffered and variable soil environment and physical limitations on the incorporation of nematodes into the soil.

Crop management activities can be directed toward creating a range of nematode densities at a research site. However, most nematodes have a patchy spatial pattern in the field, and this pattern often can be exploited to provide an appropriate range of population densities. A systematic nematode assay of the proposed research site is necessary to determine if a suitable range of densities is present, and to indicate the proper location and orientation of plots (Noe and Campbell 1985). Where the desired nematode densities are not present in the sampled area, remedial crop management practices may be implemented. Although these preliminary steps are laborious, any study in nematode epidemiology must be viewed as a long-term project. Initial investments in setting up an appropriate site are essential for deriving the most meaningful data.

The fact that nematodes occur typically in polyspecific communities often adds to the complexity of characterizing related epidemics. The relative role of given nematode species in such communities may be weighed initially through correlation and multiple regression analyses (Sasser et al. 1975). Models for such interactions also have been developed for experimental purposes (Duncan and Ferris 1983). Unfortunately, there are numerous antagonistic and synergistic interactions among species in polyspecific nematode communities that remain to be characterized in the field situation (Eisenback 1985).

5.1 Exploitation of Contagious Spatial Patterns

Where plant-parasitic nematodes occur in relatively small clumps, experimental sites can be aligned so that some plots will have very high levels of infestation and other plots will have relatively few nematodes. Proper alignment of the site depends on a careful, systematic sampling of the field. The area to be sampled should be divided into quadrats which are two to four times the size of the anticipated research plots. A soil sample consisting of 20 or more individual cores (2.5-cm diameter by 20-cm depth) should be collected in a zigzag pattern from each quadrat and assayed for plant-parasitic nematodes. The resulting densities can be plotted by hand or with the aid of a computer onto a map of the site (Barker and Campbell 1981). Proposed grids of research plots then can be superimposed on the area maps to select alignments which include the desired range of nematode densities (e.g., Fig. 15.2).

Related approaches have been used to secure an appropriate range of densities of *Meloidogyne incognita* (Kinloch 1982), *Heterodera glycines* (Alston 1985), *M. hapla* (Noe 1986), and *Hoplolaimus columbus* (Noe, unpublished) in plots for generating mathematical models representing nematode impact on crop yields. Small-plot methods have proven successful in cases where the target nematode species exhibited highly clumped spatial patterns. However, in cases where the nematodes have more uniform patterns, alternative methods must be used to derive the desired range of densities.

Soil parameters also should be assayed in individual experimental plots. The same small-plot methods can be employed for the determination of the impact of edaphic parameters on nematode populations as were used in yield-impact studies. Many of the soil parameters exhibit a clustered spatial pattern, which may overlap the spatial pattern of nematodes (Fig. 15.1) (Noe and Barker 1985), so that properly aligned plots will also provide a wide range of values for selected soil variables. Resulting soil data can be utilized to improve delineation of nematode-damage functions through multivariate analyses with nematode densities and selected edaphic parameters as independent variables.

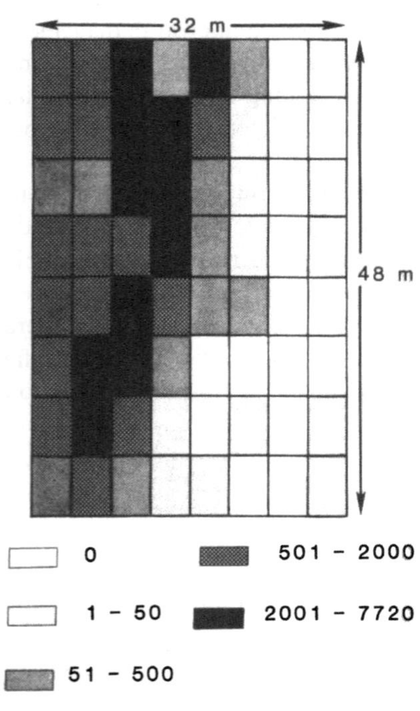

Fig. 15.2 Ranges of preplant population densities of *Meloidogyne hapla* on peanut per plot or quadrat (4 × 6 m). Numbers of nematodes per 500 cm³ soil. (After Noe 1986)

Evaluations of chemical soil treatments and/or the use of resistant and susceptible cultivars can be interfaced with measurement of density-dependent effects through the use of split-plot techniques. After plots have been identified with the desired range of nematode densities, they may be planted in two different cultivars or treated with differing chemicals, according to a statistically valid split-plot design. Utilization of this design should facilitate more precise characterizations of nematode epidemics under varying management practices. However, care should be taken when using a split-plot design with natural field infestations; if the plots are too large, different portions of each plot may have widely varying nematode densities. Plot size ideally should approximate nematode cluster size for the target species (Noe and Campbell 1985).

5.2 Means of Establishing Ranges of Population Densities

Crop-rotation systems can be used to enhance the natural variability of nematode spatial patterns, so that plots with ranges of densities can be established. Plots with varying crop-rotation histories and combinations of susceptible and resistant cultivars can be used in studies for determination of the relative aggressiveness of various nematode species (Oostenbrink 1966; Nusbaum and Ferris 1973; Jones and Kempton 1978). However, maximum differentiation of nematode densities may require 5–10 years. Also, only a single target nematode species may be manipulated successfully, since other nematodes and pests in the community may respond differently to the crop rotations. Differences in soil fertility, structure, and water potential among plots also must be considered as these variables may greatly alter the density, composition, and impact of nematode populations on crops (Vrain 1986).

Comparisons of population dynamics and yield losses with cultivars of varying susceptibility are essential to the characterization of nematode epidemics. For some highly aggressive nematodes, such as *H. glycines* on soybean, the use of a resistant cultivar, which has the same yield as the susceptible cultivar in the absence of the nematode, provides a better standard of reference for estimates of crop damage than does the use of nematicides. The maximum yield obtained with chemical soil treatment in the presence of *H. glycines* in North Carolina may reach only 70% of that of resistant cultivars (Schmitt et al. 1987).

Other management practices, such as destruction and tillage of roots at harvest may be utilized to enhance differentiation of nematode population levels across plots (Barker et al. 1985). Nematicides may aid in establishing ranges of initial numbers of nematodes, especially in studies designed to determine crop damages caused by these pests. Fumigant and multipurpose fumigant nematicides are most effective for these purposes, whereas nonfumigant materials often may fail to alter nematode density levels sufficiently. Various combinations of nematicides will often provide a wide range of effective initial nematode levels (Barker et al. 1981). Special care must be taken when applying chemicals in experimental plots for the determination of yield-loss relationships, since these relationships may be altered by the impact of the chemicals on nontarget pests, or by activity against the target pest which does not result in death of the organism. In addition, some nematicides may have slight to moderate phytotoxicity, whereas others can enhance plant growth directly (Garabedian and Van Gundy 1982).

5.3 Following an Epidemic over Prolonged Periods

Perhaps the most important aspect of the implementation of field plots in epidemiology is the length of time which the researcher will have access to the plots. Significant changes in nematode-induced crop stress and in nematode population dynamic proceed relatively slowly in the complex soil environment. Densities of some nematode species may fluctuate widely in a single growing season, but changes which affect levels of crop damage may take up to 5–10 years to become evident. These studies may involve an epidemic in a single field or over a geographic region for many years.

Nematodes interact strongly with edaphic characteristics and other soilborne pathogens in the expression of disease. These effects will have to be isolated and quantified over a number of years before a clear picture of the role of plant-parasitic nematodes can be drawn.

Where a single nematode species can be related to crop damage in a given field over a number of years, appropriate mathematical models can be developed to represent the related epidemics. This approach is similar to traditional techniques of disease-progress-curve analysis, with crop yield or nematode numbers as the dependent variable and time, expressed over a number of years, as the independent variable. These isolated, multiyear epidemics have been characterized for a number of nematodes, including *Meloidogyne arenaria* on peanut (Rodriguez-Kabana and Ivey 1986) and *G. rostochiensis* on potato (Jones and Kempton 1978).

Changes in the incidence of certain nematode species over large geographical areas also can be analyzed by using selected mathematical models. Again, the emphasis is on changes taking place over a relatively long period of time. Usually experimental units consist of individual fields, or farms, which are categorized as either infested or noninfested with the target nematode species. Mathematical functions representing the number of infested units over time then are parameterized to provide information on the rate of spread of the nematode. This approach has been used to quantify the spread of a number of important nematode species, including *Globodera tabacum solanacearum* on tobacco in Virginia (Komm et al. 1985), and *Heterodera schachtii* on sugar beet in California (Caswell and Thomason 1985).

The composition of nematode communities at the species and biotype or race taxonomic levels changes with shifts in management and cultivar usage. For example, *H. glycines* in North Carolina has evolved from populations of largely race 1, for which resistance was available, to races 2 and 5 for which no resistance was available (Schmitt, D. P. personal communication). Similarly, race 2 and 4 of *Meloidogyne incognita* as well as *M. arenaria* and *M. javanica* have increased greatly in North Carolina over the last 18 years as a result of the continuous use of cultivars of tobacco resistant to races 1 and 3 of *M. incognita*. Methods of characterizing these genetic shifts in communities have been developed only recently (Esbenshade and Triantaphyllou 1985; Ferris et al. 1985; Reddigari et al. 1985). The analysis of host-race specific nematode infestations will have to proceed similarly to the analysis of field epidemics of foliar diseases where the pathogens have known major gene interactions with their hosts. Genotypic components of the field populations need to be quantified and each pathotype assessed for its impact on host cultivars. Epidemiological models then can be formulated representing genetic components of the nematode population dynamics and the resulting impact of population shifts on crop-loss functions.

6 Modeling Approaches for Nematode-Induced Epidemics

Mathematical models of varying complexity have been used for different purposes in the representation of nematode-host relationships. As in any modeling endeavor, the selection of experimental designs, and eventual construction and formulation of the models must be controlled by the objectives of the researcher and ultimate purpose of the model (see Campbell et al. Chap. 18). In nematode management, critical-point models have been used to predict crop yield, crop loss, or changes in nematode densities, based on assays of population densities at a given time (Ferris 1985). Extensions to multiple-point models are possible in these systems, based on monitoring of nematode population dynamics over time. These methods are usually more correlative than explanatory.

Modeling has been used most extensively for descriptive and explanatory analyses of population dynamics in host-parasite systems. An equation representing population changes over a growing season as related to nematode reproductive rates and equilibrium densities has been proposed (Seinhorst 1967), and modified to allow for root damage (Jones and Kempton 1978). Simulation models of population changes over a growing season have been derived from field or greenhouse experiments in which densities were assayed and related to time and various environmental variables (Ferris 1976; McSorley and Ferris 1979). A more complex population model was formulated that incorporated life-stage categories into a discrete simulation of total population dynamics, interacting with plant-root growth at various depths (Caswell et al. 1986). Also, an empirical model has been proposed (Jeger and Starr 1985) and applied (Starr and Jeger 1985) to describe overwinter population changes of plant-parasitic nematodes.

Management-optimization models combine simulation of multiple growing-season population dynamics with crop-yield/nematode-density relationships into a comprehensive farming system framework (Jones and Kempton 1978; Noe 1986). Optimization models are relatively complex to assemble, athough mathematically they resemble their simpler counterparts for analysis of yield-loss relationships and population dynamics in that usually they are empirically derived from data collected in small experimental field plots. Nematode models are usually based on population densities and dynamics. Unlike many pathogen systems, the nematodes themselves are the most obvious and countable component of the host-parasite relationship.

Models conceivably can range in scale from multi-year, multi-crop total field simulations, to detailed biological models of single-nematode, single-plant systems. Models constructed at the broader ranges are becoming commonplace, and soon should take their place as essential management and research tools. Regardless of the scale or purpose of a proposed model, certain aspects of the modeling process are similar.

6.1 Simple Population Models

Population models can be used to represent single-season dynamics under a single crop, to compare single-season dynamics across several crops, or to link multiple-crop dynamics over many growing seasons. A simple approach is found in relating P_f to P_i after log-transformation through least squares parameterization of a linear function.

No other variables are monitored in the system, and the derived relationship represents single-season population changes only for the specific crop and environment. Elaborations of this design should involve incorporation of selected edaphic variables, such as soil structure, soil fertility, and moisture, into crop-specific functions. These more complex models then can be linked together into simulations of population dynamics under varying crops and conditions.

The experimental approach for the construction of a population dynamics simulation should begin with a description of the system, including elucidation of perceived output and utilization, followed by the determination of important variables. The success of a simulation will depend on the researcher's ability to select the appropriate variables for study, and to correctly elucidate the relationships among these variables. The output of nematode population models usually consists of nematode density at a given time. Calculation of this density may be based on a synthesis of several other components, including the beginning population density, a measure of host suitability and availability of feeding sites, and environmental variables such as temperature and moisture. A more ambitious approach would include a coupling of nematode population models with plant-growth simulators through determination of interactions between the systems.

6.2 Collecting Data to Derive Models

Experimental data to parameterize the selected models can be collected in greenhouse pots, microplots, or small field plots as previously described. Greenhouse research is most efficient for determining the effects of edaphic variables on population changes. Soil structure, moisture, and temperature can be manipulated with relative ease in the greenhouse environment or growth chambers. Most nematode species take readily to culture in pots. When determining the effects of different levels of several variables, however, factorially arranged experiments may rapidly become unwieldy. These studies can be made more practical by adroit use of preliminary research to limit numbers and levels of variables and to select appropriate incomplete factorials which cover critical biological ranges.

Microplots also can be used effectively in studies to determine population dynamics under different host crops (Barker 1985b). The microplot environment is much closer to a field situation than the greenhouse, but large numbers of replications are still possible in relatively small areas. Individual host plants can be established in a randomized block design, which would be impractical in full-scale field plots. However, containment within an artificial barrier may change native soil characteristics, and thus alter the nematode environment. This change may or may not alter the measured population dynamics. Also, removing soil for repeated nematode assays may alter the amount of soil and soil structure significantly within the limited volume of small microplots.

The field environment integrates may interacting components of the agroecosystem into a single window through which population dynamics can be viewed. Derivation of models using data from field studies inevitably becomes a massive exercise in data collection. It is impossible to collect information on every interacting biological and environmental component in the complex soil matrix, so the researcher must rely on previous work and informed intuition to select only the most critical variables to measure. This procedure is rarely successful in the first attempt.

Establishment of well-designed field plots for use in nematode modeling is a matter of using adequate numbers of replicates and selecting appropriate sizes and alignments for the plots (Noe and Campbell 1985). After the plots are established, data collection consists primarily of measuring nematode densities at various times, along with continuous or discrete monitoring of selected environmental variables. Measuring nematode population densities is an art in itself (Barker 1985 a); the sampling approach must be carefully designed and executed, or variation and sampling error inherent in nematode counts may make the entire research effort unproductive. Nematologists also have been hampered by the difficulty of monitoring important environmental variables within the soil, but improved methods for this purpose are becoming available (see Sutton et al. Chap. 7).

Data collected for nematode population modeling may require conditioning before analysis. Nematode counts often are transformed with a $\log(X+1)$ transformation to alter the variance structure so that it comes closer to fulfilling the assumptions of parametric statistical procedures. This transformation also may linearize certain population density functions, such as relationship between P_f and P_i. Soil moisture measurements, obtained via neutron probes or other means, should be adjusted to represent the matric potential of each soil type.

6.3 Formulating the Models

When sufficient data are available to begin the modeling process, mathematical formulations which accurately represent relationships among the variables measured must be selected by the researcher. Simple plotting of dependent versus independent variables will provide valuable insights. The most suitable function may be no more elaborate than a linear least-squares model, whereas more mechanistic formulations may involve elaborate equations based on exponential or logistic rates of population changes (Jones and Kempton 1978; Jeger and Starr 1985). Usually, the best model is the simplest formulation which will accurately represent the data and still satisfy the goals of the researcher.

Assembling the derived mathematical equations into a complex simulation of nematode population dynamics will follow easily from a comprehensive description of the nematode-host-environment system. The correct linkages and time sequences for the model components should be obvious, but unforeseen interactions may still cause unreliable results (Wallace 1983). For instance, soil moisture levels and soil texture or structure can interact to restrict nematode mobility or to limit the ability of infective stages to penetrate host roots. These limitations become critical at levels of the interacting variables which should not be significant when each is considered individually.

Temperature is often used as a primary independent variable in simulation models, but effects attributed to variation in this variable may actually be reflections of other system components which are correlated with temperature. When one of these other components, such as soil moisture, aeration, or host-root growth reaches a critical point, the model will fail to respond to further changes in temperature, causing erroneous simulation of the nematode population dynamics. Interacting components are of particular concern in the soil environment where so many of the variables are strongly correlated (Noe and Barker 1985).

6.4 Validation

Validation involves comparison of simulated population changes with dynamics monitored in naturally infested fields. An example of this type of validation would be to collect data from fields with widely varying soil types, such as sand, sandy loam, and clay soils, or from fields with different crop histories, and then to compare the data to results from the model, with inputs representing each field situation. When simulating nematode population densities, wide latitude must be allowed in the validation process to allow for the typically large sampling error in nematode counts. For further details on modeling and validation of models, see Campbell et al. (Chap. 18) and Teng and Johnson (Chap. 12).

7 Epilogue

Much progress has been made in developing methods required for investigating nematode epidemics. Still, greater emphasis should be placed on using improved techniques in statistics, biomathematics, and computer science. Also, increasing emphasis on techniques in biotechnology should provide new tools for the assessment of detailed, biological host-parasite interactions, leading to better models of these systems (Lindow and Webb 1983; Long 1984; Reddigari et al. 1985). More precise descriptions and models of nematode epidemics are needed to improve cropping systems and other management tools for limiting losses associated with these pests. This aspect is paticularly critical today in light of the loss of a number of the more effective nematicides.

Soil-inhabiting nematodes are small, have a contagious spatial pattern, and characteristically change in population levels during the year. Present methods of assessing nematode populations are time-consuming, expensive, and need improvement. Nevertheless, current methodology is adequate to fulfill meaningful research and epidemic-management objectives. Undoubtedly, refinements of methodology such as the utilization of computerized image systems and remote sensing will facilitate investigation of large-scale epidemics. Rapid biochemical analyses also may revolutionize nematode-epidemic studies as specific diseases and associated nematodes are better characterized and understood. Quantitative studies on the progress of nematode-induced plant diseases should become feasible, allowing nematode epidemiologists to use analyses similar to those used for aboveground, foliar epidemics.

8 References

Alston DG (1985) Population dynamics and development of *Heterodera glycines* as related to soybean phenology. MS Thesis, North Carolina State University, Raleigh North Carolina

Ball DA, Ferris H (1982) A technique for inoculating field sites with *Meloidogyne* eggs. J Nematol 14: 420–422

Barker KR (1985a) Sampling nematode communities. In: Barker KR, Carter CC, Sasser JN (eds) An advanced treatise on *Meloidogyne,* vol 2. Methodology. Cooperative Publication Department of Plant Pathology, North Carolina State University, and the United States Agency for International Development, Raleigh North Carolina, pp 3–17

Barker KR (1985b) The application of microplot techniques in nematological research. In: Barker KR, Carter CC, Sasser JN (eds) An advanced treatise on *Meloidogyne,* vol 2. Methodology. Cooperative Publication Department of Plant Pathology, North Carolina State University, and the United States Agency for International Development, Raleigh North Carolina, pp 127–134

Barker KR, Campbell CL (1981) Sampling nematode populations. In: Zuckerman BM, Rohde RA (eds) Plant parasitic nematodes, vol 3. Academic Press, London, pp 451–473

Barker KR, Todd FA, Shane WW, Nelson LA (1981) Interrelationships of *Meloidogyne* species with flue-cured tobacco. J Nematol 13: 67–79

Barker KR, Schmitt DP, Imbriani JL (1985) Nematode population dynamics with emphasis on determining damage potential to crops. In: Barker KR, Carter CC, Sasser JN (eds) An advanced treatise on *Meloidogyne,* vol 2. Methodology. Cooperative Publication of the Department of Plant Pathology, North Carolina State University, and the United States Agency for International Development, Raleigh North Carolina, pp 135–148

Campbell CL, Noe JP (1985) The spatial analysis of soilborne pathogens and root diseases. Annu Rev Phytopathol 23: 129–148

Caswell EP, Thomason IJ (1985) Geographic distribution of *Heterodera schachtii* in the Imperial Valley of California 1961 to 1983. Plant Dis 69: 1073–1077

Caswell EP, MacGuidwin AE, Milne K, Nelsen CE, Thomason IJ, Bird GW (1986) A simulation model of *Heterodera schachtii* infecting *Beta vulgaris.* J Nematol 18: 512–519

Duncan LW, Ferris H (1983) Validation of a model for prediction of host damage by two nematode species. J Nematol 15: 227–234

Eisenback JD (1985) Interactions among concomitant populations of nematodes. In: Sasser JN, Carter CC (eds) An advanced treatise on *Meloidogyne,* vol 1, Biology and control. Cooperative Publication Plant Pathology Department, North Carolina State University and the United States Agency for International Development, Raleigh North Carolina, pp 193–213

Esbenshade PR, Triantaphyllou AC (1985) Use of enzyme phenotypes for identification of *Meloidogyne* species. J Nematol 17: 6–20

Ferris H (1976) Development of a computer-simulation model for a plant-nematode system. J Nematol 8: 255–263

Ferris H (1985) Basic modeling strategies for nematode management. In: Barker KR, Carter CC, Sasser JN (eds) An advanced treatise on *Meloidogyne,* vol 2, Methodology. Cooperative Publication Plant Pathology Department, North Carolina State University and the United States Agency for International Development, Raleigh North Carolina, pp 205–213

Ferris VR, Ferris JM, Murdock LL (1985) Two-dimensional protein patterns in *Heterodera glycines.* J Nematol 17: 422–427

Freckman DW, Caswell EP (1985) The ecology of nematodes in agroecosystems. Annu Rev Phytopathol 23: 275–296

Garabedian S, Van Gundy SD (1982) Growth enhancement and inhibiting effect of insecticide/nematicides on the growth of plants in the absence of nematodes. J Nematol 14: 440–441 (abstr.)

Goodell P, Ferris H (1980) Plant parasitic nematode distributions in an alfalfa field. J Nematol 12: 136–141

Jeger MJ, Starr JL (1985) A theoretical model of the winter survival dynamics of *Meloidogyne* spp. eggs and juveniles. J. Nematol 17: 257–260

Jones FGW, Kempton RA (1978) Population dynamics, population models and integrated control. In: Southey JF (ed) Plant nematology. Minist Agric, Fisheries and Food Tech Bull 7. Her Majesty's Stationery Office, London, pp. 333–361

Karandinos MG (1976) Optimum sample size and comments on some published formulae. Bull Entomol Soc Am 22: 417–421

Kinloch RA (1982) The relationship between soil populations of *Meloidogyne incognita* and yield reduction of soybean in the coastal plain. J Nematol 14: 162–167

Komm DA, Riley JJ, Elliott AP (1985) Epidemiology of a tobacco cyst nematode *(Globodera solanacearum)* in Virginia. Plant Dis 67: 1249–1251

Lindow SE, Webb RR (1983) Quantification of foliar plant disease symptoms by microcomputer-digitized video image analysis. Phytopathology 73: 520–524

Long FL (1984) A field system for automatically measuring soil water potential. Soil Sci 137: 227–230

Marks RJ, McKenna LA (1981) A dispenser for the rapid delivery of standard inocula of the cyst nematode *Globodera rostochiensis*. Ann Appl Biol 97: 119–122

Martin MJ, Riedel RM, Rowe RC (1982) *Verticillium dahliae* and *Pratylenchus penetrans*: interactions in the early dying complex of potato in Ohio. Phytopathology 72: 640–644

Martin MJ, Riedel RM, Rowe RC (1983) A technique for quantitative use of nematodes from monoxenic tissue culture as soil inoculum. Plant Dis 67: 275–277

McSorley R, Ferris JM (1979) PHEX: a simulation of lesion nematodes in corn roots. Indiana Agric Exp Stn (Purdue) Res Bull 959, 22 pp

Noe JP (1986) Cropping systems analysis for limiting losses due to plant-parasitic nematodes: guide to research methodology. Cooperative Publication Plant Pathology Department, North Carolina State University and the United States Agency for International Development, Raleigh North Carolina

Noe JP, Barker KR (1985) Relation of within-field variation of plant-parasitic nematode population densities to edaphic factors. Phytopathology 75: 247–252

Noe JP, Campbell CL (1985) Spatial pattern analysis of plant-parasitic nematodes. J Nematol 17: 86–93

Norton DC (1978) Ecology of plant-parasitic nematodes. Wiley, New York

Nusbaum CJ, Ferris H (1973) The role of cropping systems in nematode population management. Annu Rev Phytopathol 11: 423–440

Oostenbrink M (1966) Major characteristics of the relation between nematode and plants. Meded Landbouwhogesch Wageningen, 4–66

Prot JC, Netscher C (1979) Influence of movement of juveniles on detection of fields infected with *Meloidogyne*. In: Lamberti F, Taylor CE (eds) Root-knot nematodes (*Meloidogyne* species) systematics, biology and control. Academic Press, London, pp 193–203

Reddigari SR, Sundermann CA, Hussey RS (1985) Isolation of subcellular granules from second-stage juveniles of *Meloidogyne incognita*. J Nematol 17: 482–488

Rodriguez-Kabana R, Ivey H (1986) Crop rotation system for the management of *Meloidogyne arenaria* in peanut. Nematropica 16: 53–63

Sasser JN, Barker KR, Nelson LA (1975) Correlations of field populations of nematodes with crop growth responses for determining relative involvement of species. J Nematol 7: 193–198

Schmitt DP, Ferris H, Barker KR (1987) The response of soybeans to *Heterodera glycines* in different soil types. J Nematol 18: 240–250

Seinhorst JW (1965) The relation between population density and damage to plants. Nematologica 11: 137–154

Seinhorst JW (1967) The relationships between population increase and population density in plant parasitic nematodes. III. Definition of the terms host, host status and resistance. IV. The influence of external conditions on the regulation of population density. Nematologica 13: 429–442

Starr JL, Jeger MG (1985) Dynamics of winter survival of eggs and juveniles of *Meloidogyne incognita* and *M. arenaria*. J Nematol 17: 252–256

Taylor LR (1971) Aggregation as a species characteristic. In: Patil GP, Pielou EC, Waters WE (eds) Statistical ecology, vol 1: Spatial patterns and distribution patterns. Pennsylvania State University Press, University Park Pennsylvania, pp 357–372

Vrain TC (1986) Role of soil water in population dynamics of nematodes. In: Leonard KJ, Fry WE (eds) Plant disease epidemiology. Macmillian, New York; Collier Macmillian, London, pp 101–128

Wallace HR (1973) Nematode ecology and plant disease. Edward Arnold, London

Wallace HR (1983) Interactions between nematodes and other factors on plants. J Nematol 15: 221–227

Chapter 16

Use of Geophytopathological Information

Heinrich C. Weltzien[1]

1 Introduction: Definitions — Objectives

Data collection cannot be justified by itself. The vast volume of data assembled in plant disease surveys needs generalization and interpretation. There is a clear demand to condense the information and to transfer numerical figures into a more directly comprehensible documentation. Appropriate methods have to be developed and applied. They should open the way for an ecological interpretation of plant disease data, and relate them to climatic zones and plant production areas. This may lead to a better understanding of disease occurrence and spread and even allow forecasting of the potential geographic distribution of plant diseases.

It is this geographic aspect in plant pathology which we will try to elucidate in this chapter. The term geophytopathology was first introduced by Weltzien (1967). The definition proposed calls for the "Description of plant disease distribution and its causal interpretation". It was clearly understood that geophytopathology is a part of phytomedicine or phytiatry (Weltzien 1972), and as such plant pests or weeds are also subject to geographic analysis, though we do not deal with them here. The orientation towards plant health and plant protection (Weltzien 1978) also goes beyond the concept of biogeography as first introduced into plant pathology by Reichert (1953).

The objectives of geophytopathological studies can differ widely. Most frequently one will find descriptive approaches. In this case the documentation of the results of field survey work is the primary goal. The picture here is usually static showing a disease situation at a given time and a defined location.

The study of disease spread is another objective frequently followed. More complex objectives are the search for disease-environment interactions and the analysis of complex ecological situations. Studies on forecasting may be considered as the most complicated objectives in geophytopathology.

2 Data Collection

Plant disease surveys are the first steps to obtain relevant information. Especially new records of plant diseases within continents, countries or regions are of importance. Updated plant disease lists are of high value to those who work on geographic distribution patterns. The need to identify and publish the exact geographic location of such findings cannot be over emphasized. The first infection foci may allow conclusions on

[1] Institut für Pflanzenkrankheiten, Nußallee 9, Bonn, FRG

the way of dispersal and spread of the pathogen (see also Nagurajan and Ajari, Chap. 17). Also, the early strategy on disease control is facilitated, if the medium and route of transport are known with some certainty. Geographic longitude and latitude, though the most precise information, may not be the most practical for our purpose. In most cases it will be easier to identify the locality on road maps, using distances in kilometers or miles from villages or towns or other permanent landmarks.

Frequently, first plant disease records are made by incidental findings. Thus, the probability is low that the locality identified is the place of first occurrence and the only infection focus. A more systematic survey must, therefore, follow, if a true picture of a disease situation is to be described. For geographic studies the surveys should be based on the distribution pattern of the host plant. For cultivated crops national statistics on land use frequently offer the necessary background information and will be sufficient in most cases. If this land-use pattern is transferred onto a road map, a practical survey route can be designed and followed. We used one of the following procedures to arrive at randomized data with a high degree of objectivity, if larger areas had to be surveyed by car:

1. Stop at the next host plant field visible from the road and survey it, after a prefixed distance has been driven from the last field, neglecting all other fields in between.
2. Calculate the acreage of the host plant within a given area and decide on the number of survey stops per hectare.

Methods 1 and 2 can frequently be combined by estimating the driving distance between two stops on a basis of the proposed number of samples/100 ha. Under practical conditions, some subjective decisions may be difficult to avoid. Frequently, the next field encountered in using method 1 may be atypical, e.g. extremely small or large, very poorly managed or even neglected. Also, access may be difficult or impossible. In these cases, on the spot adjustments are necessary and should be made, but they should always be the exception.

Method 1 is certainly preferable, if the area of cultivation is rather continuous and the crop is generally found everywhere. A typical example is the area of cereal cultivation in central Europe. It is also necessary to follow this method, if agricultural statistics are incomplete and difficult to obtain. This may be the case in many less developed regions.

Method 2 is advantageous whenever the crop is cultivated in well-defined and often separated regions of known acreage. As such regions may be very different in size and crop acreage, the sampling density may need to be adjusted to these parameters. As a general rule, the sampling should be more frequent in small rather than in large areas, to allow a minimum number of data for statistical analysis.

There is no rule on the absolute numbers of samples using method 1 or 2. The less frequent a pathogen is expected, the larger the number should be. Pathogens of epidemic character can be sampled in rather small numbers as they occur with a high degree of uniformity. For the actual sampling within each field, the rules explained by Cochran (1963) or Yates (1971) apply. Both disease frequency and disease intensity (see Kranz, Chap. 3) are useful information. While the frequency can usually be counted directly in the field, disease intensity is often preferably assessed on a collected sample. Thus, the time spent in each field can be short. This is necessary if large areas have to be surveyed. Other methods of data collection are based on aerial photography from either planes or satellites. In both cases, ground verification is absolutely necessary to avoid misinterpretation. These methods are, therefore, less suitable for geographic studies in larger areas, as ground travel is still necessary. They are, however, highly suitable for data collection on well-defined disease poblems in areas of cultivation, where the disease identity is well known.

3 Data Documentation

Maps are the most appropriate data bases for information with geographic relevance. They can combine qualitative and quantitative information, define localities, where these data were collected, and allow new conclusions beyond the primary information. Their use in phytopathology is wide spread, as they are comprehensive and easy to read. Their variations seem to be endless, but several basic types can be singled out.

Most common is the *location map* (Fig. 16.1). It shows the actual geographic location of an observation. The precision is of course limited by the map scale. In most cases it will not be possible to indicate the actual spot in such a way that it can be found again by other observers. If this is necessary, a written description must be added in most cases.

Area maps are also very common. Here, a line enclosing all relevant locations identifies an area, which represents some degree of abstraction and generalization (Fig. 16.2). Often shading of an area is used to avoid the rather arbitrary line. One may leave the original locations marked within the area, though this is often not very practical (Fig. 16.3). Disease incidences are also easy to map by areas, if their distribution patterns allow this. Both, lines or shadings, can be used and the actual disease intensity can be included (Figs. 16.4 and 16.5).

Isoline maps connect all points of equal values for the parameter measured. These can be temperatures, other climatic data or areas colonized at a given time, etc. They are commonly used to document the spread of diseases (Fig. 16.6) from a given location or focus.

Vector maps serve the same purpose but can indicate more and less important pathways by arrow strength (Fig. 16.7).

Diagram maps offer many potentials. In principle, all kinds of geographic designs can be mapped. Circle diagrams seem to be very useful, if several parameters are to be shown in their relative proportions. Circle size itself is proportional to a number or area, and sectors of the circle may indicate other parameters, e. g. the total area of wheat planted (circle size) and the percent of fields attacked by root rots (Fig. 16.8). Also, column diagrams are often useful, e. g. to demonstrate disease frequency on one host as related to soil types (Fig. 16.9). All types of maps can be combined into truly *synthetic maps* combining vast amounts of information and allowing new conclusions. However, one should not encourage overloading of maps beyond capacity, as their readability may than suffer.

4 Ecological Interrelationships

Environmental factors decide on disease incidence and severity, if pathogen and host are present. From the multitude of factors involved, usually only a few are overruling,

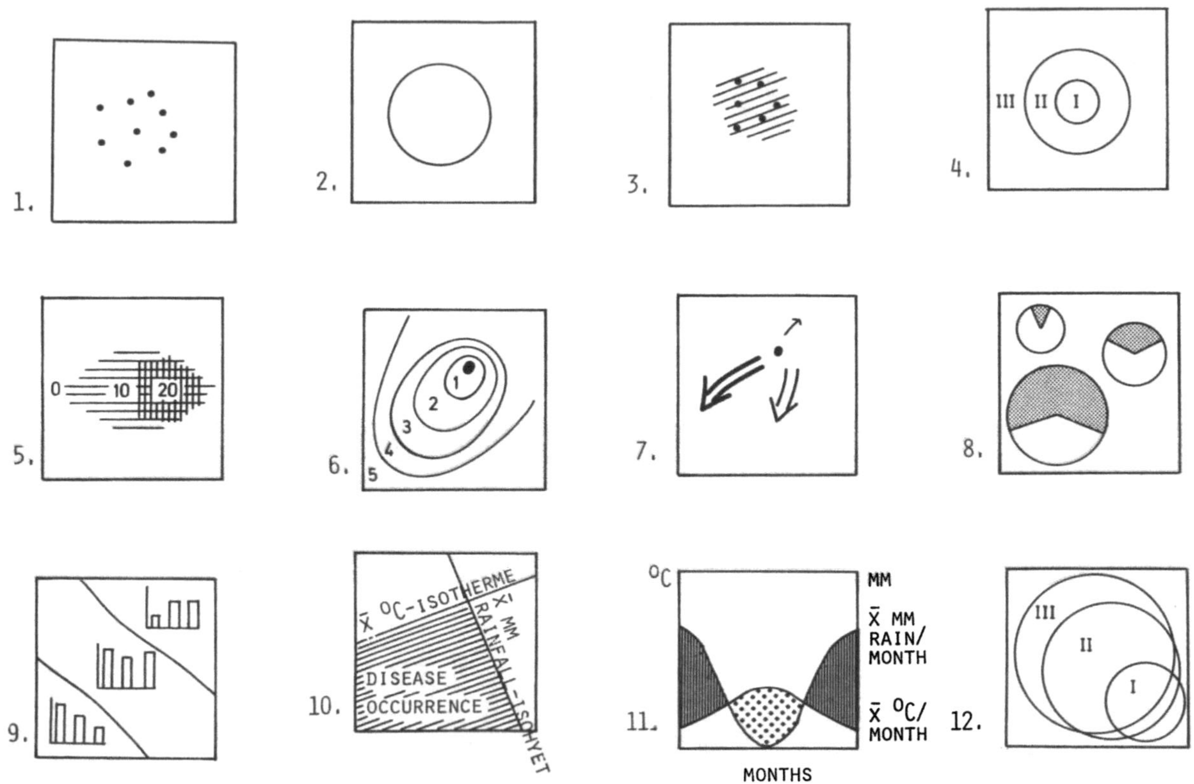

Fig. 16.1 Location map
Fig. 16.2 Area map
Fig. 16.3 Area map with locations
Fig. 16.4 Area map, e.g. three different pathogen races
Fig. 16.5 Area map, e.g. three disease frequencies
Fig. 16.6 Isoline map, e.g. disease spread from time *1* to *5*
Fig. 16.7 Vector map, e.g. intensity of disease spread from a focus
Fig. 16.8 Diagram map, circle diagram
Fig. 16.9 Diagram map, column diagram
Fig. 16.10 Temperature-rainfall limits of disease occurrence
Fig. 16.11 Simplified climatogram indicating arid ▒ and humid ▥ months
Fig. 16.12 Three zones of different infection potential, e.g. *I* main damage; *II* marginal damage; *III* sporadic attack

while many others cause only minor variations. Maximum temperatures in summer or minimum temperatures in winter can limit pathogen dispersal and survival (see Rotem, Chap. 5). In such a case mapping is the most practical way of documenting infection zones, which may be further limited by rainfall requirements (Fig. 16.10).

Leaf wetness within a given temperature range is another climatic factor combination, essential for the occurrence of many diseases, such as downy mildews and rusts. As leaf wetness is not generally measured by standard meteorological stations, rainfall data must be used, if direct data collection is not possible (see Friesland and Schrödter, Chap. 8). Often it proved to be practical to use the climatograms published by Walter and Lieth (1966) (Fig. 16.11). Periods of humidity at tolerated temperatures may correspond with potential infection periods. Periods of aridity may correspond with non-infection periods.

However, actual disease situations in the field are more differentiated. Most diseases are also found under unfavourable ecological conditions. But their occurrence is scattered within the population of host plants and they almost never reach economic threshold levels. Also, under optimal environmental conditions not all plants are heavily infected, but economic thresholds are generally passed. It is evident that between these two zones a third area may exist with intermediate disease intensities. In some years they will pass economic thresholds, in other they will not. For mapping of plant disease distribution it is therefore advisable to distinguish these three intensities of disease occurrence (Fig. 16.12). For their identification they were called:

1. Area of main damage;
2. Area of marginal damage;
3. Area of sporadic attack (Weltzien 1972, 1983).

Their demarcation has also practical relevance, as in area 1 regular plant protection is an economical activity, while it is not so in area 3. Area 2 is the classical environment, where warning and forecasting systems will work best. Here, measuring and interpretation of environmental data will enable the extension service to encourage or discourage plant treatments on an economical basis.

5 Forecasting

If ecological interdependences as outlined in section 4 are sufficiently understood, the potential disease problem on a given crop at a given location can be subject to forecasting (see also Madden and Ellis, Chap. 13). This should also be possible, if the crop has not yet been cultivated at this location or, if the pathogen has not yet been introduced. Obviously, this is of value to many agricultural projects in developing countries, where new crops are introduced into agricultural systems. It is also significant for those institutions, which, as the International Agricultural Research Centers, send plant material, especially seeds, to all parts of the world under the risk of spreading new pathogens with them. In general, any newly introduced pathogen can be studied under this aspect of potential spread, and even existing quarantine rules may be challenged by this method.

We have worked out the system for sugar beet leaf diseases (Drandarewski 1969; Bleiholder and Weltzien 1972; Weltzien and Mamluk 1982) on a worldwide basis. It was possible to correlate all sugar beet growing areas with one or more of the three disease zones as outlined in section 4, by evaluating macroclimatic data from the climatograms published by Walter and Lieth (1966). In two cases the prognosis offered for areas at this time still free from powdery mildew *(Erysiphe betae)* was verified by later disease events. This was most striking for the western and northwestern sugar beet areas in the USA (Ruppel et al. 1975) and for Chile (Steudel 1983, personal communication).

There is little doubt that a similar worldwide prognosis is possible for many more pathogens and pests, if the ecological factors controlling the development and spread are carefully studied and evaluated. The possibilities offered by careful generalization of environmental or field data, so essential to all geographic studies, are not yet fully explored in plant pathology and epidemiology but will certainly offer new insights into our science, if employed wisely.

6 References

Bleiholder H, Weltzien HC (1972) Beiträge zur Epidemiologie von *Cercospora beticola* Sacc. an Zuckerrübe. III. Geopathologische Untersuchungen. Phytopathol Z 73: 93–114

Cochran WG (1963) Sampling techniques, Wiley, New York

Drandarewski CA (1969) Untersuchungen über den echten Rübenmehltau *Erysiphe betae* (Vanha) Weltzien. III. Geophytopathologische Untersuchungen. Phytopathol Z 65: 201–218

Reichert I (1953) A biogeographical approach to phytopathology. Proc 12 Intern Biol Congr Stockholm 1950, 730–731

Ruppel EG, Hills FJ, Mumford DL (1975) Epidemiological observations on the sugar beet powdery mildew epiphytotic in western USA in 1979. Plant Dis Rep 59: 283–286

Walter H, Lieth H (1966) Klimadiagramm — Weltatlas. Fischer, Jena

Weltzien HC (1967) Geopathologie der Pflanzen. Z Pflanzenkr Pflanzenschutz 74: 175–189

Weltzien HC (1972) Geophytopathology. Ann Rev Phytopathol 10: 277–298

Weltzien HC (1978) Geophytopathology. In: Horsfall JG, Cowling EB (eds) Plant disease — an advanced treatise. Academic Press, London, pp 339–358

Weltzien HC (1983) Climatic zoning and plant disease potential — examples from the Near and Middle East. Eppo Bull 13: 69–73

Weltzien HC, Mamluk O (1982) Survey on the geographical distribution of sugar beet pests and diseases in the Near and Middle East. Printed as Manuscript, Institut für Pflanzenkrankheiten Bonn, p 114

Yates F (1971) Sampling methodes for censuses and surveys, 3rd edn. Griffin London

Chapter 17

Monitoring and Mapping Long-Distance Spread of Plant Pathogens

S. Nagarajan[1] and Ajai[2]

1 Introduction

Plant pathogens may spread to widely separated areas. Longdistance spread implicitly means dispersal over hundreds of kilometers. In such a dispersal most propagules die but the chance of success permits the occupation of a new habitat with a vast potential for multiplication. Proof for such disseminiation is the incontrovertible association between the pathogen and the resulting disease in the source, from which the spores dispersed, and the invaded habitat; or the interception of migrant spores.

Long-distance dispersal can be classified into (1) spread in waves and (2) spread in a single jump. Spread in waves occurs when a plant pathogen introduced into a new ecological area occupies the new habitat by successively advancing waves. Such patterns were observed for *Puccina polysora* in Africa (Rhind et al. 1952). Spread in a single jump is the only means to success when there are no susceptible plants between the source and the target area. Such was the case of spread of *Melampsora medusae* and *M. larici-populina* from Australia to New Zealand (Wilkinson and Spiers 1976). To this class also belongs the seasonal spread of *P. graminis tritici* uredospores from south to central India (Nagarajan et al. 1976).

2 Factors that Facilitate Long-Distance Dispersal and Deposition

Long-distance dispersal may be facilitated by birds, floating plants, ships, aircraft, etc. Here, we deal mainly with long-distance dispersal by wind. An actively sporulating plant at the source, and efficient transport, bulk deposition over vulnerable host and congenial ground-level conditions at the target should coincide for a sucessful long-distance spread of the pathogen. The sporulating lesions release propagules, most of which land closer to source. Only about 6% of them become windborne and only a fraction of these is carried to greater heights that enable a long-distance transport. The presence of propagules in the upper air is a prerequisite for long-distance dispersal and is facilitated by several means. Gushing wind sweeping over a tall mountain with agricultural vegetation transports spores into the upper air without difficulty. Also cyclonic circulation of wind lifts the spores to high altitudes. Lifting of spores also occurs when strong thermals prevail due to differential heating of the land surface.

[1] Indian Agricultural Research Institute, Division of Mycology and Plant Pathology New Delhi 110012, India
[2] Space Applications Center, Indian Space Research Organization, Ahmedabad 380053, India

On a hot sunny day, air over a dark surface becomes hotter and carries pollen, spores and insects to a height of a kilometer or more. The horizontal spread of spores in the air by either means depends on mid-to-low-level wind currents.

Spore viability during dispersal is affected by temperature, radiation and time. The quality and dosage of absorbed radiation depends on spore pigmentation, thickness and hydration of spore walls, duration of exposure to radiation and the dispersal distance involved (Leach and Anderson 1982; and Rotem, Chap. 5). The mechanisms that secure deposition of spores travelling close to the ground (impaction, sedimentation, attraction due to electrical charges and aggregation) have a negligible effect on the deposition of spores carried by wind currents to higher altitudes. In these cases deposition invariably follows impaction with falling rain drops. The finer the water droplets, the greater is the impaction efficiency. The small droplets then aggregate into larger drops. In addition to impaction, rain drops also act as electrostatic collectors due to coulombic attraction between oppositely charged spores and rain drops.

In a thunder shower, coupled with induced attraction, about 20% of spores have been deposited by these mechanisms (Gregory 1973).

3 Methodology of Monitoring

Spread of plant pathogens is monitored in ground surveys, weather and climatic analyses and remote sensing. Direct evidence of spread is obtained by virulence analysis and insect or spore catching.

3.1 Ground Surveys

Geophytopathological maps based on historical records reveal the nature and rate of spread (Weltzien, Chap. 16). Such surveys enabled re-examination of theories on the nature and recurrence of wheat rust in India (Joshi et al. 1971) and the development of epidemiological concepts and models for efficient disease management (Nagarajan and Joshi 1985). Collection of data from the ground with well-planned field surveys with appropriate sampling of the pathogen in question is thus a fundamental prerequisite for a study of a long-distance dispersal.

Various spore-trapping devices have been developed for usage in aerobiological studies on the ground, which include trap plants (Gregory 1973; Rotem, Chap. 5). Also aircrafts have been used to conduct vertical and horizontal monitoring of spore clouds (Hirst et al. 1967), but this method is expensive and not easily accessible. Cheaper and easier is to study the problem by a combination of meteorological analyses and ground surveys. An inexpensive air sampling is made with live traps. For instance, barley plants acting as live traps monitored the spread of *Erysiphe graminis f. sp. hordei* across the North Sea (Hermansen et al. 1976). Spore-catch studies showed that *Helminthosporium maydis* appeared up to a height of 1.5–2.0 km. The severe epidemic of southern corn leaf blight in the entire corn-growing USA in 1970 was facilitated by favourable wind currents (Frease and Yarger 1974). *Nilaparvata lugens,* the vector of grassy stunt in rice, does not overwinter in Japan and arrives each year from the Chinese mainland under the influence of typhoon/cyclone and associated cold fronts. Kisimoto (1976) predicted the arrival of these viruliferous insects to Japan according to trap catches and weather patterns. *Spiroplasma citri* from Citrus in the southwestern part of the USA spread to new habitats (1660 km away) in which Citrus are absent. The vector fed on horseradish (*Armoracia rusticana*) and induced symptoms different from those present on Citrus. However, effects of isolates from the source and target area on an array of hosts indicated their common source (Fletcher 1983). The similarities in host reactions confirmed the spread of barley rust virulence 24 from Europe to South America, and strains of wheat stem rust from the Angola-Mozambique belt to Australia. Better evidence for a long-distance dispersal of pathogens is obtained through proper genotype pattern analysis (see Welz, Chap. 11) and their comparison with the pattern at the source assumed. Spores as inoculum for these tests can be obtained from trap plants or plots, leaf samples from fields surveyed or devices which collect the rain in a target area.

Circumstantial evidence may help to validate the spread of plant pathogens. For instance, airborne soil and dust particles from West Africa, trapped in Barbados (Carlson and Prospero 1972) indicated that uredospores of *Hemileia vastatrix* from Africa can also spread to the continent of South America (Bowden et al. 1971).

3.2 Monitoring Dispersal from the Air

It is possible to monitor the probable arrival and the subsequent spread of inoculum from synoptic situations by means of weather and/or earth resource satellites. Integrating information received from ground surveys with remote sensing information is a proper approach to monitoring long-distance dispersal.

3.2.1 Monitoring the Synoptic Situation

Wind is the main carrier of plant pathogens. Isobar patterns are good guide to wind flow except in areas close to the equator. Deposition of spores invariable occurs with rain which can be often predicted or recorded. Under Indian conditions primary inoculum of *P. graminis tritici* in the upper air is scrubbled by rain long before its presence in ground-level air. Such a rain-associated deposition of spores is linked with the definite air circulation pattern perceivable in 700 mb synoptic weather charts. In these situations, collectively called "the Indian stem rust rules", a tropical cyclone in November is the most important factor (Nagarajan and Singh 1975). A cyclonic storm is a large vortex in the atmosphere with counterclockwise wind direction in the northern hemisphere. The size of the spiral can vary from 150 to 1000 km, its height may exceed 15 km, the wind speed may reach 150–200 km h^{-1} and the system may cover a distance of 500 km day^{-1}. Such a weather system or front can be easily perceived in sequential synoptic weather charts drawn twice a day. Under Indian conditions we can use synoptic charts to monitor the weather system called "western disturbance" which influences the epidemics of brown and yellow rusts of wheat over northwestern India (Nagarajan et al. 1982). The outbreak of barley yellow dwarf in central USA is facilitated by the low-level jet winds in April to mid-May that link Oklahoma and Iowa. This disease can be forecasted according to the synoptic situations that favour the long-distance dispersal of the vector (Wallin et al. 1967).

A trajectory is a curve in space, which traces the positions successively occupied by a moving air parcel. According to tracts of wind in the synoptic charts we can draw a "trajectory" of a spore cloud, i.e. the probable path of transport of the airborne organisms. Calculated trajectories move either forward or backward. The common error in trajectory construction derives from incorrect wind data and a wrong assessment of the initiation of spore transport. For instance, an error of 500 m in the supposed altitude of the spore cloud can result in a totally different trajectory pathway (Durst and Davis 1957).

3.2.2 Use of Weather Satellites

Timely detection and efficient monitoring of the weather system by polar orbiting and geostationary satellites have added a new dimension to studies on long-distance dispersal. In addition to high temporal resolution, geostationary satellites provide consistent viewing geometry to any given location and synoptic coverage over a large area. A small cumulus, stratus and cirrus clouds manifest themselves clearly even to an untrained person. According to the type of clouds, one can locate fronts, squall lines, jet streams and ridges. Cloud movement can be accurately monitored using half-hourly images. Low level ($\sim$ 850 mb) and high level ($\sim$ 200 mb) wind records are obtained from satellites by tracking the cloud system in three, half-hourly images. Genesis and growth of tropical cyclones can be monitored by observing the cloud formation associated with low pressure systems. Empirical methods have been established by Dvorak (1975) to calculate maximum surface winds associated with a cyclone by observing its extent, spiral bands, tightness, etc.

It was found that tracks of cyclones and the paths of the *P. graminis tritici* uredospore trajectory are nearly identical (Nagarajan and Joshi 1980) and can be substituted. The chance of getting disease is very high in the area where the cyclone dissipates. Over a span of time the accumulated cloud cover influences the micrometeorological situation in the field and the subsequent disease development; both can be estimated from records made by the meteorological satellite (Nagarajan et al. 1982).

3.2.3 Validation of Disease by LANDSAT

The high resolution multispectral scanner (MSS) in the resource sattellite LANDSAT can be used to monitor the growth and health of a crop over a large area, thus validating the arrival of diseases. LANDSAT in a near-polar, sun-synchronous orbit covers the earth's surface every 18 days and scans an area of 185 km^2. The MSS measures and records light reflected from the earth's surface in four broad spectral bands (0.5–0.6, 0.6–0.7, 0.7–0.8 and 0.8–1.1 µm) in the visible and infrared spectrum. A LANDSAT scene consists of approximately 7.5 million pixels, each with a ground-level resolution of 57 × 79 m. The records can be converted into images for manual interpretation, or can be digitally processed and analyzed by computer.

Annual crop profiles have a characteristic form that is related to the sequence of development changes in the crop. The profile features include maximum amplitude or peak, rise and decline of the peak, half-maximum width and profile duration and shape. These features can be used to discriminate between crops that are grown in the same season (Ajai et al. 1985). The near-infrared region (0.75–1.35µm) is a region of high reflectance caused by multiple scattering which takes place between intercellular air spaces of spongy mesophyll (Gausman 1974). Healthy plants give a higher reflectance in the near-infrared region and a lower one in the visible region. Diseased plants show a higher reflectance in the visible spectrum and a lower one in infrared region (Teng and Close 1977). Thus, reflectance may indicate plant growth and health. For describing these phenomena we use a vegetation index that measures near-infrared radiance relative to that in the visible bands: the infrared to red radiance ratio (IR/red), the normalized difference

$$ND = \frac{(IR - Red)}{(IR + Red)},$$

the transferred vegetation index ($TVI = ND + 0.5$) or the tasselled cap greenness (Kauth and Thomas 1976). The tasselled cap Greenness Vegetation Index (GVI) is obtained by transforming LANDSAT MMS information into four-dimensional data space using the four MSS bands (Thompson and Wehmanen 1980):

$$GVI = -0.28\,MSS4 - 0.66\,MSS5 + 0.58\,MSS6 + 0.39\,MSS7, \tag{17.1}$$

where MSS4, MSS5, MSS6 and MSS7 are LANDSAT data expressed in digital numbers in the four bands. Greenness (Eq. 1) is basically a contrast between the high reflectance in the near-infrared and the high absorption in the visible bands, and so is a measure of the vegetation vigour and health.

LANDSAT data in near-infrared bands (MSS6 and MSS7) have been used to discriminate the rust-affected wheat from the healthy wheat (Nagarajan et al. 1984a). The greeness vegetation index derived from LANDSAT MSS digital data in the four bands has been sucessfully used to distinguish between diseased wheat in the Pakistani Punjab and the healthy wheat in the Indian Punjab (Nagarajan et al. 1984b).

3.3 Future Trends in Aerial Reconnaissance

As mentioned above, the monitoring of sequential weather patterns by weather satellite can be used for prediction of long-distance dispersal: and the high resolution LANDSAT MSS data may be utilized for assessing disease severity and crop loss. A combination of both methods will improve the system. The following procedure can be adapted for a "Crop Health Watch System" that would form part of a larger, food resource management system:

1. Use meteorological satellite data to signal potential areas of disease occurence.
2. Use resource satellite data to determine vegetation greenness of the sample segments within and outside potential areas and evaluate against a normal year.
3. If potential damage is indicated, use full-frame imagery in conjunction with segment greenness to outline the extent of damage.
4. Compare crop vigour over years for indication of yield reduction, and the extent of cut in the food resources.

The spatial resolution of LANDSAT MSS restricts its use to homogeneous and monocropped areas only. However, we may use the very high spatial resolution data provided by the LANDSAT Thermatic Mapper and the French satellite SPOT. Microwave remote sensing can be used for all weather conditions as the scans can also penetrate the atmosphere under cloudy conditions. The Indian Remote Sensing Satellite (IRS) scheduled for launch in 1987 will provide high resolution (36.5 m) data in four spectral bands that can be used for a routine health watch.

4 References

Ajai, Gopalan AKS, Kamat DS (1985) Discrimination of winter crops using temporal spectral profiles. Int J Remote Sensing 6: 1575–1580
Bowden J, Gregory PH, Johnson GG (1971) Possible wind transport of coffee leaf rust across the Atlantic Ocean. Nature 229: 500–501
Carlson TN, Prospero JM (1972) The large scale movement of Saharan air outbreaks over the northern equational Atlantic. Appl Meteorol 11: 283–297
Durst CS, Davis NE (1957) Accuracy of geotrophic trajectories. Met Magazine (London) 86: 138–141
Dvorak VF (1975) Tropical Cyclone intensity analysis using enhanced infrared satellite data. Eleventh Technical Conference of Hurricanes and Tropical Meteorology, Miami, American Meteorological Soc, pp 268–273
Fletcher J (1983) Brittle root of horseradish in Illinois and the distribution of *Spiroplasma citri* in the United States. Phytopathology 73: 354–357
Frease HM, Yarger ND (1974) Movement of southern corn leaf blight spores. Iowa State J Res 48: 231–245
Gausman HW (1974) Leaf reflectance of near infrared. Photogrammetric Engeneering 40: 183–191
Gregory PH (1973) Microbiology of the atmosphere, Leonard Hill, UK, pp 377
Hermansen JE, Torp U, Prahm L (1976) Evidence of long distance dispersal of live spores of *Erysiphe graminis* f. sp. *hordei*. Roy Vet Agric Univ Copenhagen, 1975 Year Book, pp 17–30
Hirst JM, Stedman OJ, Hogg WH (1967) Long distance spore transport: Methods of measurement, vertical spore profiles and the detection of immigrant spores. J Gen. Microbiol 48: 329–355
Joshi LM, Saari EE, Gera SD (1971) Epidemiology of wheat rusts in India. INSA Bull 37: 449–453
Kauth RJ, Thomas GS (1976). The tasseled cap a graphic description of spectral temporal development of agricultural crops as seen by Landsat. Proc Symp Machine Processing of remotely sensed data. West Lafayette, Indiana, 4N–41–51
Kisimoto R (1976) Synoptic weather conditions inducing long distance immigration of planthoppers. *Sogatella furcifera* and *Nilaparvata lugens*. Ecol Entomol 1: 95–109
Leach CM, Anderson AJ (1982) Radiation quality and plant diseases. In Hatfield JL, Thomason IJ (eds) Biometeorology in integrated pest management. Academic Press, London, pp 267–306
Nagarajan S, Joshi LM (1980) Further investigations on predicting wheat rusts appearance in central and penisular India. Phytopathol Z 98: 84–90
Nagarajan S, Joshi LM (1985) Epidemiology in the Indian subcontinent. In: Roelfs AP, Bushnell WR (eds). The Cereal rusts, vol 2. Academic Press, London pp 371–402
Nagarajan S, Singh H (1975) Indian stem rust rules. An epidemiological concept on the spread of wheat stem rust. Plant Dis Rep 59: 670–672
Nagarajan S, Singh H, Joshi LM, Saari EE (1976) Meteorological conditions associated with long distance dissemination and deposition of *Puccinia graminis* f. sp. *tritici* uredospores in India. Phytopathology 66: 198–203
Nagarajan S, Kranz J, Saari EE, Joshi LM (1982) Utility of weather satellites in monitoring cereal rust epidemics. Z Pflanzenkr Pflanzenschutz 89: 276–281
Nagarajan S, Sashikumar MN, Ajai, Kamat DS (1984a) Detection of wheat rust disease from Landsat Multispectral data. Proc Seminar on Crop Growth Conditions and Remote Sensing June 1984, New Delhi, India, pp 7–1–1 to 7–1–21
Nagarajan S, Seiboldt G, Kranz J, Saari EE, Joshi LM (1984b) Monitoring wheat rust epidemics with Landsat–2 satellite. Phytopathology 74: 276–281
Rhind D, Waterston JM, Deighton FC (1952) Occurrence of *Puccinia polysora* underw in West Africa. Nature 169: 631
Teng PS, Close RC (1977) Spectral reflectance of healthy and leaf rust infected barley leaves. Aust Plant Pathol Soc. Newsl 6: 7–9
Thompson DR, Wehmann OA (1980) Using Landsat digital data to detect moisture stress in corn-soybean growing regions. Photogrammetric Engeneering Remote Sensing 47: 1087–1093
Wallin JR, Peters D, Johnson LC (1967) Low level jet winds, early cereal aphid and barley yellow dwarf detection in Iowa. Plant Dis Rep 51: 528–530

Wilkinson AG, Spiers AG (1976) Introduction of the popular rusts *Melamspora larici populina* and *M. medusae* to New Zealand and their subsequent distribution. NZ J Sci 19: 195–198

IV The Synopsis of Experimental Results

Chapter 18

Modeling Epidemics of Root Diseases and Development of Simulators

C. Lee Campbell[1], Keith M. Reynolds[2] and L. V. Madden[3]

1 Introduction

Plant pathologists are engaging in more quantitative research as the need is increasing to understand complex pathosystems rather than to simply describe them qualitatively. This quantitative research has been facilitated by the availability of computers, other electronic technology, and computer software. The generation of large data bases for plant pathosystems has led to an increase in modeling efforts and to a heightened awareness of the importance of such efforts.

Models have become an integral part of epidemiology as epidemiology has become an integral discipline of plant pathology. It is essential, therefore, to understand not only the modeling process in plant pathosystems but also the purposes, uses, and abuses of models. Our intent is to briefly explore the rationale for modeling in plant disease epidemiology, to survey the types of models that may be used with an emphasis on simulation models, to present a systems analytic approach in simulation modeling of root diseases, and to suggest the need for coupling simulators of root diseases with other disease simulators and root and crop growth models. This chapter will be complemented by the following one on foliar diseases.

2 Scope, Purpose, and Types of Models in Plant Disease Epidemiology

System simplification and modeling are essential to the scientific process. Scientists observe and describe natural phenomena, synthesize, summarize, and attempt to explain their observations; and use their observations not only to understand how a system works but also to predict the outcome of other still unknown situations. Scientists deal with nature directly only during discrete observations which subsequently have to be translated into processes and measurements usually with some simplification.

Models represent our perception (or imagination) of reality in a symbolic and simplified form. Ideally, a model retains the essential aspects of a real sytem that are pertinent to the problem at hand. The complexity or simplicity of the model must be in

[1] Department of Plant Pathology, North Carolina State University, Raleigh, NC 27695, USA
[2] Research Associate Department of Plant Pathology, Ohio State University, Wooster, OH 44691, USA
[3] Department of Plant Pathology, Ohio State University, Wooster, OH 44691, USA

accord with its purpose. While simplicity allows for easier understanding of the model, complexity may allow a greater accuracy in the description of the system.

The modeling approach is a valuable tool, especially if the intention of the research is to increase our understanding of a pathosystem. Components of pathosystems are too complex to be understood and controlled without abstraction. Modeling enables us to obtain a holistic understanding of a system by placing the interrelated factors in perspective at perhaps a fraction of the cost in time and resources of prolonged experimentation.

Models, by their nature, are abstractions of reality but must not become too abstract. As Skellam (1972) states: "Science demands more from its constructs than mathematical meaning; it also demands relevance." Thus, without an explicit statement of purpose and an adherence to that purpose, models lose their relevance and may not be worthwhile. The degree to which a model fulfills its purpose is measured with respect to how well that purpose could have been achieved without the model (Shoemaker 1980)

Models can be classified in various ways based on their properties and use. Teng (1985) should be consultated for a further discussion of some model classifications. We briefly describe the following more common ones. A *linear* model can be represented as:

$$Y = b_j X_j, \tag{18.1}$$

in which Y and the X's (i.e., X_1, X_2, ... X_j, ... X_n) are variables and the b's are parameters. A parameter is an unknown constant estimated from data. Equation (18.1) can be described more precisely as being linear in the parameters. The X's can correspond to time, environmental or biological factors, or even transformations of some measured variable, e.g., $X_2 = X_1^2$. In a *nonlinear* model, parameters appear as exponents or are multiplied or divided by other parameters. Two examples are:

$$Y = b_0 X^{b_1} \tag{18.2}$$

and

$$Y = b_0 - \exp(-b_1 X). \tag{18.3}$$

It is much harder to estimate parameters of nonlinear models than of linear ones. However, nonlinear models frequently arise and are appropriate in biology.

The models in Eqs. (18.1–18.3) are all *mathematical.* A special group of these models are *statistical,* in which the random variability or probabilities are represented. For instance, a statistical version of $Y = b_0 + b_1 X$ can be written as:

$$Y_i = b_0 + b_i X_1 + e_i, \tag{18.4}$$

in which the i subscript corresponds to the i-th value of X. For example, there may be five observations of $X = 2$, with each corresponding to a different Y. The e_i term represents the "error" or "unexplained variability", and equals the difference between the observed Y_i and that *expected* from the model, i.e., $b_0 + b_1 X_i$. Implicit in statistical models is the realization that for each unique X, there is a whole population of Y's and not just a unique Y. These types of statistical models arise frequently in epidemiology because exact relationships without variability are seldom found. Assuming a specific model, least squares regression can be used to estimate the parameters.

The other type of statistical model describe the *probability* of an event. One example is the following model that represents a Poisson distribution:

$$P(x) = \frac{\mu^x e^{-\mu}}{x!}, \tag{18.5}$$

in which $P(x)$ is the probability of x (e.g., infections) events occurring; μ is a parameter and equals the mean number of X (Pielou 1977). Probability models are extremely important in describing events at the root surface and have been important throughout plant disease epidemiology (Bald 1970; Gregory 1948).

The method of model development leads to additional classifications. *Empirical* or *correlative* models are derived by fitting data with an acceptable model. No theory regarding underlying mechanisms is required. Linear models are popular for this approach because of their ease of handling, although many simple nonlinear models may also be used. The number of infections in relation to inoculum levels is often well described by an empirically derived model as in (18.2). A *mechanistic* or *explanatory* model is derived from the *tentative* understanding of reality. This understanding could be obtained from previous experiments or biological principles. After a mechanistic model is formulated, data are collected to test the model's predictions. Mechanistic models can be of many simple or complicated forms.

3 Systems Analysis in Plant Pathology

Models are part and parcel of systems ecology, a branch of ecology directly applicable to plant disease epidemics (Zadoks 1971). A *system* is a part of reality that contains interrelated elements characterized by many reciprocal cause-effect pathways (deWit and Rabbinge 1979; Watt 1966). Systems analysis in ecology is based on the assumption that the state of an ecosytem can be expressed quantitatively and that changes in that system can be expressed in mathematical terms (deWit and Goudriaan 1974). The totality of the relationships within a system is the structure of the system and while a model is itself a system, the reverse generally is not true. Plant disease epidemics have structures composed of at least three subsystems which interact over time: the pathogen, the host, and the disease (Kranz 1978).

Systems analysis provides an outline for the organization of experimentation and modeling and takes into account the system complexity and variabilty. As Kranz and Hau (1980) point out, systems analysis serves as an aid in determining what is biologically relevant in a disease system, what is feasible or needed from a practical viewpoint, what the priorities should be, and what methods of measurement of epidemic characteristics are most suitable for the stated objective. The systems analytic approach also helps to refine ideas and identify previously unknown, important interactions and to reveal weaknesses in hypotheses and assumptions (Jeffers 1978). Systems analysis may also assist researchers in modifying experimental procedures and modeling efforts to obtain unifying concepts and operational models rather than trivial or irrelevant models or mere catalogs of fractional processes. Systems analysis can also serve as a guide toward the optimal allocation of resources for fulfilling a stated research objective.

4 Simulation Modeling and General System Theory

Simulation modeling is an integral part of systems analysis and plant disease epidemiology. *Simulation* is the art and science of constructing composite models and mimicking one system with another. The study of system properties exhibited by these models is also a concern of simulation. Historically, many of the concepts of simulation have been derived from general systems theory, so it is worthwhile to briefly consider some of the relevant, fundamental concepts.

Biological models are necessarily simplifications of the systems they represent and, consequently, are abstractions. However, a system itself is an abstraction in a sense, since in viewing it as a segment of reality, we are treating it as an object. If a system is an object, then it is bounded, and everything that is not a part of the system belongs to the environment of the system. Furthermore, the definition of a system boundary should be such that quantities that belong to the environment are not influenced by quantities that belong to the system (Klir 1969). For example, soil moisture may be viewed as an environmental quantity; however, plants influence soil moisture status through transpiration. Thus, except in unusual circumstances, soil moisture is properly a system quantity, whereas rainfall would commonly be treated as an environmental quality. A plant community would not affect rainfall activity if the size of the community is suitably small. However, if the system being modeled were subcontinental in spatial scale, then transpiration could certainly influence the volume of moisture held in the atmosphere and would , consequently, influence rainfall activity. In general, spatial and temporal resolution may determine whether a given quantity belongs to the system or its environment.

A system can be defined in terms of: (1) its input and output quantities; (2) activity; (3) behaviour; and (4) two alternative types of structure. These definitions have been thoroughly discussed by Klir (1969). Klir also provides a detailed discussion of necessary and sufficient subsets of these definitions that allow a complete system specification in several alternative forms.

Simulation models have been developed for both pathosystems involving airborne and soilborne pathogens. The simulators for foliar diseases (see Hau, Chap. 19) have usually simulated entire disease epidemics and have generally been temporal. For details see the review by Teng (1985). The simulations performed for systems involving soilborne pathogens (Bloomberg 1983; Gutierrez et al. 1983; Jones et al. 1978; Reynolds et al. 1986) have tended to concentrate on specific aspects of a system rather than on the system as a whole.

A simulation model is amenable to manipulation which would be impossible to experience or impractical to perform on the system for which it was developed (Naylor et al. 1966). Also, precise replication of experiments can be performed using a simulator $I-M$ a situation hardly, if ever, allowed in the real world.

The degree of difficulty encountered in developing a simulation model will depend on the complexity of the planned simulation, the availability of real world data for development of component models, and the researcher's understanding of the system. Data collected from published literature may or may not be appropriate for use and large initial gaps may exist in the data base. Experimentation with biological components of the system and simulation modeling may, per necessity, proceed simultaneously.

The planning and development of simulation experiments occur in a series of steps identified in Fig. 18.1 (Naylor et al. 1966). Steps 1 and 2 provide the organizational framework for the research project. Problem formulation is essential and frequently does not get the attention it deserves. At this stage, both the modeling objectives and

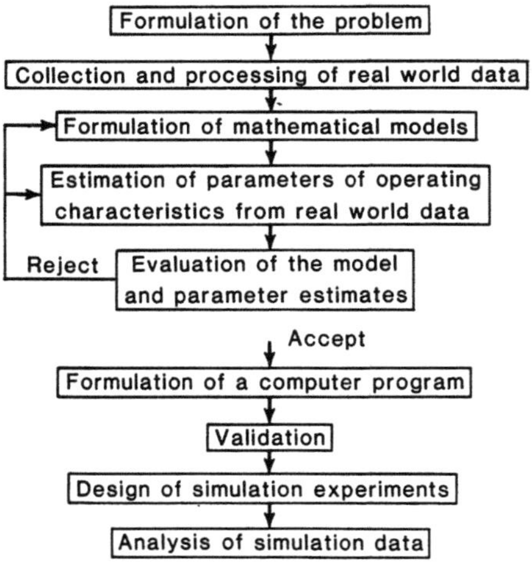

Fig. 18.1 The planning and development of simulation experiments. (After Naylor et al. 1966)

purpose(s) of model development need to be set out as explicity as possible. An important aspect of objective specification is the development of an object criterion which is used to assess the degree of correspondence between the model and the system being modeled (Overton 1977). Since model construction is an iterative process, the objective criterion also provides a basis for deciding when to terminate the process.

Detailed explication of the model purpose is valuable in two respects. First, it helps insure that the necessary input and output quantities, as well as internal mechanisms, are present in the model to answer the questions that the model is intended to address. Second, a clear and detailed statement of purpose may help insure that the end user of the model (who frequently is not the model developer) understands what types of questions the model can and cannot be used to answer.

Steps 3–5 help to identify gaps in our knowledge. The importance of item 7 cannot be overemphasized and has been discussed (Teng 1981). Steps 8 and 9 are the rewards of the entire process and should be viewed as an essential but logical consequence of simulation development.

5 Simulation Models for Root Diseases

Although a rationale exists for simulating epidemiological systems that involve soilborne plant pathogens, comparatively few simulators for these systems have been developed (Bloomberg 1979, 1983; Reynolds et al. 1986; Gutierrez et al. 1983; Gilligan, 1983; Barker and Noe, Chap. 15). At least part of the explanation for this scarcity of root disease simulation models is the lack of detailed information on the structure and behavior of nearly all root disease systems. In addition, much of the knowledge that exists is not available in a form which is readily integrated into simulation models. Systems involving *Phytophthora cinnamomi* are a good case in point; this pathogen is one of the most well studied of all root disease-causing organisms and yet it has only been possible to formulate a tentative simulation model for *P. cinnamomi* (Reynolds et al. 1986). These general problems are precisely the reason that the application of systems analysis to root disease systems may be so beneficial. In particular, a systems approach could help to optimize the progress of research by providing a structural framework within which to organize current knowledge of a system and by formalizing the purpose and objectives of research into a given system.

Before entering into a discussion of specific modeling approaches, it is necessary to clarify the meaning of certain terms. The term primary infection is reserved for infection which originates from pathogen propagules in the soil. All such infections, whether they are the first or 101st, are considered to be primary. In contrast to the meaning attached to secondary infection by Gilligan (1985, and Chap. 6, this Vol), we will use this term only to refer to spread of the pathogen between an infected plant or plant and another plant or plant root. Gilligan (1985) included autoinfection (*sensu* Robinson, 1976) under the term secondary infection, but since autoinfection is almost always a consequence of systemic infection in root pathosystems, we prefer the term colonization to autoinfection.

The root disease simulation models for fungal pathogens are each quite distinctive. Bloomberg's (1979) model for infection of Douglas fir seedlings by *Fusarium oxysporum* deals only with primary infection of seedlings, since secondary infection, i.e. spread between an infected plant or plant root and another plant or plant root, was found to be a very minor component of the epidemic process in this system. A stand model for the spread of *Phellinus weirii* in young-growth Douglas fir stands, which is based on the rate of infection center enlargement and intensification of disease within centers, has been developed by Bloomberg (1983). The model of Gutierrez et al. (1983) for cotton production in response to verticillium wilt is relatively unique among the models thus far developed in that the physiological effects of disease on plant metabolism are treated in some detail. Like Bloomberg's earlier model, the cotton-verticillium model does not deal with secondary infection. The latter model is basically a single plant model, although it does incorporate plant density effects and related compensation phenomena. Finally, Reynolds et al. (1986) have developed a model for the spread of *Phytophthora cinnamomi* in Fraser fir nursery beds. This model deals with primary infection from soilborne inoculum, seedling colonization, and interplant spread of disease.

In order to illustrate approaches to developing simulation models for root diseases (for nematodes, see Sect. 6.0 by Barker and Noe, Chap. 15), we will examine specific methods that have been employed for modeling the basic elements of the infection cycle (primary and secondary infection, and colonization) as well as host-pathogen interactions (such as effect of colonization on growth and yield). In the course of our discussion, particular emphasis will be placed on the usefulness and appropriateness of system simplification. An important consideration in the development of root disease simulation models is the potential for interfacing such models with current or future crop or individual plant growth models. Consequently, we will also examine how present models deal with this problem.

5.1 Primary Infection

A variety of approaches to modeling primary infection events have been utilized. One approach, which has been used quite successfully, is to simply not model primary infection explicity (Gutierrez et al. 1983). In this approach, only the observed effects of inoculum density on plant growth and yield are actually modeled. Discussion of this approach is provided in Sect. 5.3.

Many soilborne plant pathogens survive in soil either within or on the surface of decaying host tissues that were infected and colonized at some earlier point in time. If the colonized tissue is woody, a pathogen may persist for long periods and its inoculum may, therefore, be consistently associated with residual organic matter. In such cases, it may be appropriate to consider the colonized organic residue as the unit of inoculum. Bloomberg's (1979) model for infection of Douglas fir seedlings by *Fusarium oxysporum* and the model of Reynolds et al. (1986) for infection of Fraser fir seedlings by *Phytophthora cinnamomi* both treat colonized plant material as the unit of inoculum in soil.

Two techniques have been developed for determining when a root contacts organic material containing pathogen propagules. Both methods start with the assumption that inoculum-bearing organic matter is randomly distributed within small volumes of soil occupied by a tree seedling's root system (Bloomberg 1979; Reynolds et al. 1986). The validity of this assumption has been demonstrated for the *Fusarium* root rot model which assumes that the inoculum is distributed over a lattice. The size of the lattice depends on inoculum density, while the number of cells in a lattice element is determined by the size of the organic material (which is assumed to be cylindrical in shape and located at the center of a lattice element). When a root tip grows into a new lattice element, its precise location within that element is determined stochastically with the coordinates of the tip being independent, uniform deviates. Infection occurs when contact or "near contact" between the inoculum and root tips occurs (Chi et al. 1964; Griffin 1969). Bloomberg (1979) found that the infection efficiency of this form of inoculum varied with the size of organic matter fragments.

As a second technique, calculus and analytical geometry also can be used to determine whether a root and a piece of colonized organic matter make contact or come within some specified radial distance of one another (Reynolds et al. 1985a). In this case, both the root and pieces of organic matter have arbitrary orientation in space. Details of this approach are considered in Sect. 5.4.

Other methods of simulating primary infection are more appropriate when inoculum in soil is not associated with large organic fragments. Equation (18.6) presents a slightly modified version of Ferris' (1981) formula for calculating rhizosphere width:

$$W = \left[\frac{H}{LMIE} - r\right]^{1/2} - r, \tag{18.6}$$

in which: W = rhizosphere width (mm);
r = root radius (mm);
H = number of infections;
M = number of roots;
L = mean length of susceptible roots (mm);
I = inoculum density propagule (mm^{-3});
E = inoculum efficiency (infections per propagule).

Equation (18.6) can be used to calculate the volume of soil traversed by a root during a simulation time step. Given values for I and E, the probability of root tip infection can then be calculated. Drury et al. (1983) have discussed a number of problems related to the use of Ferris' model (Eq. 18.6); however, Eq. (18.6) has been successfully used to model changes in the rhizosphere width of Fraser fir seedlings (Reynolds et al. 1985a).

One potential difficulty with this equation is the need to estimate E, a quantity that is not constant over the radius of the rhizosphere (Drury et al. 1983; Gilligan 1985). This aspect of the problem can be considered in two ways. First, an expected (average) value for E can be used. Second, if the distribution of E over rhizosphere radius is known, propagule location within the soil volume traversed by the root can be specified stochastically and a value for E can be computed as a function of radial distance from the root. In general, however, values for E will vary not only with propagule to root distance, but also with the past and current states of the environment of the propagule. No root disease simulation models to date have considered the effect of environmental history on inoculum, yet this is clearly an aspect of fungal infection biology that deserves attention. For short-term simulations, it is probably safe to assume that changes in E are not significant.

Equation (18.6) can be applied equally well to individual root tips, a population of root tips on a plant, or a population of root tips in a population of plants. In the latter two cases, one might rearrange it to give an expression for H. The total quantity H could then be distributed among the population of root tips assuming a Poisson distribution, for instance. A similar approach has been suggested by Gilligan (1985), who gives a well-developed account of three alternative probability models.

5.2 Colonization and Inoculum Production

The process of colonization may be important not only with respect to the growth and development of an individual infected plant, but also in influencing progress of the epidemic when secondary infection is a possibility. Detailed submodels for colonization have only been developed for nursery diseases of two tree species (Bloomberg 1979; Reynolds et al. 1986). In both cases, it was assumed that variability in factors other than soil temperature were relatively unimportant. Validation studies in the *Fusarium* root rot of the Douglas fir system demonstrated this assumption to be acceptable.

An important feature of the effects of soil temperature on the rate of colonization relates to the temperature gradient that will always be present in a natural system. Consequently, the rate of lesion expansion will differ with soil depth as it may be exposed to very different temperature regimes at a single point in time (Reynolds et al. 1986). The latter differences will be most pronounced near the soil surface where temperature gradients tend to be steepest.

The physiological condition of an infected root may also influence the rate of lesion elongation. Wallis (1976) has observed that acropetal colonization of roots (direction is with respect to a root's point of attachment to a lower order root) may be more rapid than basipetal colonization. Since this effect was demonstrated for large, woody roots, it should certainly be considered as a factor in small root systems.

Detailed simulation of colonization is not feasible when the plant population is relatively large, or when the root structure of large plants is very complicated. As an alternative, one might describe the progress of colonization of an individual plant by specifying the areal coverage of colonized root mass. Suppose, for instance, that the root mass occupies a circular area when projected into a horizontal plane. If the distribution of root volume over the plane is known, then a model that describes the change in the area colonized could also be used to calculate the quantity of root mass affected. This model should function well if infection is initiated at the root crown. Since the root mass is distributed vertically as well as horizontally, exclusive use of this method for modeling colonization would only give very approximate results. A significant improvement in the above approach could, however, be realized by combining the areal description of colonization with the more detailed method discussed previously. For instance, following initial infection of an individual root, colonization could be modeled first at the level of the individual root. Once infection has progressed to the root crown, an areal colonization algorithm could then be used.

The only detailed simulation model for fungal colonization of roots to deal with production of pathogen inoculum on roots is that of Reynolds et al. (1986). Studies of chlamydospore and sporangium production by *Phytophthora cinnamomi* on infected Fraser fir roots showed that sporangium production by a vegetative thallus occupying colonized roots was both sparse and restricted spatially (Reynolds et al 1985 b). In contrast, large numbers of chlamydospores were observed to form on the root surface. Because of the relative rates of production of these two sources of inoculum, and because sporangia of *Phytophthora* are usually less persistent in the soil environment than chlamydospores, it was assumed for simplicity in modeling this system that chlamydospores formed at the root surface were the sole source of secondary inoculum. Under conditions of both warm soil temperature and high soil moisture, rapid changes in propagule numbers can be expected even when inoculum has been newly produced (Reynolds et al. 1985 b).

5.3 Effects of Colonization on Plant Growth and Yield

Gutierrez et al. (1983) have developed the most physiologically sophisticated root disease simulation model to date. Their model for verticillium wilt was interfaced with a model developed earlier for cotton production (Gutierrez et al. 1975). The cotton growth model consists of a system of differential equations that describe changes in plant number, leaf, stem, and root mass as well as mass and number of fruit. The model is described as a continuous implementation of the Leslie matrix approach (Leslie 1945). Predicted values for the systems state variables are based on submodels that prescribe priority allocation of photosynthetic resources which are available from a metabolic pool.

The effect of *Verticillium dahliae* on cotton production is mediated through a scalar quantity, Z, which is a function of physiological time and pathotype virulence. Inoculum density and the quantity, Z, modify both the rate of photosynthate accumulation in the metabolic pool and rates of carbohydrate flux from this pool to the four possible sinks (leaf, stem, root, and fruit). Changes in cotton growth and fruit production are empirically related to inoculum density and Z; however, root system infection is not modeled explicity. Gutierrez et al (1983) note that model validations show good agreement between the model and the field data up through 70 days and poorer agreement thereafter. It is possible that incorporating a root infection algorithm into the model structure would improve matters since it would then be possible to include submodels to account for changes in inoculum efficiency and/or root resistance to infection.

Reynolds et al. (1986) also used a scale factor to adjust an expression for seedling shoot mass increment. In this case, the scale factor, S, was calculated as a function of root length as affected by fungal colonization. "Affected" root length refers not only to that length of root which has actually been colonized, but also to any additional length of root that lies distal to the lesion(s), where distal refers to the point of attachment to a lower order root.

The rationale for assessing the effect of root colonization by a pathogen on shoot mass accumulation can best be illustrated with an extreme example. Consider a subject seedling which has no previous infections and which has recently become infected at or near the root crown. If radial colonization of root tissue proceeds at a rate similar to longitudinal colonization, a girdling infection will develop before the lesion has extended very far down the root. Consequently, the total length or mass of root actually colonized is quite small, but the root and shoot may be effectively isolated from one another by the infection. Clearly, in such a case, there is no longer any functional root mass (with respect to the shoot) and cessation of shoot growth would be expected. Use of loss of root length as opposed to root mass, to predict effects on shoot growth is justifiable if root systems are small and absorptive roots are approximately uniformly distributed along the structural roots. More information on sampling of soilborne diseases can be found in Chapters 3, 6 and 15 in this Volume.

5.4 Secondary Infection

Secondary infection can be modeled at three levels. The simplest method is to model spread in terms of change in areal coverage over time. This approach has been used very successfully in the development of a simulation model for the impact of *Phellinus weirii* on stands of Douglas fir trees (Bloomberg 1983).

The next level involves coupling a simple interplant infection submodel to a submodel for areal spread of infection on an individual plant. Reynolds and Bloomberg (1982) developed a series of regression models and a discriminant model to predict degrees of root association. The regression models predict the number of root contracts (for three different levels of root association) in a volume of soil based on attributes such as total root volume, number, and cross-sectional area. If the contributions to the latter quantities by each of two individuals can be predicted, then it is possible to predict the degree of root association in that volume of soil and, hence, when and if cross-infection will occur. The discriminant model predicts the number of root contacts based on a variety of tree, stand, and site attributes.

At the highest level of realism, secondary infection can be modeled by coupling a detailed model of colonization with a detailed model of root association. The model of Reynolds et al. (1986) provides a complete spatial specification of a Fraser fir seedling root structure for 2-year-old plants. The colonization algorithm predicts the limits of lesion development for all lesions on all seedlings in the simulated population of plants. The submodel for secondary infection treats an infected root as a cylinder whose radius is the maximum radial distance at which infection of a new root can occur (Reynolds et al. 1985b; Reynolds et al. 1986). The length of the cylinder is equal to the length of the infected root at the time that infection was established. Each time a new root is added to a seedling during simulation, a root contact alogrithm is called to check whether the longitudinal projection of this new root will intercept any of the cylindrical volumes associated with infected roots that currently exist. Both roots are represented parametrically in a set of simultaneous equations. Interception is determined by evaluating the determinant of the Jacobian matrix derived from the set of simultaneous equations. Details of the algorithm can be found in Reynolds (1985, Appendix A, p. 203–205).

6 Interfacing Epidemiological Models with Crop and Single Plant Growth Models

Simulation models can provide valuable insight into the development and epidemiology of root diseases. These models provide an opportunity to assess, organize, and further expand our knowledge about epidemics of root diseases and numerous interactions among soilborne pathogens, host roots, and the soil environment. Often, however, such models are limited to a single disease due to the acknowledged complexities involved in modeling even a single host-pathogen interaction. If we are to gain the maximum theoretical and practical benefits from simulation modeling of root diseases, these models must be interfaced with other models to provide information on the total agroecosystem (Kranz 1977).

Ideally, a simulation model for an agroecosystem could be developed by a team of biological, biomathematical, and computer science specialists. If, however, one biological specialist is unavailable or so little is known about one component system that modeling of the component is intractable, other components of the agroecosystem may still be modeled and the problem system treated as a black box. The team approach to simulation modeling is advantageous in that synergistic interaction among the participants can provide previously unforeseen insights into the system (Gutierrez et al. 1979) and the process can stimulate a heightened awareness by each participant of the interdependencies that exist among the modelers and scientists of the various disciplines.

Alternatively, if a plant or crop growth model exists, it may be possible to interface a root disease model with it. This requires that a submodel for root growth be present in the plant or crop model or that the root component is, at least, accounted for in a way that allows the addition of the disease model. If such a root growth model exists and is compatible with the requirements for development of the disease, it can then serve as a framework for developing a root disease model. If biological components or environmental variables important to the modeling of the root disease are not present in the model, it may be essential to first modify the relevant submodel(s). Before serious effort is devoted to this alteration and development, however, it is essential to assess the potential for compatibility between the existing and proposed models.

The interfacing of simulations for root diseases with other disease and pest models requires an understanding of the biology, ecology, and interactions among these systems that is lacking for most agroecosystems and forest ecosystems. This lack of information should not be viewed as a limitation for the development of simulation models for root disease, but as a challenge to acquire the requisite information, to concentrate on the synecological as well as autecological nature of systems involving root diseases, and to collaborate with scientists in disciplines outside of phytopathology.

The ultimate success in our development of interfaces among appropriate models and, in fact, in our development of simulation models for root pathosystems will depend on our ability to conceptualize root diseases from the organismal level to the level of the agroecosystem. System simplification is inevitable and essential but must be done with reference to the other ecosystem components and on an empirical and rational basis. New methodologies and new ways of thinking about root pathosystems are also essential to insure success in development and utilization of simulation models for root diseases.

7 References

Bald JG (1970) Measurements of host reaction to soilborne inoculum. In: Toussoun TA, Bega RV, Nelson PE (eds) Root disease and soilborne pathogens. Univ CA Press, Berkeley

Bloomberg WJ (1979) A model of damping-off and root rot of Douglas-fir seedlings caused by *Fusarium oxysporum*. Phytopathology 69: 74–81

Bloomberg WJ (1983) A ground survey method for estimating losses caused by *Phellinus weirii* root rot. III. Simulation of disease spread and impact. Can For Serv Pacific For Res Center Rpt BC-R-7, pp 25

Chi CC, Childers WR, Hanson EW (1964) Penetration and subsequent development of three *Fusarium* species in alfalfa and red clover. Phytopathology 54: 434–437

de Wit CT, Goudriaan J (1974) Simulation of ecological processes. Centre for Agricultural Publishing and Documentation, Wageningen, The Netherlands, pp 195

de Wit CT, Rabbinge R (1979) Systems analysis and dynamic simulation. EPPO Bull 9: 149–153

Drury RE, Baker R, Griffin GJ (1983) Calculating the dimensions of the rhizosphere. Phytopathology 73:1351–1354

Ferris RS (1981) Calculating rhizosphere size. Phytopathology 71: 1229–1231

Gilligan CA (1983) Modeling of soilborne pathogens. Annu Rev Phytopathol 17: 431–460

Gilligan CA (1985) Construction of temporal models. III. Disease progress of soilborne pathogens. In: Gilligan CA (ed) Advances in plant pathology, vol 3, Mathematical modeling of crop disease. Academic Press, London, pp 255

Gregory PH (1948) The multiple-infection transformation. Ann Appl Biol 35: 412–417

Griffin GJ (1969) *Fusarium oxysporum* and *Aspergillus flavus* spore germination in the rhizosphere of peanut. Phytopathology 59: 1214–1218

Gutierrez AP, Falcon LA, Loew W, Leipzig PA (1975) An analysis of cotton production in California: a model for Acala cotton and the effects of defoliators on its yield. Environ Entomol 4: 125–136

Gutierrez AP, Wang Y, Jones RE (1979) Systems analysis applied to crop protection. EPPO Bull 9: 133–148

Gutierrez AP, De Vay JE, Pullman GS, Friebertshauser FE (1983) A model of verticillium wilt in relation to cotton growth and development. Phytopathology 73: 89–95

Jeffers JNR (1978) An introduction to systems analysis: with ecological applications. Arnold, London

Jones FGW, Kempton RA, Perry JN (1978) Computer simulation and population models for cyst-nematodes (Heteroderidae: Nematoda). Nematropica 8: 36–56

Klir GJ (1969) An approach to general systems theory. Van Nostrand, New York

Kranz J (1977) Die Entwicklung von Pflanzenschutzsystemen. Mitt Biol Bundesanst Land Forstwirtsch Berl Dahlem 175: 85–100

Kranz J (1978) Comparative anatomy of epidemics. In: Horsfall J G, Cowling E B (eds) Plant disease vol 2 Academic Press, London

Kranz J, Hau B (1980) Systems analysis in epidemiology. Annu Rev Phytopathol 18: 67–83

Leslie PH (1945) On the use of matrices in certain population mathematics. Biometrika 35: 213–245

Naylor TH, Bolintfy JL, Burdick DS, Chu K (1966) Computer simulation techniques. Wiley, New York

Overton WS (1977) A strategy of model construction. In: Hall CAS, Day JW Jr (eds). Ecosystem modeling in theory and practice. Wiley, New York
Pielou EC (1977) Mathematical ecology. Wiley, New York
Reynolds KM, Bloomberg WJ (1982) Estimating the probability of intertree root contract in second-growth Douglas-fir. Can J For Res 12: 493–498
Reynolds KM, Benson DM, Bruck RI (1985a) The epidemiology of *Phytophthora* root rot of Fraser fir: Estimates of rhizosphere width and inoculum efficiency. Phytopathology 75: 1010–1014
Reynolds KM, Benson DM, Bruck RI (1985b) The epidemiology of *Phytophthora* root rot of Fraser fir: root colonization and inoculum production. Phytopathology 75: 1004–1009
Reynolds KM, Gold HJ, Bruck RI, Benson DM, Campbell CL (1986) Simulation of the spread of *Phytophthora cinnamomi* causing a root rot of Fraser fir in nursery beds. Phytopathology 76: 1190–1201
Shoemaker CA (1980) The role of systems analysis in integrated pest management. In: Huffaker CB (ed) New technology of pest control. Wiley, New York
Skellam JG (1972) Some philosophical aspects of mathematical modelling in empirical science with special reference to ecology. In: Jeffers JNR (ed) Mathematical models in ecology. Blackwell, Oxford
Teng PS (1981) Validation of computer models of plant disease epidemics: Z Pflanzenkr Pflanzenschutz 88: 49–63
Teng PS (1985) A comparison of simulation approaches to epidemic modeling. Annu Rev Phytopathol 23: 351–379
Wallis GW (1976) Growth characteristics of *Phellinus (Poria) weirii* in soil and on root and other surfaces. Can J For Res 6: 229–232
Watt KEF (1966) The nature of system analysis. In: Watt KEF (ed) Systems analysis in ecology. Academic Press, London
Zadoks JC (1971) Systems analysis and the dynamics of epidemics. Phytopathology 61: 600–610

Chapter 19

Modelling Epidemics of Polycyclic Foliar Diseases and Development of Simulators

B. Hau[1]

1 Introduction

In the previous chapter (see Campbell et al., Chap. 18), the scope, purpose and types of models in plant disease epidemiology as well as the role of systems analysis in plant pathology are outlined and it is not necessary to repeat them here. The terms, e.g. model or simulation, are used in this chapter in the same sense as defined in the previous one.

This chapter describes modelling of polycyclic epidemics on plant foliage. These epidemics start when primary inoculum, coming from outside the field or originating from the overwintered pathogen, infects the host and causes disease symptoms on leaves. The leaf spots produce secondary inoculum which again attacks the host and the pathogen multiplies through successive generations in the course of an epidemic.

2 Modelling of Polycyclic Epidemics

Before we describe the different steps in a modelling procedure we shall define the objectives of modelling epidemics.

2.1 Objectives of Modelling Epidemics

As in all other research fields, modelling epidemics aims at several objectives which in turn influence the complexity of the models to be constructed. The three objectives generally connected with modelling are: description, prediction and explanation (Kranz and Royle 1978; Campbell et al., Chap. 18). However, in our case only the ability of models to explain the system and to predict its behaviour is important.

In epidemiology one of the main objects is to understand the course of epidemics and to gain insight in their structures. This means especially to comprehend the interlocking effect of the different parts of the infection chain and the effects of external factors. This aspect of modelling demands detailed knowledge of biological processes. On the other hand, models can assist in experimentation and guide research projects.

Besides the application of simulation models in research, simulators can be used in the education of students and the instruction of farmers. Models can be extended to

[1] Phytopathologie und Angewandte Entomologie, Wissenschaftliches Zentrum Tropeninstitut, Justus-Liebig-Universität, Gießen, FRG

management games which enable one to test different tactical or strategical decisions and to compare them in order to find an optimal solution under given circumstances.

Epidemiological models and simulators are also needed for practical applications, such as decision making. Together with weather forecasts, models may predict further disease progress. In combination with the present extent of observed damage, the forecast of future disease development can help to decide whether a fungicide application is needed. Moreover, specific models simulating disease development over several seasons may detect strategies to delay fungicide resistance or race adaptation.

2.2 Steps in Modelling Epidemics

The construction of models according to a systems analysis approach proceeds in several stages and involves some recycling (Kranz and Hau 1980). Description of these stages is given in further sections of this chapter. In all of them the researcher should keep in mind that certain steps will be repeated several times during the improvement of the model.

2.2.1 Restriction of Models

The first step in modelling epidemics is the recognition and bounding of the problem. One has to decide which components of an epidemic are essential and how they interact. Obviously the complexity of models depends on the purpose for which they are developed and the model should be as simple as possible to serve the desired purpose. This requirement can be achieved by raising the complexity of simple models or by reducing the number of elements included in sophisticated models. This is the problem of synthesis versus analysis or, in other words, the bottom-up versus the top-down approach.

Epidemics develop in time and space. The researcher has to decide whether the time or space factor, or both are the subject of modelling. However, the spatial spread of diseases within a field is integrated in only a few simulators, for instance EPIDEMIC (Shrum 1975) or EPIMUL (Kampmeijer and Zadoks 1977). Mathematical models based on physical processes of dispersal (see McCartney and Fitt 1985) have not been used in published simulation models. However, the combination of spatial and temporal characteristics of epidemics has been carried out using simple theoretical equations (Minogue and Fry 1983a, b; Jeger 1983).

Considering the bounding one should clarify in advance whether a deterministic or a stochastic model is needed. The latter models are more realistic but their construction is more laborious and experimentation more extensive. Up to now only few of the published simulators, e.g. Teng et al. (1980) and Sall (1980), include stochastic components together with Monte Carlo simulation techniques.

Every epidemic consists of the two subsystems, viz. pathogen and host, which interact over time under the influence of environment and cause the disease. The environment is not part of the model but effects pathogen and host by external variables like temperature, humidity, radiation, etc. Though the influence of these driving variables on host growth is evident, the "most sophisticated epidemiological models have gone no further than to include plant growth as a single variable that increase according to an empirical relationship with time" (Rouse 1983), for instance as a leaf area index curve like in EPIDEM (Waggoner and Horsfall 1969) or as a function of available leaf surface in cm^2 like in BARSIM-I (Teng et al. 1980). In EPIGRAM (Aust et al. 1983) host growth given as total leaf area is divided into the proportions of the different leaf layers to account for the adult plant resistance of the upper leaves.

In any case one has to decide on the manner in which the host should be represented in the epidemiological model. The alternative to simple functions is a detailed crop growth simulator that accounts for the interactions with the pathogen such as reduction of host growth after disease outbreak. We will come back to this point later when dealing with the combination of disease and host simulators.

2.2.2 Approaches to Modelling

The next step in systems analysis is the design of a conceptual model, "which summarizes available information, experience and imagination" (Kranz and Hau 1980). The conceptual model of a polycyclic epidemic can be formed (1) according to detailed biological processes like the infection chain or (2) by using general equations for epidemics which incorporate biological meaning like the paralogistic equation of Vanderplank (1963). The first approach is called analytic, while the second one is called synthetic (Vanderplank 1982). We follow here the terminology of Vanderplank, but one should be aware that in the literature a number of different terms are used to classify these models. Recently, Teng (1985) compared and discussed some of the different modelling approaches (see also Campbell et al., Chap. 18).

2.2.2.1 The Analytic Approach

Models resulting from the analytic approach (*sensu* Vanderplank 1982) are strictly oriented on the biological processes of the system. Every phase of the fungal life cycle is modelled separately starting from infection, incubation and lesion growth and proceeds to sporulation and dispersal. Details of the modelled phases depend on the available information and the purpose of modelling. Each phase may be split into subphases, for instance the infection process into the subphases germination, hyphae growth, penetration and so on. On the other hand, certain phases of the life cycle may not be accounted for and may be replaced by data actually measured in the field. For instance in modelling powdery mildew of barley, Aust et al. (1983) fed the model with spore-catch data from the field instead of using a simulated amount of dispersed spores. Incidentally, simulating spore dispersal is always the weak link in the simulation of the infection chain. In some models it may be desirable to combine several phases of the pathogen life cycle into simple equations which is adequate in management systems designed to supervise epidemics in commercial fields, e.g. EPIPRE (Zadoks 1981). Independent of the complexity, epidemics are simulated by linking the components and running through the life cycle again and again.

This analytic approach was applied by almost all simulators in the literature starting 18 years ago with EPIDEM (Waggoner and Horsfall 1969). Details of the development of simulation models since then have been reviewed extensively (Waggoner 1978; Zadoks 1979; Hau 1985; Teng 1985). Their particular characteristics have been elucidated by comparison of various models (Hau 1985; Teng 1985).

To model epidemics with the analytic approach, the influence of external factors (like temperature or host resistance) on specific phases of the pathogen life cycle and their linkage must be known in such completeness that quantitative relationships between driving variables and the pathogen can be expressed in mathematical terms.

Vanderplank (1982) criticized this approach by pointing out the accumulation of possible errors due to the often great number of variables involved. He instead recommended to use the synthetic approach based on simple equations consisting of only a few variables, not error prone.

2.2.2.2 The Synthetic Approach

In the synthetic approach (*sensu* Vanderplank 1982), an epidemic is treated as an entirety without splitting it up into details. The models consist of one or several equations, normally differential equations, based on parameters with a biological meaning. The advantages and disadvantages of these models are discussed elsewhere, e.g. Madden (1980), Hau et al. (1985) and Rouse (1985). We will not treat simple biological equations like the logistic growth curve and statistical equations, for instance polynomial models. However, some models will be considered which include more biological facts than the growth rate and initial inoculum to start numerical integration.

Vanderplank (1963) proposed a synthetic model which describes the diseased proportion of the leaf area in time. The parameters of the equation are an infection rate, the latent period and the infectious period. Based on his model Vanderplank (1982) calculated some tables which "give clearer illustration of the structure of disease progress than anything else yet published".

Jeger (1982) recommended a system of differential equations linking by biologically meaningful parameters the rate of change of total, infectious and removed diseased plant tissue. By means of an analytical solution of this synthetic model he felt that valuable insights into the area of epidemic theory can be achieved. A model developed by Fleming (1983) contains a mean infection rate and again the length of the latent period. Berger and Jones (1985) combined an infection model with a host-growth model. In the system of differential equations, variable latency was achieved by a distributed-delay submodel.

Normally, methods of numerical integration are applied for the solution of the differential equation and the systems of equations. As mentioned above, starting values in the numerical procedure can have biological meaning, for instance the level of primary inoculum.

For most examples in the literature, the parameters of the equations remain constant. However, it is also possible to assume a time dependence of the parameters under the influence of the temporal development of meteorological variables and to describe the parameters as mathematical functions or tables influenced by external factors. The actual changes of the parameter values and of the infection chain processes can then be calculated at each time during the vegetation period.

2.2.3 Experimentation

The data base generated from the literature or from one's own experiments is the prerequisite for establishing the mathematical relationships between driving variables and pathogen development. In most cases, published data can only be used in the first step of modelling because details of the experiments are often not described accurately and cannot be reproduced, *inter alia,* this holds for experiments in which the races have not been determined or the definition of variables is not clearly described (e.g. the length of the incubation time). In modelling this implies that one must conduct experiments either to bridge gaps in knowledge or to clarify processes which have been reported with contradictory results.

For the development of simulation models, experiments were often conducted under controlled conditions in growth chambers. If, for instance, temperature is the main driving variable, the experiments are carried out at different, but constant temperatures and other external factors are held constant (e.g. relative humidity). It is consequently assumed that the effect of temperatures not examined can be assessed from the temperature levels tested. But one should be aware that extrapolation outside the temperature range investigated in the experiments can be misleading, especially when the cardinal temperatures are subject to compensation phenomena (Aust et al. 1980). Because of these phenomena, the usual approach in experimentation to test the effect of single meteorological factors does not suffice for modelling purposes, since several variables may affect disease by interaction. Therefore, besides various levels of each meteorological factor, the combination of factors should be tested in experiments. This dictates an experimental design which can guarantee an appropriate evaluation of data.

Even if the influence of constant conditions can experimentally be deduced, it may not always be possible to forecast what will happen if these conditions change. Hence, it is necessary to perform experiments in growth chambers under changing conditions (see Rotem, Chap. 2).

Aust and Bretschneider-Herrmann (1982) compared experiments in growth cabinets and in the field and claimed that results in a controlled environment are useful for explaining causal relations, but not for determining quantitative relations as observed in field experiments (see also Chaps. 1 and 2). However, the "highly variable and uncontrolled field conditions make results difficult to interpret", while the conditions in growth cabinets "can be consistently repeated and varied at the will of the researcher" (Coakley 1983). Part of the imponderable variability can be excluded by conducting a mixed-field/growth-chamber experiment. For instance, one may use spores produced under controlled conditions to inoculate leaves of different ages in the field as Aust (1981) did in his experiments to quantify adult plant resistance of barley powdery mildew.

In all experiments one should realize that the macro- and micrometeorological variables can differ greatly. For instance, the temperature on the leaf surface may be 8–10°C greater or 6–8°C lower than ambient air temperature (Rouse 1983). Therefore, it should be clear which meteorological variables have been measured and are to be used in the model (see Sutton et al., Chap. 7; Friesland and Schrödter, Chap. 8).

2.2.4 Development of Models

After the experimentation is completed the main step in systems analysis starts, viz. the construction of the model. First, the general conclusions of the previous results are drawn. For instance, we determine the incubation time at 18°C when the experimental work was done at 16° and 20°C. This can be achieved by linear interpolation using the two neighbouring data points. The effect of temperature would then be described by a polygon curve and implemented in the computer programs by the construction of a table.

In another method, not only the two closest temperature levels are used, but all temperature levels tested in the preceding experiments. By means of regression analysis, a function is fitted to the complete data set. This method is more difficult, especially when non-linear regression procedures are applied.

The general procedures to formulate mathematical models, to choose appropriate functions and to estimate parameters are explained by Hau et al. (1985) and include cases of several influencing variables in the model.

As mentioned before it is not enough to grasp the effect of constant conditions, one must be able to mimic the processes in the changing environment. Waggoner and Parlange (1974a, b; 1975) described possible effects on germination of spores due to frequent changes of temperature and presented a mathematical model which can be generalized to other environmental variations. Hau et al. (1985) discussed and compared some methods developed to account for the influence of changing temperatures.

Though the temporal processes within an epidemic are continuous, the simulation in the analytic approach is carried out in discrete time steps using the Euler method or the rectangular formula for the integration. With specialized simulation languages, other integration routines can be applied, for instance the trapezoidal rule or the Runge-Kutta-integration. The same holds for the use of differential equations or systems of these equations in the synthetic approach.

In both modelling approaches, initial values of the state variables are needed to start simulation. It is difficult, however, to determine the initial values of the many variables under field conditions (Teng 1985). Accurate estimations of initial states are important because they regulate the subsequent epidemic. This is especially crucial with the status of the disease because actual field observations can only reflect visible disease symptoms, but not infections which are still latent. This means that instead of a single initial value for the disease, a function is needed which describes its development before the starting point. In some cases this can be achieved by means of daily spore catches. The same problem arises with the synthetic approach when the course of disease development up to the starting point must be known for the solution of the delay differential equations.

The permanent repetition of the processes within the infection chain in the complex models and the necessary numerical integration of differential equations is conditioned by the use of computers with their high speed in computation and huge storage capacity. Hence, state and rate equations should be formulated in computer language. Most of the simulation models have been programmed in FORTRAN, but some are written in special simulation software like CSMP, e.g. EPISIM (Rijsdijk 1975) and recently the model of Berger and Jones (1985). This kind of simulation language is particularly adequate for models designed as differential equations because different integration methods are already included.

2.2.5 Testing of Models

The philosophy and methodology of testing epidemiological models has been reviewed by Teng (1981) and will not be repeated here. Therefore, we will comment only on model validation, which means to test the agreement between predicted and observed epidemics. Subjective and statistical procedures for the validity are described by Teng (1981), while a reliability index for the choice between available models is proposed by

Legett and Williams (1981). Feldman et al. (1984) proposed a statistical procedure for validation instead of standard statistical tests which are not applicable to time series of dynamic processes.

However, it should be clear that "even when model output is consistent with observation, one cannot conclude that assumed biological mechanisms on which the model is based are realistic" (Fleming and Bruhn 1983). Hence, it is necessary to test the components of a model separately and to make sure that the linking of submodels reflects the biological process correctly. This procedure has been called verification (Teng 1981).

On the other hand, it may be possible that a model is verified correctly but does not reasonably mimic epidemics in the field. This can always happen in the early phase of an epidemic because then the initial state of variables in the field cannot be determined accurately. Even if the state variables are known exactly, deviations between predicted and observed epidemics can occur because in the model the development is only determined by the initial state of the pathogen. In reality, incoming inoculum from outside the experimental field can influence disease progress, especially at the beginning of an epidemic.

Generally, one should not forget that the observations of the disease used for the validation of a model cannot be measured or assessed without errors and in some cases these errors "may be more significant than those arising from the model itself" (Legett and Williams 1981). This statement may especially be true when disease intensities are low and unevenly distributed.

Other problems connected with the testing of models include the transformation of data (Van der Plank 1975), the normalization of model outputs (Kranz 1979) and the robustness of models (Fleming 1980).

In the tests of simulators, sensitivity analysis of models is rarely executed. The model of Zadoks (1971) can be regarded as an example in which the influences of the latent period and of other parameters on the dynamics of epidemics were examined by means of the paralogistic equation of Van der Plank (1963). Sensitivity tests are essential for models designed for practical applications where the attainable accuracy of parameter assessment is limited. It can also lead to a reduction of the number of variables in the model.

2.2.6 Implementation and Application of Models

The implementation of models is related to the objectives for which they were constructed. Models designed as a research tool can be applied as they are. Restricted to a certain problem, this kind of model is in most cases only used by the person who developed it. It often happens that these models are hardly documented, never published and easily forgotten when the designer, mostly graduate students, have left the institution. With the exception of EPIDEMIC (Shrum 1975), which was designed as a flexible simulator of many diseases, all published simulators have been prepared to deal with a special disease. To the best of my knowledge the approach of Shrum was not adopted by other workers.

The modification of models is necessary when they are used in education. The simulator should then be interactive, self-explanatory, easily to handle and well documented. Besides its ability to simulate epidemics, other features must be added e.g. the establishment of disease-loss relationships, damage thresholds and effects of fungicides. Such a management tool enables the user to compare model outputs influenced by different strategic or tactical decisions. APPLESCAB (Arneson et al. 1978), LATEBLITE (Bruhn et al. 1980) and SPUD CROP (Krawczak and MacKenzie 1981) are some of the pest management programs used in teaching with the aid of microcomputers. One should be aware that the development of computer-assisted learning systems requires both teaching experience and knowledge of computer programming (McLaren and Craignon 1981).

Only few research models are implemented for practical pest management, for example SYMPHYT in the German Democratic Republic (Spaar and Ebert 1985). In some cases simplified versions of sophisticated computer models are used in forecasting systems, for instance in EPIPRE (Zadoks 1981). Besides other reasons which were discussed by Butt and Jeger (1985), the restricted use of models in forecasting is due to the difficulty to measure the variables required for application of the original models in commercial fields. Thus, the monitoring aspect determines the acceptance of simulators by farmers or extension agents (Kranz et al. 1984). Therefore, an adaptation phase is indispensable in which monitoring and sampling is adjusted to restraints in farming practice (Kranz et al. 1984).

3 Combination of Models

3.1 Combination of Crop Growth and Plant Disease Models

Following the different steps of modelling, one may end up with a model which can simulate plant disease epidemics under the influence of external factors. On the other hand, models exist which explain the growth of the host plant in dependence on similar driving variables (see Seem, Chap. 4). Therefore, it is desirable to combine crop growth with plant disease models to overcome the gap between the detailed simulation of the pathogen and the simple functions for the host. Moreover, the crop growth models provide a framework for integrating effects of pathogens on crop growth and yield (Boote et al. 1983).

Generally, some effects of the host on the epidemics are included in plant disease simulators. For example, even in the simple synthetic models corrections for host growth have been proposed (Van der Plank 1963; Kushalappa and Ludwig 1982). More complicated models incorporate the influence of adult plant resistance in the various leaf layers, as in EPIGRAM (Aust et al. 1983). However, the physiological effects of pathogenesis on plant development are largely ignored by epidemiologists due to the fact that little quantitative data are available (Rouse 1983). Plant pathologists and agronomists should concentrate more on the linkage aspect of models so that epidemics can really be modelled as a process in the disease triangle, host-pathogen-environment.

3.2 Combination of Epidemiological Models

The usual plant disease simulators are valid only for a certain cultivar of the crop plant and the mixture of races of the pathogen tested in the experiments. To generalize a model it is necessary to adapt the simulator to different cultivars as Rapilly (1977) did with EPISEPT and to extend it to allow for the intraspecific competition of races and their dynamics.

So far, no analytical simulation model is available to describe the dynamics of different races of a pathogen. However, simple synthetic models based on differential equations of the Lotka-Volterra type have been proposed by Barrett (1983) who included the approaches of other authors as special cases.

In reality, the situation is even more complex because apart from the different races of one pathogen, also different pathogens interact. Therefore, a combination of plant disease simulation models is needed to account for the interspecific competition of plant pathogens; this is more important than the presence of competing races. To the best of my knowledge no models combine the simulation of two or more interacting pathogens. Even the easiest case, i.e. the occurrence of two pathogens on the same plant organ (e.g. mildew and rust on the same leaf), has never been tackled. Surveillance systems, like EPIPRE (Zadoks 1981), treat the observed pests and pathogens independently, assuming that fungicides can lower the density of agents so that interactions can be neglected. For the coupling of models on epidemiological facets and models on crop growth see Chap. 18.

4 Concluding Remarks

Simulation models are undoubtedly necessary tools in disease management in the future. However, today, 18 years after the publication of the first plant disease simulator, only a few epidemiological models have passed the status of being a research instrument and are used in practical decision making. The reason for this is that almost all simulation models in plant pathology are problem-oriented and only valid for a certain mixture of the pathogen and only one host cultivar. But, even with this simplified assumption, some parts of the infection chain, as a biological basic model, are treated rather loosely. This is especially relevant to data on spore dispersal and landing.

Moreover, the representation of the host plant in epidemiological models is poor (see Chap. 4, Sect. 5.3). This weakness of the models can only be overcome when crop growth simulators are combined with conventional plant disease simulators.

The next step must be the extension of such simulators by modelling intra- and intersprecific competition of plant pathogens, for instance to explain the buildup of fungicide resistance or to understand the dynamics of a pest complex. However, this goal can only be reached when an adequate documentation of the models enables the non-experts to understand and use the models as guidelines. Interdisciplinary cooperation remains a prerequisite in modelling of epidemiological models.

5 References

Arneson, PA, Oren TR, Loria R, Jenkins JJ, Goodman ED, Cooper WE (1978) APPLESCAB: a pest management game. Michigan State Univ, East Lansing (MI)

Aust HJ (1981) Über den Verlauf von Mehltauepidemien innerhalb des Agro-Ökosystems Gerstenfeld. Parey, Berlin (Acta Phytomedica, vol 7)

Aust HJ, Bretschneider-Herrmann B (1982) Nutzen und Grenzen biologischer Forschung unter kontrollierten Umweltbedingungen mit besonderer Berücksichtigung phytopathologischer Fragestellungen. Angew Bot 56: 201–217

Aust HJ, Bashi E, Rotem J (1980) Flexibility of plant pathogens in exploiting ecological and biotic conditions in the development of epidemics. In: Palti J, Kranz J (eds) Comparative epidemiology. A tool for better disease management. Pudoc, Wageningen, pp 46–56

Aust HJ, Hau B, Kranz J (1983) EPIGRAM — a simulator of barley powdery mildew. Z Pflanzenkr Pflanzenschutz 90: 244–250

Barrett JA (1983) Estimating relative fitness in plant parasites: some general problems. Phytopathology 73: 510–512

Berger RD, Jones JW (1985) A general model for disease progress with functions for variable latency and lesion expansion on growing host plants. Phytopathology 75: 792–797

Boote KJ, Jones JW, Mishoe JW, Berger RD (1983) Coupling pests to crop growth simulators to predict yield reductions. Phytopathology 73: 1581–1587

Bruhn JA, Bruck RI, Fry WE, Arneson PA, Keokosky EV (1980) LATEBLIGHT: a plant disease management game. User's manual. Cornell Univ. Ithaca (NY), Department of Plant Pathology, Mimeo 80–1

Butt DJ, Jeger MJ (1985) The practical implementation of models in crop disease management. In: Gilligan CA (ed) Mathematical modelling of crop disease. Academic Press, London, pp 207–230 (Advances in Plant Pathology, vol 3)

Coakley SM (1983) Ambient meteorological factors — light, temperature, and moisture. In: Kommedahl T, Williams PH (eds) Challenging problems in plant health. American Phytopathological Society, St Paul, USA, pp 154–167

Feldmann RM, Curry GL, Wehrly TE (1984) Statistical procedure for validating a simple population model. Environ Entomol 13: 1446–1451

Fleming RA (1980) Selection pressures and plant pathogens: robustness of the model. Phytopathology 70: 175–178, 71:268

Fleming RA (1983) Development of a simple mechanistic model of cereal rust progress. Phytopathology 73:308–312

Fleming RA, Bruhn JA (1983) The role of mathematical models in plant health management. In: Kommedahl T, Williams PH (eds) Challenging problems in plant health. American Phytopathological Society, St Paul, USA, pp 368–378

Hau B (1985) Epidemiologische Simulatoren als Instrumente der Systemanalyse mit besonderer Berücksichtigung eines Modells des Gerstenmehltaus. Parey, Berlin (Acta Phytomedica, vol 9)

Hau B, Eisensmith SP, Kranz J (1985) Construction of temporal models. II. Simulation of aerial epidemics. In: Gilligan CA (ed) Mathematical modelling of crop disease. Academic Press, London, pp 31–65 (Advances in Plant Pathology, vol 3)

Jeger MJ (1982) The relation between total, infectious and postinfectious diseased plant tissue. Phytopathology 72:1185–1189

Jeger MJ (1983) Analysing epidemics in time and space. Plant Pathology 32:5–11

Kampmeijer P, Zadoks JC (1977) EPIMUL, a simulator of foci and epidemics in mixtures, mosaics and multilines. Pudoc, Wageningen

Kranz J (1979) Simulation of epidemics caused by *Venturia inaequalis* (Cooke) Aderh. EPPO Bull 9:235–242

Kranz J, Hau B (1980) Systems analysis in epidemiology. Annu Rev Phytopathol 18:67–83

Kranz J, Royle DJ (1978) Perspectives in mathematical modelling of plant disease epidemics. In: Scott PR, Bainbridge A (eds) Plant disease epidemiology. Blackwell, Oxford, pp 111–119

Kranz J, Hau B, Aust HJ (1984) Monitoring in crop protection modelling. In: Conway GR (ed) Pest and pathogen control: strategic, tactical, and policy models. Wiley-IIASA, Chichester, pp 233–241 (International series on applied systems analysis, vol 13)

Krawczak DA, MacKenzie DR (1981) SPUD CROP, the potato pest management game. Pennsylvania State Univ, University Park (PA)

Kushalappa AC, Ludwig A (1982) Calculation of apparent infection rate in plant diseases: development of a method to correct for host growth. Phytopathology 72:1373–1377

Legett RW, Williams LR (1981) A reliability index for models. Ecol Model 13:303–312

Madden LV (1980) Quantification of disease progression. Prot Ecol 23:159–176

McCartney HA, Fitt BDL (1985) Construction of dispersal models. In: Gilligan CA (ed) Mathematical modelling of crop disease. Academic Press, London, pp 197–143 (Advances in Plant Pathology, vol 3)
McLaren JS,Craignon J (1981) A computer assisted learning system for use in teaching of crop production. Agric Prog 56:69–83
Minogue KP, Fry WE (1983a) Models for the spread of disease: model description. Phytopathology 73:1168–1173
Minogue KP, Fry WE (1983b) Models for the spread of plant disease: some experimental results. Phytopathology 73:1173–1176
Rapilly F (1977) Recherche des facteurs de résistance horizontale à la Septoriose du Blé (*Septoria nodorum* Berk.). Résultats obtenus par la simulation (1). Ann Phytopathol 9:1–19
Rijsdijk FH (1975) A simulator of yellow rust of wheat. C R Semaine d'étude agriculture et hygiène des plantes. Bull Rech Agron Gembloux: 411–418
Rouse DI (1983) Plant growth models and plant disease epidemiology. In: Kommedahl T, Williams PH (eds) Challenging problems in plant health. American Phytopathological Society, St Paul, USA, pp 387–398
Rouse DI (1985) Construction of temporal models. I. Disease progress of air-borne pathogens. In: Gilligan CA (ed) Mathematical modelling of crop disease. Academic Press, London, pp 11–29 (Advances in Plant Pathology, vol 3)
Sall MA (1980) Uses of stochastic simulation: grape powdery mildew example. Z Pflanzenkr Pflanzenschutz 87:397–403
Shrum R (1975) Simulation of wheat stripe rust (*Puccinia striiformis* West) using EPIDEMIC, a flexible plant disease simulator. Penn State Univ Agri Exp Sta, Progress report 347
Spaar D, Ebert W (1985) Monitoring and forecasting in plant protection in the German Democratic Republic. EPPO Bull 15: 299–310
Teng PS (1981) Validation of computer models of plant disease epidemics: a review of philosophy and methodology. Z Pflanzenkr Pflanzenschutz 88:49–63
Teng PS (1985) A comparison of simulation approaches to epidemic modeling. Annu Rev Phytopathol 23:351–379
Teng PS, Blackie MJ, Close RC (1980) Simulation of the barley leaf rust epidemic: structure and validation of BARISM-I. Agric Syst 5:55–73
Vanderplank JE (1963) Plant diseases: epidemics and control. Academic Press, London
Vanderplank JE (1975) Principles of plant infection. Academic Press, London
Vanderplank JE (1982) Host-pathogen interactions in plant disease. Academic Press, London
Waggoner PE (1978) Computer simulation of epidemics. In: Horsfall JC, Cowling EB (eds) Plant disease, vol 2. Academic Press, London, pp 203–222
Waggoner PE, Horsfall JG (1969) EPIDEM, a simulator of plant disease written for a computer. Conn Agric Exp Stn Bull 698, pp 80
Waggoner PE, Parlange JY (1974a) Mathematical model for spore germination at changing temperature. Phytopathology 64:605–610
Waggoner PE, Parlange JY (1974b) Verification of a model of spore germination at variable, moderate temperatures. Phytopathology 64:1192–1196
Waggoner PE, Parlange JY (1975) Slowing of spore germination with changes between moderately warm and cool temperatures. Phytopathology 65:551–553
Zadoks JC (1971) Systems analysis and the dynamics of epidemics. Phytopathology 61:600–610
Zadoks JC (1979) Simulation of epidemics: problems and applications. EPPO Bull 9:227–234
Zakoks JC (1981) EPIPRE: a disease and pest management system for winter wheat developed in the Netherlands. EPPO Bull 11:365–69

Chapter 20

The Methodology of Comparative Epidemiology

J. Kranz[1]

1 Introduction

Comparison is an important cognitive and research tool, and is thus widely used. In fact, most statistical tests are designed just for it. Using comparisons, plant pathologists commonly draw conclusions on the effects on their various experimental treatments, as do epidemiologist. Comparative epidemiology, however, in a narrower sense is a research approach with both analytic and synoptic objectives. It aims at establishing or sustaining general epidemiological principles. Against the background of accepted theories, comparative epidemiology gauges the validity of hypotheses and theories, and abolishes them if they fail the test (Kranz 1980). Experiments, or a critical scrutiny of published evidence, are intended to help to reduce the diversity of various epidemics, their components and facets to a set of basis principles, systems or types. Another application of comparative epidemiology is the provision of generally valid information for teaching and textbooks.

Comparative epidemiology also serves immediate applications. From an understanding of the different behaviour of epidemics, affected by factors acting on their components and structures, we may assess, for instance, the relative importance of certain weather factors, various treatments or control tractics and strategies. In the case of integrated pest management, the comparison of techniques may help in selecting the most promising combinations of control measures. In all cases, comparative epidemiology should provide the tools with which speculations about disease management strategies can be transformed into quantified and testable hypotheses (Zadoks and Schein 1980).

Against the background of the experimental methods dealt with in the preceding chapters, this chapter will attempt to describe some of these tools. Concepts, scope, methods and applications of comparative epidemiology have been described in the earlier literature (Kranz 1974, 1978; Palti and Kranz 1980; Vanderplank 1982). Some chapters in this book implicitly also refer to comparative epidemiology in addition to the methods described. This chapter does not summarize these relevant points in detail, but rather outlines the principles and approaches of methodology aimed at the specific goals of comparative epidemiology.

[1] Phytopathologie und Angewandte Entomologie, Wissenschaftliches Zentrum Tropeninstitut, der Justus-Liebig-Universität, Gießen, FRG

2 Some Basic Aspects of Comparative Epidemiology

2.1 The General Framework

Comparison in all branches of science is based on some common procedures. For a certain topic, a working hypothesis evolves and material (data, facts, information) is collected and ordered to achieve some generalization, rule or hypothesis. Thereafter, more material will be needed to test the hypothesis in order to confirm or refute it. If confirmed for all possibilities, a theory might eventually emerge and sufficiently explain the phenomenon under study. Each suggested theory is subject to further modification and, finally to acceptance or rejection.

Comparison needs well-defined objectives, adequate inventories and full comprehension of the pathosystems and their epidemics. Whenever possible, comparative epidemiology should be both quantitative and experimental. Design of experiments should ensure comparative measurements of relevant variables and features (e.g. structural elements, progress curves, interactions) and pertinent statistical evaluation. Ideal is the simultaneous and consistent measurements of the relevant factors under exactly the same conditions of the studied pathosystems.

Comparisons of epidemics, present at different sites, climates and years and in fields variously treated, are usually limited to a given level of integration. For comparisons of entire systems, e.g. the complexity of a pathosystem for a computer simulator (Sect. 4.3), one has to bear in mind that they are always influenced in many different facets by external factors or man. Therefore, an adequate understanding is required of the system, its elements and structures (Fig. 20.1), and their intrinsic features, e.g. recurrence of events, limits and thresholds. All effects on these facets lead to processes which result in a diversity of epidemic behaviour (e.g. slow rusting, explosive or slow epidemic) epidemic and patterns (e.g. shape of the disease progress curve). Systems analysis often will be the best research approach to cope with all these complexities.

2.2 Terminology

It is imperative that of comparative epidemiology adopts a language which avoids "remoteness and ambiguity of terms" as discussed by Butt and Royle (1980). These authors have made proposals for a proper terminology, and for definitions to be used in comparative epidemiology. Various terms have to be defined operationally as a sort of recipe written for a specific purpose to ensure appropriate measurements. They also should be explicit as they bound and characterize a concept. Butt and Royle (1980) also distinghuish between "measurement terms" (e.g. spore number, disease efficiency, incubation period) and "descriptive terms" which have no connotation of measurement (e.g. dispersal, infection process, incubation).

2.3 Published Evidence as Data Base for Comparison

Comparative epidemiology based on published evidence can be fruitful as Gäumann (1951) und Van der Plank (1963) have shown. But use and interpretation of such information requires caution. Essential is the use of results which emanate from

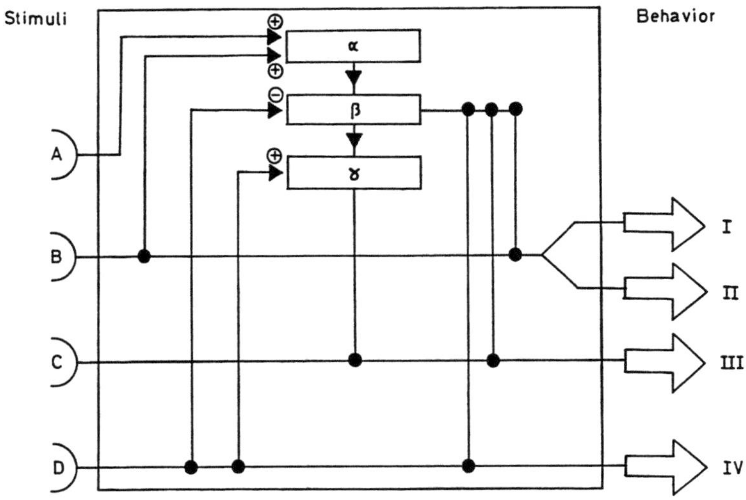

Fig. 20.1 Schematic representation of a system with inputs (stimuli) *A–D* and outputs (behaviour) *I–IV*, and the elements α–γ with structures (indicated by *lines* and *dots*) and their effects (⊕, ⊖)

commensurate methods, parameters and measurements. Unless such information is amenable to experimental tests or other forms of verification, e.g. repeatability it can only be considered as inferences or tentative deductions. Kranz (1980) described the four types of published evidence to be used, and the restrictions which should guide the researcher. These data should at least be critically scrutinized for the experimental techniques from which they result, as well as for their evaluation and interpretation by the authors. Obviously, the evaluation of publications must be subject to the same standards that apply to the evaluation of laboratory or field eperiments. Data present in old experiments must be reconsidered, recalculated and verified. Such an approach can save time and money, and it can help us to avoid spurious strategies (Zadoks and Schein 1980).

Hypotheses and theories inferred from published evidence should, whenever possible be experimentally verified.

3 Comparative Experiments

3.1 The Scope

Experiments in comparative epidemiology should be conducted at the same level of biological integration, viz. the levels of pathogens, epidemics and communities (Zadoks and Schein 1980). They may be done in the field, greenhouse, growth chamber or with computer simulation. The latter method is particularly suited for comparison of elements, structures etc. and factors acting on them within pathosystems. For simulation amongst pathosystems each of them has to be modelled to practically the same degree of realism.

For comparison amongst these levels (e.g. going from spore production to spore dispersal as pathogen and epidemic levels respectively; see also Sect. 3.2.1) it is obvious that prerequisites of quantitative analysis have to be observed. Scaling effects (Zadoks and Schein 1980) may require more adequate attention, as in the case of spore dispersal at different scales of distance, e.g. from plant to plant, or between fields.

Comparative field trials are most easily performed when carried out on several diseases in the same crop (e.g. ear blotch, powdery mildew and stripe rust of wheat), or a disease affecting several crops which can be grown side by side in the same field. (For experimental requirements which are valid also for proper comparative field trials, see Aust and Kranz, Chap. 1; Kranz, Chap. 3; Sutton et al., Chap. 7; Friesland and Schrödter, Chap. 8.) For comparisons involving soilborne diseases, we refer to Gilligan's and Barker and Noe's contributions in this volume (Chaps. 6 and 15). Raccah and Irwin (Chap. 14) may be consulted for peculiarities of virus diseases and Welz (Chap. 11) and Dekker (Chap. 10) for aspects of disease and fungicide resistance.

Problems may arise with comparisons between epidemics of field diseases versus forests, field versus glasshouse diseases, etc., or between diseases in a temperate versus tropical environment (unless done by the same person). Experimental verification of such comparisons are often not possible, and inference from such comparisons should be drawn with caution.

3.2 The Criteria

The choice of criteria of comparison depends on the objective of comparison. In general, comparisons should be based on features which for any given disease have the highest degree of invariance, or consistency. These are usually the criteria of primary importance (see Sect. 4). Those of secondary or tertiary importance may be distinguished according to their degree or range of variability (see Williams 1976, for details and examples). For a comparison based on differences, redundant characteristics should be excluded. But there are also objectives for which the frequency of identical or redundant features can be used to determine similarity.

3.2.1. Comparison of the Pathosystems Level and Population Level

For comparison of pathosystems Zadoks and Schein (1980) suggest component analysis to compare their behaviour whithin or amongst pathosystems. Measurement terms are infection efficiency, latent period, lesion growth, infectious period and number of spores per lesion per day. Infact, this level of comparison has been widely employed, e.g. in studies of slow rusting. Even details like the sporulation curves within the infectious period can be compared. One may relate increase in population to key parameters of the organism's fecundity curve, A, T, W, S, which in turn are related to r_l (Leonard and Mundt 1984). These terms denote the time at which sporulation starts (A) after the latent period, when sporulation V_t reaches its maximum S at time T, and the time when sporulation ceases (W) in the lesions.

In the transition from the pathosystem to the population level of monocyclic diseases, Zadoks and Schein (1980) propose the use of the life-table technique. This technique permits comparison of the intrisic growth rate of the population (r_{max} viz. the maximum relation growth rate of population), mean length of a generation (Tg, viz. the latent period p) and the net reproduction rate R per generation. For polycyclic diseases the same authors suggest the determination of the initial disease intensity (y_0), the latent period (p), the infectious period (i), the daily spore production per lesion (N) and the proportion of spores that will effectively lead to new sporulating spores (E). According to these authors, the product of the latter two parameters can also be expressed in Vanderplank's (1982) equation:

$$\frac{d(yt)}{dt} = [y(t-p) - y(t-i-p)]\,[1-y(t)], \tag{20.1}$$

where R is the corrected infection rate, y is the disease intensity ($0 < y \leqq 1$), i is the infectious period in days, p is the latent period in days and t is the time (day of recording).

Vanderplank (1982) has also added implicitly the progeny/parent ratio $\alpha = iR$ with i as the unit of time, level of their asymptote and the dimensionless products pr and pR as further criteria for comparison.

3.2.2 Comparison of Epidemics, Their Behaviour and Patterns

Structures (Fig. 20.1) of epidemic aspects can be compared by using measurement and descriptive terms for the similarity or diversity of:

1. The structure itself or how the elements interact, and their role and function in the system; (e.g. does spore production, spore dispersal or surviving inoculum have a greater impact on disease progress; which is more sensitive for compensation of external factors, or more easily controlled?).
2. Structural elements and their reaction to external factors (e.g. mechanism of spore dispersal) and their resulting output-input relationship (e.g. in latent and infectious periods).

Similarity of epidemiologically relevant functions of structural elements may also be judged by their intrinsic features (e.g. limits, thresholds); or their output-input relationships between preceding and succeeding state variables (or elements), as in the temperature effect on infection efficiency and on length of the incubation periods (Kranz 1980). For practical purposes, it would suffice to compare only components which consistently have shown (e.g. in sensitivity analyses) an important effect on the systems behaviour. In many areas of biology, research on homologues has led to far-reaching conclusions (see Kranz 1980). In the comparison amongst pathosystems, criteria might be adopted from taxonomy.

Comparison of patterns, e.g. different shapes of disease process curves, that express the interaction of systems structures stimulated by external factors can either be made for entire epidemics, or for their specific parts. For instance, disease progress curve patterns allow one to compare: (1) the starting point of curves, (2) the growth rates, (3) the slopes, (4) the existence and duration of an asymptote and (5) the maximum incidence, y_{max}. In a comparison of structures in bean root rot epidemics, Campbell et al. (1980) analyzed eight disease progress curve elements, viz. the WEIBULL distribution-function-shape parameter, WEIBULL distribution-function-scale parameter; the area under disease progress curve (AUDPC, see also Berger, Chap. 9), y_{max}, first-difference regression linear coefficient, first-difference regression quadratic coefficient, disease incidence (%) and time of disease onset, and related them to four "unobservable" common factors. The first is the overall description of disease progression, including curve shape, rate of disease progression and y_{max} after 50 days;

the second factor related the area under disease progress curve and the disease intensity and final disease intensity (y_{max}). As third factor, they identified the location of an epidemic in time through the time of onset. Their fourth factor is the relationship between disease incidence and severity.

The shape of disease progress curves or disease gradients often carries useful biological information. Consequently, basic patterns can be expected, to which observed patterns may be assigned to reduce chance variation. Basic patterns of complete disease progress curves, e.g. untreated with fungicide, may also be bilateral, bimodal, multimodal or oscillating. These patterns often reflect disease cycles, discontinuities in infection, variation in incubation periods and changes in host susceptibility due to new growth or other factors. Kranz (1974) also proposed various quantitative curve elements. Thes curve elements represent certain biological facets that are integrated in the curve itself. Therefore, curve elements, in addition to the entire disease progress curve, may indicate the effects of independent variables and serve as a basis for quantitative comparison of epidemics. If linearization is required, e.g. for statistical purposes, slope and position of the lines as well as the transformation equations used to obtain the best fit can be employed as criteria for comparison.

Rates of epidemic increase express the speed of response of the system to external or internal factors in relation ot previous states. Rates are usually reactive to the effects of weather, treatments, degree of resistance, etc. This makes them useful criteria in the comparison of entire systems behaviour. Rates that suit our purpose best are those which can be subjected to experimental or statistical tests. Hence, when the data base permits, the regression coefficient b should be preferred over the infection rate r. The latter is justified if the available data consist only of measurements at time t_1 and t_2, e.g. at intervals of recording or between growth stages of the host. Since rates are variable, it might be preferable to assign them to meaningful classes (e.g. between 0.1—0.2, 0.21—0.35) or give them a relatively low weight as a secondary or tertiary criterion for comparison in a set of other criteria.

3.2.3 Comparison of Soilborne Diseases and Their Epidemics

For comparison of soilborne diseases and their epidemics, the variance to mean ratio V/m of various frequency distributions (Poisson, negative binomial, Neyman type A, etc.) is commonly used. However, these comparisons fail to provide information on the spatial pattern of a disease in a field. The same also applies to David and Moore's indices for mean crowding $m^* = mt\,(V/m-)$, and patchiness, m^*/m (Nicot et al. 1984). Random distribution (Poisson) applies when $V/m = 1$, IC $= 0$ and $m^*/m = 1$. For a comparison of spatial distribution, these authors refer to the nearest neighbour or Greig-Smith methods, or to spatial autocorrelation analysis and to two-dimensional spectral analysis (see Nicot et al. 1984 and Chaps. 6 and 15).

3.2.4 Comparison at the Community Level

Epidemiology at the community level is not yet well developed. One approach is a comparison of the occurrence of the pathogens in various types of natural vegetation and localities for which Kranz and Knapp (1973) used Venn diagrames (Fig. 20.2). Another example is the study of virulence/race dynamics (Chap. 11). Race patterns, frequencies and diversity coefficients are used as measurement terms. Zadoks and Schein (1980) suggest $r_d = ln(1-s)$ as a relative rate (with s as the selection coefficient)

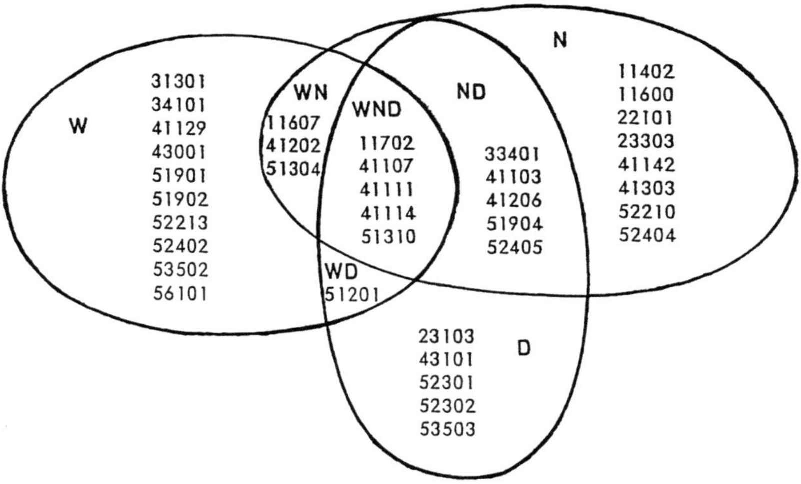

Fig. 20.2 Venn diagrams with pathogens (*code numbers*) existing in various plant communities (*D*, *N*, *W*, etc.). Similarities and differences in the pattern of plant pathogens are thus shown. (After Kranz and Knapp 1973)

at which a weaker race is replaced by a stronger race. In relation to fungicide resistance Skylakakis (1982) defined a standard selection time s_d for the resistant pathogen strain (see Dekker, Chap. 10). For comparisons on wider scales than at the community level, see Weltzien (Chap. 16) and Nagarajan and Ajai (Chap. 17).

4 Tools and Procedures in Comparative Epidemiology

4.1 Data Reduction and Curve Approximation

Results obtained from experiments and other sources can be compared by means of simple plotting, mapping or tabulation of quantitative data (e.g. life tables) using statistical, mathematical and computer methods designed for this purpose. Although versatile, plotting essentially yields indications or permits inferences, especially when the curves are smoothed by moving averages to reduce accidental variations. The original curves will also be needed when linearization is required for the statistical verification of, for instance, differences between various progress curves, as biological implications are much better understood from the non-linearized curves. Curvilinear functions approximate disease progress curves more naturally than linear models, which may be compared by means of such functions (see also Sect. 3.2).Among the functions that fit an observed disease progress curve, the simplest will be the best as long as its accuracy suffices (see also Madden et al., Chap. 13). In many cases the choice of the function for approximation (of which there are many in appropriate books on mathematics) may be arbitrary. But even then these functions would contribute to the comparison of epidemics as they reveal which function fits the epidemics of which diseases best or which other diseases have a similar set of best fitting functions.

Nearly all functions used so far (e.g. growth or growth rate functions; Madden 1980) have only a few constants or variables which may be useful criteria, e.g. the upper asymptote value k, the position factor b and the skewness n of the disease progress curve. The latter could be a good criterion for comparison, as it may describe the natural coincidence between susceptible stages of the host, the amount of inoculum present in the course of an epidemic, or both. Hau et al. (1985) discuss the curvilinear approximation of various components for simulators, including disease progress curves. But also linearized relationships may well be used to derive principles, as shown by Gregory (1968) with disease gradients.

4.2 Use of Coefficients and Multivariate Statistical Methods

Similarity of elements, structures and behaviour (i.e. patterns and rates) can also be measured by similarity or matching coefficients (Sokal and Sneath 1963; Williams 1976). Ranking correlations, where appropriate, may be of similar use. Also, analyses of variance with suitable data can be used in comparative epidemiology, as Van der Plank has shown by distinguishing vertical and horizontal resistance. The possible use of diversity coefficients as used by ecologists needs to be explored in comparative epidemiology.

Factor, cluster and discriminant analyses (Mardia et al. 1979) and their various test criteria can be helpful. All thes methods require the measurement of several characteristics for each pathosystem, epidmemic, etc. They all help to avoid conclusions resulting from a superficial resemblance. Factor, discriminant and cluster analyses are tools borrowed from applied mathematics, which would also permit to store, sort and reduce the overwhelming number of seemingly different epidemics, and to filter their essential communalities. Principal Axis Factor Analysis (PAFA) was used to reveal some intercorrelations between elements (see Kranz 1974). Even if this information is only indicative, it may be connected with factors inherent in a disease and in ecological requirements of host and/or pathogen.

Various types of cluster analysis have proven useful for comparison of entire epidemics (Kranz 1974). Elements that exhibit the least variation in the disease progress curves are most useful and rank high in their importance. However, a realistic evaluation of the importance of various elements might sometimes cause problems as selection of the test parameters can be crucial. The choice of very stringent test parameters may eventually lead to clusters of two components with the closest resemblence. These are useful when one looks for the most similar epidemic among those available for comparison in the computer. However, it is not certain that the same disease will always appear in the same cluster. With improved knowledge of measuring and weighting, the importance of various elements, clustering methods and cluster determination is expected to yield better results. In these cases chance fluctuation of a disease between clusters of epidemics can be reduced. However, we should not aim at fixing a disease to only one "epidemic taxon". For instance, we should expect a given disease to have an epidemic of one cluster (class) in a wet year and of another class in a dry year.

Also promising for some aspects of the comparison of epidemics are the multiple and stepwise discriminant analyses available in statistical packages of computer software. With these analyses, the relations of the observed process to known factors or aspects are quantitatively established, thus making their result more objective (Mardia et al. 1979).

4.3 Use of Computers in Comparison

It is obvious that methods like factor, cluster or discriminant analysis would hardly be possilbe without using computers. Computers are thus useful in comparison of epidemics, particularly if data become too bulky and confounded with the many elements involved. The computer in such instances serves research retrieval and identification. Information, e.g. known patterns of disease progress curves stored in the computer, could become a yardstick against which the incoming information is tested. It would then be easier to check the different shapes, which a disease progress curve of a given disease may attain, and with which other diseases and under which conditions they tend to be associated.

We also cannot do without a computer when dealing with more complex models. This applies in particular to what is now known as a simulator. Simulators may be used as synoptic instruments for comparison by testing sensitivity and validity of assumptions, conclusions or models. In comparative epidemiology simulators may replace filed and laboratory experiments. As simulators, at present, are developed for particular diseases only, the technique can only be used within such a pathosystem. Hence, only the relative importance of structural and program elements and the effect of external factors (weather, treatments or practices) can be compared. Comparisons between pathosystems would require a general simulator with all relevant phase of the infection chain as components. With sensitivity analyses one could then test the relative importance of components/pathosystems and arrive at general conclusions in addition to the particulars of a given disease. Also, the type of approximation fitting best a given component of the epidemic may be used as a criterion for comparison.

Similar comparisons may be done with less complex models and without a computer. But the variables, and the constants included in such a model, may limit the scope of comparison. Nevertheless, these parameters (y, i, p) may generate useful conclusions by comparisons between pathosystems as shown by Vanderplank (1982). The advantages of R (see Equations 20.1.) and the effect of the progeny/parent ratio L make them generally adequate to be employed in comparison between various pathosystems.

5 Conclusions

It is obvious that comparison of epidemics should be handled with flexibility and in an object-oriented way. This requires good comprehension of the biology of the pathosystems involved. Purely mechanistic comparison of variables, constants and parameters alone may not ensure successful comparative epidemiology. A given criterion may have a different weight in another context or another systems level. Criteria preferably should have a biological background, which should describe each pathosystem with its particular epidemic competence and the reactions of this pathosystem to influences exerted by host, environment, agricultural and control practices.

In an earlier paper (Kranz 1974) we suggested the classification of epidemics. Many of the measurement terms and descriptive terms referred to in this chapter may be used as descriptors of diseases with regard to their epidemic competence. An epidemiological more competent pathogen, for instance, incites a disease that produces more rapidly great amounts of an easily dispersed inoculum of a high disease efficiency, and has a shorter latent period and requires a less stringent temperature range than another disease which lacks all or some of these advantages. Table 20.1 suggests some possible epidemiological descriptors. They eventually could enhance the possibilities of automatic comparison, which can be useful with large bodies of data. But, more importantly, such descriptors will facilitate the application of known epidemiological principles to more and more plant disease and thus assist the improvement of systems control and systems design.

Table 20.1 Some suggested epidemiological descriptors[a]

Life cycle: Mono- or polycyclic

Host range: Affected host species, *forma specialis,* races; competitiveness (soilborne fungi)

Range of environmental conditions required:
Specific, narrow, indifferent, wide for temperature, dew, humidity, rain, degree days for major facets on an epidemic (e.g. first outbreak, latent period), etc.; major compensation effects

Dispersal and spread:
Means (wind, water, vector, etc.)
Gradients of spread (steep, flat, etc.)
Flight distance and time
Distribution of loci
Ease of spread

Epidemic competence:
Mean sporulation capacity
Mean infectious period
Mean disease efficiency
Mean progeny/parent ratio
Mean latency
Mean infection rate
Mean max. disease intensity
Longevity and infectiousness of inoculum
Yield loss mean

Epidemic patterns:
Types of disease progress curves that disease is known to attain
Mean delay between crop and disease appearance (in days); early or late disease
Mean duration of epidemic (in days)
AUDPC (if standardized)
Duration of progressive phase (in days)

[a] Under optimum conditions for disease progress.

6 References

Butt DJ, Royle DJ (1980) The importance of terms and definitions for a conceptually unified epidemiology. In: Palti J, Kranz J (eds), pp 29–45 (see below)

Campbell CL, Madden LV, Pennypacker SP (1980) Structural characterization of bean root rot epidemics. Phytopathology 70:152–155

Gäumann E (1951) Pflanzliche Infektionslehre, 2. Aufl. Birkhäuser, Basel

Gregory PH (1968) Interpreting plant disease dispersal gradients. Annu Rev Phytopathol 6:189–212

Hau B, Eisensmith S, Kranz J (1985) Construction of temporal models. II. Simulation of aerial epidemics. In: Gilligan CA (ed) Advances in plant pathology, vol 3. Mathematical modelling of crop disease. Academic Press, London

Kranz J (1974) Comparison of epidemics. Annu Rev Phytopathol 12:355–374

Kranz J (1978) Comparative anatomy of epidemics. In: Horsfall JG, Cowling EB (eds)Plant disease, vol 2. Academic Press, London, pp 33–62

Kranz J (1980) Comparative epidemiology: an evaluation of scope, concepts and methods. In: Palti J, Kranz J (eds) pp 18–28 (see below)

Kranz J, Knapp R (1973) Qualitative and quantitative Unterschiede der Vergesellschaftung parasitischer Pilzarten in verschiedenen Vegetationsformen. Phytopathol Z 77:235–251

Leonard KJ, Mundt CC (1984) Methods for estimating epidemiological effects of quantitative resistance to plant diseases. Theor Appl Genet 67:219–230

Madden LV (1980) Quantification of disease progression. Prot Ecol 2:159–176

Mardia KV, Kent JT, Bibby JM (1979) Multivariate analysis. Academic Press, London

Nicot PC, Rouse DI, Yandell BS (1984) Comparison of statistical methods for studying spatial patterns of soilborne plant pathogens in the field. Phytopathology 74:1399–1402

Palti J, Kranz J (eds) (1980) Comparative epidemiology. A tool for better disease management. Pudoc, Wageningen

Skylakakis G (1982) The development and use of models describing outbreaks of resistance to fungicides. Crop Protection 1:244–262

Sokal RS, Sneath PH (1963) Principles of numerical taxonomy. WH Freeman, San Francisco

Vanderplank JE (1963) Plant diseases: epidemics and control. Academic Press, London

Vanderplank JE (1982) Host pathogens interactions in plant disease. Academic Press, London

Williams WT (ed) (1976) Pattern analysis in agricultural sciences. Elsevier, Amsterdam

Zadoks JC, Schein RD (1980) Epidemiology and plant-disease management, the known and the needed. In: Palti J, Kranz J (eds), (see above), pp 1–17

Subject Index

Accuracy 36–40, 43, 47
Acyrthosiphum pisum 212
Abundance 219
Aerial reconnaissance 247
Aerial spore pool 167
Aerobiological studies 244
Age stratification of lesions 42, 46, 48
Aggregation parameter (*k*): see negative bionomial distribution
Aggressiveness 165, 170
Agro-ecosystem 263
Analysis of
- Anova 194
- mapped data 93
- regression 194
- virulence 165
Alternaria macrospora 73
A. porri 70
A. porri f. sp. *solani:* see *A. solani*
A. solani 30–33, 74, 75, 80, 83
Alternaria sp. 28, 82
Analytic approach, 269, 272
Anemometer 109, 110
Aphids 209, 210, 211, 216, 217
- movement of 217
- as vectors 211, 215
Aphid-borne viruses 209, 220
Aphis citricola 217
Apple powdery mildew 63
Apple scab: see *Venturia inaequalis*
Area maps 239
Area of
- main damage 241
- marginal damage 241
- sporadic attack 241
Area under disease progress (AUDPC) 60, 141, 144, 149, 150, 181, 184, 193, 283, 284
- computer program 145
Areal photography 48, 238
Arena test 215
Armoracia rusticana 245
Artificial fruits 108
- leaves 107
Atmospheric humidity 105
Autecological nature of systems 263

Bait trap plants 212
Barley powdery mildew: see *Erysiphe graminis*
Barley yellow dwarf virus 212, 218, 245
BARSIM-I 269
Beet western yellow virus 212
Behavior of vectors 218
BETE function 123
Binomial coefficient 167
Biogeography 237
Biometeorological
- relations 118, 119
- variables 123, 125, 126
BLITECAST 192, 196, 201, 203, 205
Blue mold: see *Peronospora tabacina*
Botrytis cinerea 139
B. squamosa 198
"Bottom-up" approach 179
Brevicoryne brassicae 212

Canopy partitioning 46
Calibration curve 43
Cauliflower mosaic virus (CaMV) 212
Cercospora apii 139
C. arachidicola 139
C. beticola 155, 156, 157, 160
Cercospora spp 197
Chi-square test 173
Citrus tristeza virus (CTV) 210, 213
Classification 44
- of epidemics 288
- of spatial patterns 86
Climatic zones 237
Climatograms 241
Cluster analysis 286
Cobb scale 38
Coefficient of determination 37
Collecting spores from leaves 72–74
Colony counts 170
- growth 171
Combined field and growth chamber experiments 13, 15
Comparative epidemiology 13, 279, 280, 285, 287
- experiments 281

Comparative epidemiology
- objectives of 279, 282, 283
- published evidence 280
- terminology 280
- tools and procedures 285–288
Comparison
- at the community level 284
- between pathosystems 287
- criteria of 282
- of soilborne diseases 284
Compensation phenomena 14
Component analysis 180, 185, 282
Computer in comparative epidemiology 287
- use of 203
- program to fit data to mathematical functions 145
Continuous selection models 174
Control
- effects of. c. measures 141
- epidemiological principles of 136
- optimal c. 204
Controlled conditions 8, 21–23, 69, 74, 77, 78, 194, 195, 203
Coulter counter 29
Critical disease level (CDL) 198
Crop growth models 186, 263, 269, 274
Crop health watch systems 247
Crop loss assessment 25, 38, 46, 179, 180, 185, 187
Crop measurement 46, 47
Crop microclimate 99
Crop-pest loss models 186, 187
"Cryptic error" 185
Curve approximation 285
Curve elements 284
Curvilinear functions 285
Cyclone spore collector 169
Cylindrocladium crotalariae 93

Damage coefficients 48
Data acquisition 8, 11
- analysis 194
- communication 102
- base for comparison 280
- evaluation 11
- loggers 99, 100, 102
- reduction 285
Defoliation 44, 48
Degree days 56, 225
- base temperature 120, 121
- solar-thermal units 57
Delay (dt) 138
- distributed delay 64
Density-dependent selection of virulences 174
Deposition of spores 126, 244
Design of virulence tests 168
Detection of
- races 166
- viruses 210, 211
Destruction 36, 38, 48
Dew 23, 28–30, 71, 72, 76, 78, 81, 106, 124, 131
Diagram maps 239
Differentials, set of 169
Disease assessment 35, 38, 42, 46, 49, 181, 210
- leaf shedding 36, 38, 48
- precision 36–40, 47
- psychological responses 39
- psysiological stimuli 39
- several diseases 47
- surveys 49
- through spore catches 46
see also: accuracy and errors
Disease assessment aids: see standard diagrams
Disease assessment in classes 43–45
- class means 44
- scales 42
- automatic measurement 42, 45
- descriptive keys 56
- direct estimates 44
Disease development
- asysmptotic 54, 62, 63
- carrying capacity 62, 63
- and host growth 62
- negative 55
- simulation of 63
Disease distributions
- binomial 168
- compound Poisson 89
- contagious 89
- negative binomial 40, 48, 89–92, 225, 284, 286
- Neyman Type A 89–91
- normal 94–96
- Poisson 40, 87, 91, 175, 196, 254, 260
- random 284
- Weibull 184
Disease efficiency 170, 171, 283
Disease frequency 35, 238
Disease forecasting systems 191
Disease gradients 284
Disease incidence 35, 36, 40, 41, 42, 45, 47, 179, 212, 239
Disease increments 183
Disease indices 37
Disease intensity 35, 40, 41, 42, 46, 148, 149, 202, 238
- final 284
- initial 283
- scaling of disease intensities 43
Disease management 244
Disease measurement: see disease assessment
Disease mosaics 95

Disease occurrence 237
Disease progress 40, 143, 149, 212
Disease progress curves 142, 179, 184, 283–286
- actual 147
- analysis of 142, 180
- theoretical 147
Disease severity 35–38, 42–49, 69, 179, 183, 185
- preferred values 38, 43
Disease simulation model, see: simulation
Disease triangle 118
Disease cycle 196
Disease zones 242
Discriminant analyses 200, 286
Dispersal 238, 241
- mechanisms 209
see also spore dispersal
Dispersion parameter 40
Dominance index 175
Dosage-response curve 154
Dot blot immunosorbent assay (DBIA) 211
Doublet analysis 94, 96
Downy mildews 28, 30, 31, 72, 202, 241

ED_{50} ratio of strains 161
Effective period of fungicide 147, 149
ELANCO tables 43
Electrical impendance probe 108
Electron microscope (EM) 210
Electronic leaf area meter 46
Electronic
- monitoring 99, 102
- sensors 205
ELISA 46, 211
Endodemic, see: esodemic
Energy balance 124, 125, 131
Energy budget 116, 129
Environmental
- factors 239
- parameters 217
- water 105
EPIDEM 269
EPIDEMIC 268, 273
Epidemic behaviour 280
Epidemic competence 288
Epidemic rate 142, 150
Epidemiological components 179, 180
Epidemiological descriptors 288
EPIGRAM 269, 274
EPIMUL 268
EPIPRE 198, 269, 275
EPISEPT 274
EPISIM 272
Errors
- experimental 10, 11, 12
- individual 8, 38
- object-dependent 39
- representional 12
- sources of ... in disease assessment 38
- systematic 40
Erysiphe betae 242
E. graminis 45, 70, 82, 118, 128, 140, 160, 166, 170, 173, 244
Esodemic 168, (as endodemic) 181, 183
Eutypa sp. 65
Evaporation 124, 125
Exodemic 168, 181, 183
Experimental design 8, 195, 281

Factor Analysis 286
False colour imageries 46
Fecundity curve 282
Field experiments 7, 8, 39, 187
- holistic 7, 9, 11, 13
- meristic 7, 9, 11, 13
- natural infestation 227
- operational plan 8
- pilot experiments 8
- planning 8
- standard designs of 8
Field (assessment) keys 44, 45
Field live assays 216
Field plots
- plot sizes 167, 229
- repetitive planting 213
Field record form 8, 11, 49
Fitness 165, 173, 174
- components of 165, 171, 176
- estimation of 159
- parameters of 158, 159
- fungicide resistant strains 158
Forecasters 191, 203
- apple scab 205
- classification 191
- empirical 197, 196
- evaluation 198
- for vectors 195
- qualitative 197, 201
- quantitative 198, 201
- models 198
- regression equation 198
- single season 197, 198, 201
- validating in field trials 204
Forecasting 21, 168, 237, 241, 268
- constraints 204
- delivery and operation 204
- empirical 203
- fundamental 192, 203
- implementation 204
- multiple season 201
- systems 192, 196, 241
- within seasons 201

Free leaf moisture 23, 28, 80, 117, 118
Frequency distribution 43, 48, 120, 175
- log-transformed 174
Fungicide
- applications 192
- efficacy 147
- equivalents 142
- resistance 158, 268, 275
- - definition 153
- - fitness of resistant strains 153
- - level of 153
- - monitoring 159
- - recognition of strains 154
Fungicide sensivity
- baseline data 154
- monitoring 153, 160
Fusarium oxysporum 258, 259

Gaeumannomyces graminis 86, 124
Gene-for-gene interactions 170
Genetic shifts in communities 230
Genotype pattern analysis 245
Geographic distribution 237
Geophytopathology 237
- data documentation 239
Geophytopathological maps 239, 240
Global radiation 116, 117, 129
Globodera tabacum solanacearum 230
G. rostochiensis 227
Globodera spp. 224
Gompertz transformation 140, 142, 143, 150
Gompit 140, 142
Greenness Vegetation Index (GVI) 247
Ground verification 238
Growth, analysis of 58
- models 263
- rate (GR) 58
- chamber 8, 9, 13, 14, 39, 215
G-test of independence 172, 173, 174
Guard rows 12
Germination 123
Guignardia bidwellii 192, 193, 194

Habitat 13, 14, 22
Helicotylenchus digonicus 224
Helminthosporium maydis 140, 244
Helminthosporium sp 29
Hemileia vastatrix 245
Heterobasidion (Rigodoporus) annosum 82
Heterodera glycines 228, 229, 230
H. schachtii 230
High virulence 171
Holistic approach 7
Homologues 283
Hoplolaimus columbus 228

Horizontal resistance 142, 286
Horsfall and Barratt scale 43, 44
Host, carrying capacity 63
- phenology 56
- plant age 194
- growth 148
Host growth, analysis 57
- asysmptotic 62
- classes 62
- compensation 53
- constraints 52
- flushes 53
- perennials 64, 65, 66
- phenological key 57
- senescence 52
- simulation 63
- stages of development 46, 55, 181, 185
- and yield 261
Host-seeking activity of vectors 218
Humidity 22-24, 27-30, 72, 74, 77-79, 81, see also: rel. humidity
Hygrometer 111
Hypersensitivity 27
Hypothesis, formulation of 9
Hyoscyamus desertorum 214

Immunoelectron microscopy (IEM) 210
Incidence: see disease incidence
Incubation 27, 28, 121, 122
- period 15, 118, 122, 169, 283
Infection 21, 25, 30-32, 71, 74-78, 80, 81, 118, 120, 123
- primary 86, 258
- secondary 258, 260, 262
- systematic 258
Infection efficiency 148, 283
Infection foci 237
Infection period 192
Infection rate 16, 141, 171, 284
- apparent 55, 62, 180
- corrected 283
Infection types (IT) 170, 172
Infectious period 31, 171, 283
Infectiousness 74
Indexing 212
Index of Shannon 175
Indian stem rust rules 245
Indices of aggregation 92
Indices of dispersion 48
Inhibitors, specific-site 153
- multi-site 153
Initial level of disease 140
Inoculation 23, 25, 26, 30, 75, 76, 79, 169
Inoculators 26
Inoculum 26, 29, 30, 32, 69, 71-76, 79, 80, 196, 216

- buildup 69
- on roots 260
- potential 30, 79, 81, 82
- production 69, 75, 81
- units of 259
Irradiance 110
Irrigation 72, 76, 79, 81, 119
Isolates, bulk 168, 171
Isoline maps 239
Interception 117, 124, 125, 127, 131
Interplot interference 9, 185
Intrinsic features 283
- growth 282

Jet spore traps 166, 168
Julian days 198

"k" parameter 40

Laboratory assays 215
Lag-correlation 48
LANDSAT 246, 247
Latent heat flux 116
Latent period 28, 31, 118, 120, 121, 122, 171, 283
Lattice map, two dimensional 95
Leaf area index (LAI) 58, 60, 129, 131
- healthy 47
- total 57
Leaf temperature 124
Leaf wetness duration 117-119, 131, 241
Leptosphaeria nodorum 82
Lesion growth 118, 119, 122
- rate of 260
Lesions: see symptoms
Leslie matrix 260
Level of biological integration 281
Life-table technique 282
Light: see radiation
Linear regression analysis 174, 180, 284
Liquid nutrient medium 156
Living plant material 156
Lloyd's index of mean crowding 92
Lloyd's index of patchiness 92
Logistic growth 150
- transformation 143
Logit 136
- transformation 180
Long distance dispersal 243-247
- monitoring of 243
Loss assessment 47
- equivalents 48
- magnitudes 48
- models, empirical 182
- profiles 48
- prediction 48
Lycopodium clavatum 70, 82

Macroclimate 115, 242
Maize dwarf mosaic virus 94
Maize streak virus 212
Manipulating of the
- environment 16
- host 15
- pathogen 15
Management practices 225
Mapping 48, 93
- long distance dispersal 243
Maps, use of 239
Matching coefficient 286
Maximum likelihood 89, 92, 174
- temperature 119
Mean crowding 284
Measurement, methods of 255
Measurement of aggressiveness 171, 176
- of disease: see disease assessment
- of fungicide resistance 154-159
- of lesion sizes 42
- of pathogen diversity 174
- of race and virulence dynamics 173
Measurement terms 280, 282, 288
Mesoclimate 115
Metabolic process, inhibition of 156
Meteorological variables 117
Melampsora medusa 243
M. larici-populina 243
Meloidogyne arenaria 230
M. hapla 228
M. incognita 224, 225, 227, 228, 230
Merlinius brevidens 224
Microclimate, 21, 76, 100, 115, 116, 117, 119, 124, 131, 132
Microcomputer 43, 46
Microdensitometer 46
Microeinsteins 22, 24
Microplots 226, 232
Microprocessor 99, 205
Miniepidemics 29-31
Minimal inhibitory concentration (MIC) 161
Minimal threshold of disease 184
Mobile nurseries 166, 168, 171
Mobile spore traps 167
Model building 200
Models 47, 179, 187, 220, 244, 253
- data to drive - 232
- formulating - 233
- interfacing epidemiological - 263
- linear - 198, 200
- management optimization - 231
- multiple point - 182

Models
- nonlinear - 198, 200
- loss prediction - 183
- types of 253
- validation of 200, 272
Modeling 203, 253
Modeling for nematode-induced epidemics 231
Model verification 199
Moericke yellow pan trap 219
Moisture: see wetting
Monocyclic diseases 282
- experiments 25-29
Monilinia spp. 157, 160
Monitoring 112, 160, 205, 210, 216, 218
- aphids 217
- fungicide resistance 159-161
- soil variables 110-112
- spread of pathogens 244, 245
- weather variables 104-110
Moran's index 96
Multiple predictions 197
Multispectral scanner (MSS) 246
Multivariate
- regression analyses 9, 228
- loss assessment equations 49
- statistical methods 286
Mycosphaerella pinodes 75
Myzus persicae 212

Near-infrared light 246, 247
- -isogenic lines 171
Nearest neighbour methods 284
Negative binomial distribution 40, 48, 89-92, 225, 284, 286
- Polya-Aeppli mode 89-91
Negative interference 12
Nematicides 229
Nematodes 69, 209
Nematode community characteristics 223
Net assimilation rate (NAR) 58
- radiation 116, 129
- reproduction rate *R* 282
Neyman Type A distribution 89-91, 284
Nilaparvata lugens 244
Non-destructive electronic equipment 46
Nonlinear curve fitting 143, 285
- regression equations 174
Non-systemic diseases 38
Normal approximation 94, 95, 96

Oidium heveae 53
Oidium sp. 66
Ontogenetic predisposition 54
Overseasoning 75, 76, 118

Paperdisc bioassay 156
Partial resistance 176
Patchiness, index m^*/m 284
Pathogens diversity, measuring 174
Pathogenicity association coefficient (PAC) 173
Pathosystems 138, 140, 147, 180, 181, 183, 253, 258, 281, 282
- components of 254
Penicillium expansum 157
P. italicum 157
Peronospora destructor 82
P. tabacina 73, 82, 83
Phellinus weirii 258, 262
Phenometric measurements 10
Phenotypes, percentage of distinct 175
Photosynthetically active radiation (PAR) 116, 128
Physiological damage threshold 36
- perception of stimuli 38
Phytophthora cactorum 192, 193
P. cinnamomi 257, 258, 259, 261
P. infestans 21, 28-33, 45, 71, 73-76, 80, 82, 157, 160, 197, 201
P. phaseoli 201
Plant-pathogenic viruses 209
Plant virus epidemiology 210
Plasmopara viticola 76, 157, 160
Poisson distribution 40, 87-91, 168, 175, 196, 254, 260, 284
Polya-Aeppli mode 89-91
Polycyclic diseases 149, 267, 282
- experiments 29-30
Population
- densities 229
- dynamics 229
- level 282
- model 231
- patterns 224
Position factor *b* 286
Potato late blight: see *Phytophthora infestans*
Potato leafroll virus (PLRV) 211
Powdery mildews 26-31, 77, 118, 128: see also *Erysiphe* spp.
Precipitation 117-127, 131
Predictor(s) 149, 181, 195, 200, 201, 202, 216, 218
- criteria 197
- physical factors 195
Prediction interval 203
Predictive models 204, 212
- fundamental 204
- empirical 204
Predisposition 25, 194
Probability functions 89
- models 196
Progeny/parent ratio (L) 283, 287

Pseudomonas syringae pv. *syringae* 196
Pseudoperonospora sp. 57
Psychrometer 105, 109, 111
Puccinia graminis f. sp. *tritici* 33, 167, 169, 243, 245, 246
P. polysora 243
P. striiformis 173, 197, 198, 205
Pyranometer 110
Pyrenophora tritici-repentis 52
Pyricularia oryzae 155, 157, 160

Quantitative inoculator 172
Quantitative nematology 223

Race analysis, differential set 169, 172
Race designation 173
- dynamics 173
- frequencies 171
Radiation 21–24, 28, 30–32, 69, 72–74, 77–80
Radiation balance 116
- intensity 128
Rain 24, 29, 30, 72, 76, 79, 81, 109, 241
Randomization 94
Randomness 40, 86
Random variables 88
Ranking 44
Ranking correlations 286
Rate, developmental 120, 123
Reflectance 246, 247
Regression analysis 198
- linear 142, 143
- nonlinear 142
Relative growth rate (RGR) 58
Relative humidity (RH) 105, 117–119, 123–125, 126, 131: see also humidity
Remote sensing 42, 45, 245
Response surface 126, 128, 184
Rhizoctonia solani 86, 87
Rhizoctonia in wheat 48
Rhizosphere width, calculating of 259
Root diseases 46, 47, 253
- disease intensities of 48
- root growth model 263
- root mass 262
Run test 94
Rusts 26–29, 31, 69, 71, 72, 241: see also *Puccinia*

Sample size 40, 41, 47, 48, 142, 172, 174, 226
Samples, handling and processing of 160
Sampling 42, 166, 225, 238, 274
- dates 40, 172
- destructive 10, 11, 43, 46
- errors 142
- frequency of 42, 168
- intervals 10, 39
- method 87
- multi-stage 45
- number of 238
- plan 42
- random 48, 166, 167
- sequential 40, 42
- soilborne disease 262
- stratified 41, 167
- techniques 10, 35, 39, 40, 166
- units 39, 40, 41, 42, 43, 46, 47, 87, 88, 93, 166
- systematic 228
Sanitation, effect of 139, 140
- practices 138, 141
- ratio 138
Satellites 46, 238, 245–247
Scaling disease intensity 43
Scheduling fungicide application 203
Sclerotinia sclerotiorum 82
Screening for resistance 44
Selection coefficient 174
Selective media 70
Sensible heat flux 116
Sensitivity tests 273
Sensors 104
- positioning of 116, 117
Septoria blotch of wheat: see *Septoria nodorum*
Septoria glume blotch of wheat: see *Septoria nodorum*
Septoria nodorum 45, 198, 199
Serological methods 211, 212
Settling tower 26, 77
Shedding of leaves 36, 38, 48
Shelterbelts 128
Short-duration epidemics 182
Similarity coefficient 283
Simpson's measure of diversity 175
Simulation 21, 31, 32, 253, 255
- experiments 255–257
- planning and development of 257
Simulation models 14, 186, 256, 263
- development of 253
- for root diseases 257, 258
Simulator 47, 49, 286, 287
Single colony isolates 169
Single-plot-experiments 7, 9
Single point models 181
Skewness 286
Slide germination test 155
Small-plot methods 228
Smuts 77
Soilborne diseases 40, 48, 69, 75, 256
Soil heat flux 116
- humidity 126, 127
- moisture 117, 131
- temperature 110

Soil heat flux
- types 234
- water 110
Solar-thermal unit 56
Soybean mosaic virus (SMV) 216, 217
Space-dependent susceptibility 53
Spatial patterns 10, 48, 86
- analysis of 86
- distribution of disease 39
- of mosaics 86
- of patches 87
- population patterns 224
- contagious 227
Specific-site inhibitors 158
Spectral analysis 48
Sphaerotheca fuliginea 157, 160
Spiroplasma citri 245
Spore catches 244
- counting 69, 75
- deposition 69, 71, 72, 77–81
- dispersal 28, 30–32, 69, 70–72, 74, 76–81, 118, 119, 126, 127, 130
- germination 30, 74–78, 80, 81, 118, 123
- infectiousness 69
- release 72, 73, 76, 126
- survival 27, 31, 32, 69, 72, 74–78, 80, 82, 118
- trapping devices 28, 69–72, 81
- viability 244
Spore production per lesion, daily 283
Sporulation 25, 28, 30–32, 72–74, 76, 77, 118, 119, 171
- capacity 171
SPOT 247
SPUD CROP 274
Splash dispersal 126
Spatial autocorrelation analysis 284
Stabilizing selection 174
Standard diagrams 39, 42, 44, 46
- design of 38
State variables 283
Statistical analysis 194, 197
- models 187
- of race frequencies 173
- computer packages 199
Stemphylium botryosum f. sp. *lycopersici* 32, 33
Sterol biosynthesis inhibitors 158
Stewart's bacterial wilt 197
Stimuli 281
Storage of isolates 168
Stripe rust of winter wheat: see *Puccinia striiformis*
Structure(s) of epidemics 281, 283
Sunshine duration 128
Sulphonated polystyrene humidity sensor 105
Surveys 39, 237, 238
Survival of pathogen 241
Susceptibility of host tissue 60, 61
- time-dependent 59, 62, 64, 65
- space-dependent 59
SYMPHYT 274
Symptoms 27–30, 73, 74, 76, 78–80
Synecological nature of systems 263
Synoptic charts 245
- loss models 185
Synthetic approach 270, 272
- maps 239
Systematic-grid pattern 225
Systemic fungicide 149
Systems analysis 255, 267–271, 280
- approach 258
- behavior 283
- boundary 256
- control 288
- design 288
- theory 255

Taylor's index *b* 92
Techniques for measuring disease 42
TE function 122, 123, 125
Temperature 21–23, 28–32, 75, 78, 79, 116–119, 122, 123, 125, 129–131, 239, 241
- course of 119, 121
- daily means of 119
- dependance 122
- distribution 120, 129
- function 120
- mean 121, 122
- minimum 119
- monitoring 104
- response function 122
- sum 120–122
Terminology 280
Thermocouple 104, 109, 110
Time-dependent susceptibility 53, 54
Time-domain reflectometry (TDR) 111
Tipping-bucket gauge 109
Thermistor 104, 105, 110
Thermometer 117
Threshold value 41, 47
Tobacco etch virus 211
"Top-down" approach
Training for disease assessment 39, 42
Trajectories for spore dispersal 245, 246
Transects 94
- of diffuse clumps of diseass 95
Transformation
- arcsin$\sqrt{y}$ 149
- equations 202, 284
Trapping techniques 70, 219
Trap plants 9, 15, 16, 30, 76, 77, 80, 213, 244
Total dry weight 57
Two-dimensional spectral analysis 284

Two-way contingency tables 172
Turbulence 128, 130
Turbulent mass transfer 116, 127, 128
Types of stimulus-response 39
Turntable 26

Unmapped data 89
Uromyces phaseoli 139
Ustilago maydis 156

Validation of models 47, 234, 273
Vapor pressure 22, 78
Variables, density – dependent 93
Variance-to-mean ratio 40, 48, 92, 284
Vector abundance 217
– activity 212, 217, 218
– maps 239
– propensity 214–218
Vectors 195, 209, 210, 211, 217, 245
Venn diagrams 284
Vertical resistance 286
Verticillium wilt 53, 258, 261
Verticillium dahliae 261
Venturia inaequalis 57, 64, 76, 83, 155, 156, 157, 160, 196, 198, 274
Viability of spores 72, 74, 76
Videocamera 46
Virus epidemiology 209
Viruses 69
– nonpersistant 211
– persistent 211
Virulence 165, 166, 167
– analyses 165, 244
– associations 172
– coefficient (VAC) 173
– complexity 174
– difference 175
– dynamics 173, 284
– frequencies 167, 171, 172, 173, 176
– patterns 165, 168
– spectrum 171
– surveys 166, 172
– tests 166, 168
Virulence tests, plant material for 169
Virus sources 214
Volumetric spore traps 168

Warning systems 241
Washout 126
Water potential 105, 109, 111
Weather variables 99, 104
Weber-Fechner Law 39, 43, 44
Weibull distribution-function-shape parameter 283
Wetness 21, 27–31, 75, 81, 108, 123, 124
– duration 117, 125, 126
– sensors 107, 108, 117
Wind 21, 24, 28, 69, 72, 73, 77
– impaction spore trap (WIST) 167
– profile 130
– velocity 109, 117, 118, 127, 131
Wisconsin tank 25

Yield components 25, 182
– loss assessment 47, 179, 180, 197, 229
– potential 186

MIX
Papier aus verantwortungsvollen Quellen
Paper from responsible sources
FSC® C105338

If you have any concerns about our products,
you can contact us on
ProductSafety@springernature.com

In case Publisher is established outside the EU,
the EU authorized representative is:
Springer Nature Customer Service Center GmbH
Europaplatz 3, 69115 Heidelberg, Germany

Printed by Libri Plureos GmbH
in Hamburg, Germany